Progress in Probability and Statistics

Volume 11

Peter Huber
Murray Rosenblatt
series editors

Ernst Eberlein Murad S. Taqqu
editors

Dependence in Probability and Statistics

A Survey of Recent Results

(Oberwolfach, 1985)

1986

Springer Science+Business Media, LLC

Ernst Eberlein
Institut für Mathematische Stochastik
Universität Freiburg
7800 Freiburg i. Br.
Federal Republic of Germany

Murad S. Taqqu
Department of Mathematics
Boston University
Boston, MA 02215
U.S.A.

Library of Congress Cataloging in Publication Data
Dependence in probability and statistics.
(Progress in probability and statistics; vol. 11)
Includes bibliographies. 1. Random
1. Random variables. 2. Probabilities.
3. Mathematical statistics. I. Eberlein, Ernst.
II. Taqqu, Murad S. III. Series: Progress in
probability and statistics; v. 11.
QA273.18.D47 1986 519.2 86-12927

CIP-Kurztitelaufnahme der Deutschen Bibliothek
Dependence in probability and statistics. —
Boston ; Basel ; Stuttgart : Birkhäuser, 1986.
(Progress in probability and statistics ; Vol. 11)
ISBN 978-1-4615-8163-5
NE: GT

Originally published by Birkhäuser Boston, Inc., in 1986

ISBN 978-1-4615-8163-5 ISBN 978-1-4615-8162-8 (eBook)
DOI 10.1007/978-1-4615-8162-8

PREFACE

Practitioners and researchers in probability and statistics often work with dependent random variables. It is therefore important to understand dependence structures and in particular the limiting results that can be obtained under various dependence assumptions.

It is now well known that a great number of limit theorems which in the classical approach were always studied under the assumption that the underlying random variables were independent, continue to hold under certain dependence structures. Independence is not necessary. On the other hand, as dependence becomes stronger, at a certain point new phenomena arise. One is then led to ask: how much dependence, qualitatively and quantitatively, is allowed for the classical limit results to hold; and what are the results when dependence becomes stronger.

Our understanding of the complex relationships between dependence assumptions and limiting procedures has improved considerably during the last years. It is the purpose of this book to cover in an expository fashion a broad spectrum of recent results. Some of the theorems appear here for the first time. Because of the great variety of dependence structures, it would be difficult for a single author to treat so wide a topic. The contributors to this book participated in a conference on this subject held at the Mathematical Research Institute Oberwolfach in April 1985.

The book is aimed at an audience that is well-versed in probability and statistics but not necessarily well acquainted with the subject matter of the various papers. These papers tie together known results, describe the underlying ideas, summarize the state of the art and state some open problems.

The volume is divided in six sections. The first deals with sequences of random variables that exhibit *long-range dependence*. In that case, there is typically too much dependence for the usual central limit theorem to hold. The paper of Sun and Ho reviews results in the area of long range dependence in the case where the random variables are non-linear functions of Gaussian ones. The contribution of Giraitis and Surgailis considers more general types of random variables. Taqqu and Levy show how long-range dependence can be obtained by summing suitable renewal processes. Maejima reviews recent results on sojourns of multidimensional stationary Gaussian processes with long-range dependence.

Long-range dependence gives rise to unusual self-similar processes whose paths are fractals. Kôno studies the Hausdorff dimension of such paths. Dehling reviews recent results on U-statistics. Note the similarities in the results described by Sun and Ho and by Dehling. Papers concerning self-similarity and long-range dependence are scattered through various literatures. These separate sources are brought together in Taqqu's bibliographical guide, which is found at the end of the first section.

The ordinary central limit theorem holds when the summands are weakly dependent. There is weak dependence when the summands satisfy a suitable mixing condition, a martingale type condition, are moving averages with adequately chosen weights or satisfy various special dependence structures.

Section 2 concerns *mixing conditions*. These are measures of dependence between two σ-algebras. Bradley's paper surveys the basic properties of various strong mixing conditions, focusing on the "structural" properties of these conditions and on their relationships, but not on limit theory. The paper of Peligrad focuses on the (weak) limit theory. For the central limit theorem or for a functional central limit theorem to hold, the mixing coefficients between the past and the distant future must decay to zero sufficiently fast, but also moment conditions must be imposed on the random variables. Peligrad's paper discusses in detail the delicate interplay between rates of decay and moment conditions.

In the early 1950's, an invariance principle was a functional central limit theorem. There is now a much stronger tool, the strong invariance principle, which has been developed in the intervening years. The strong (or almost sure) invariance principle is an almost sure approximation theorem. Philipp's paper describes theorems of that type for weakly dependent random variables.

For a strongly mixing sequence, the central limit theorem holds if and only if the squares of the normalized partial sums are uniformly integrable. This result is proved in Denker's paper. Goldie and Morrow discuss the central limit theorem for weakly dependent random fields, i.e. processes with a multidimensional index set.

Various summability methods lead to various strong laws of large numbers. Bingham describes these summability methods, shows that many of the strong laws are equivalent and that the equivalence continues to hold when the random variables obey some types of weak mixing conditions.

Section 3 concerns *martingale dependence structures.* Gaenssler and Haeusler give a streamlined account of central limit theorems and functional central limit theorems for martingale difference arrays and for continuous time local martingales. Eberlein provides general conditions for a strong invariance principle and shows that the basic dependence assumption covers most of the generalizations of martingales that have been considered in the literature.

Section 4 concerns *other dependence structures* such as positive dependence (association) and Gibbs distributions. Dabrowski points to similarities in the structure of proof of the functional central limit theorem and the functional law of the iterated logarithm, and discusses these similarities in the context of the functional law of the iterated logarithm for associated sequences.

Waymire discusses Gibbs distributions such as the ferromagnetic Ising model. These distributions are of great interest in physics because they provide models for dependent systems of spins. Correlation inequalities that were previously known to hold for ferromagnetic Ising models hold also for infinitely divisible laws. Burton and Waymire describe central limit theorems for associated random measures.

Moving averages with infinite variance are studied in Section 5. When the variance is finite, the covariances provide a partial description of the dependence. When the variance is infinite, one can use the weights of the moving average to characterize the dependence. Different choice of weights may yield either short or long-range dependence. Avram and Taqqu discuss limit theorems for moving averages that belong to the stable domains of attraction. They point to the fact that for functional limit theorems to hold, it may be necessary to use an unusual topology. Davis and Resnick describe the limiting behavior of the sample correlation function for moving averages with infinite variance.

Section 6 concerns *extreme values*. O'Brien provides a convenient framework for understanding the asymptotic behavior of the maximum of dependent random variables. Hüsler considers non-stationary sequences. Vervaat characterizes all limiting processes that can appear in a functional central limit theorem for the maximum of stationary dependent random variables. These (extremal) processes are self-similar and they are stationary if one regards them as random functions of time intervals rather than time itself. An intrinsic definition of these processes involves random upper semicontinuous functions.

The papers in this volume survey various aspects of dependence. We hope that they will serve as a general reference and will inspire further research.

Ernst Eberlein
Institut für Mathematische Stochastik
Universität Freiburg
7800 Freiburg i. Br.
Germany

Murad S. Taqqu
Department of Mathematics
Boston University
Boston, MA 02215
USA

December 1985

TABLE OF CONTENTS

1. LONG-RANGE DEPENDENCE

2. MIXING CONDITIONS

3. MARTINGALE DEPENDENCE STRUCTURES

4. OTHER DEPENDENCE STRUCTURES

5. MOVING AVERAGES IN STABLE RANDOM VARIABLES

6. EXTREME VALUES

SECTION 1

LONG-RANGE DEPENDENCE

On Central and Non-Central Limit Theorems for Non-Linear Functions of a Stationary Gaussian Process

by

Tze-Chien Sun and Hwai-Chung Ho

Abstract. In this paper we shall give a brief review of recent results on both the central limit theorems and the non-central limit theorems for non-linear functions of a stationary Gaussian process.

1. Introduction

Throughout this paper, we shall always assume that the underlying process $\langle X_t, t=0,\pm1,\pm2,\ldots\rangle$ is a stationary Gaussian process with $EX_t=0$ and $EX_t^2 < \infty$ for all t. The assumption of stationarity implies that its covariance function can be written

$$r_n = EX_tX_{t+n} = \int_{-\pi}^{\pi} e^{in\lambda}\, dG(\lambda).$$

where G is the spectral distribution function of $\langle X_t\rangle$. If G is absolutely continuous, its derivative, denoted by f, is called the spectral density function and then

$$r_n = \int_{-\pi}^{\pi} e^{in\lambda}\, f(\lambda)d\lambda.$$

Let

$$\Omega = \langle \omega | \omega = (\ldots,\omega_{-1},\omega_o,\omega_1,\ldots),\omega_i \in R\rangle$$

be the coordinate space. Define

$$X_t(\omega) = \omega_t \text{ for each } t \text{ and } \omega\in\Omega.$$

Let F be the σ-field of subsets of Ω generated by the finite dimensional cylinder sets and P be the probability on F determined by the finite dimensional distributions of the X_t's. For the convenience of notation, we shall refer to this space (Ω,F,P) as the probability space on which the process $\langle X_t\rangle_{t=-\infty}^{\infty}$ is defined. We can define a shift operator $T : \Omega \to \Omega$ such that

$$(T\omega)_t = \omega_{t+1} \text{ for each } \omega\in\Omega.$$

Suppose $\langle Y_t, t=0,\pm1,\ldots\rangle$ is another process defined on (Ω,F,P) i.e., each Y_t is F-measurable or each Y_t is a function of X_t's. We shall always assume that the process $\langle Y_t\rangle$ is subordinate to the process $\langle X_t\rangle$ i.e., $Y_t\ (\omega) = Y_0$

$(T^t\omega)$ for all t and all ω. Two simple examples of such processes are

(i) instantaneous filters:

$$Y_t = H(X_t) \text{ for some function } H,$$

(ii) polynomial filters:

$$Y_t = \sum_{m=1}^{M} \sum_{k_i=-K}^{K} C_{k_1,\ldots,k_m} \cdot \prod_{j=1}^{m} X_{t+k_i}.$$

We define the renormalization processes $\langle Z_n^N \rangle_{n=-\infty}^{\infty}$ by

$$(1) \qquad Z_n^N = A_N^{-1} \sum_{t=nN}^{(n+1)N-1} Y_t, \quad n=0,\pm 1,\pm 2,\ldots$$

where A_N is a proper norming factor and N is a positive integer.

Our interest here is to study the weak limit of the renormalization process $\langle Z_n^N \rangle_{n=-\infty}^{\infty}$ as $N \to \infty$ under suitable conditions. In this paper, we shall give a brief review of the results obtained so far in this area. For the sake of clarity and simplicity of the presentation, we shall state the results only for the case when the parameter set is discrete and is one-dimensional. References to the parallel results to the case when the parameter set is continuous or multi-dimensional will be mentioned in the text.

The non-Gaussian limiting processes are self-similar processes expressed in the form of Wiener-Ito integrals. However, the property of such processes will not be discussed in this paper and the reader shall be referred to the book ([9]) by Major.

2. Some Earlier Results

In 1963 and 1965, Sun showed in [17] and [18] the

following theorem

Theorem 1

Under the conditions

(2) $\sum_{n=0}^{\infty} r_n^2 < \infty$ and $\lim_{N\to\infty} \frac{1}{N} \sum_{i=1}^{N} \sum_{j=1}^{N} r_{i-j}$ exists and is finite,

the sequence Z_0^N defined in (1), with $A_N = \sqrt{N}$, has an asympotic Gaussian distribution with mean zero and some finite variance as $N \to \infty$, when either

$$(i) \quad Y_t = \sum_{m=1}^{M} \sum_{k_i=-K}^{K} C_{k_1,\ldots,k_m} \left[\prod_{j=1}^{m} X_{t+k_j} - E\left[\prod_{j=1}^{m} X_{t+k_j} \right] \right]$$

for some finite M and K and constants $C_{k_1,\ldots,k_m}$, or

$$(ii) \quad Y_t = H(X_t) \text{ with } EY_t = 0 \text{ and } EY_t^2 < \infty \text{ for all } t.$$

The above theorem was proved originally under the conditions

(3) $f \in L^2(-\pi,\pi)$ and $\lim_{n\to\infty} \frac{1}{N} \int_{-\pi}^{\pi} \frac{\sin^2(\frac{N}{2}\lambda)}{\sin^2(\frac{1}{2}\lambda)} f(\lambda)d\lambda$ exists and is finite.

It is easy to see that (3) is equivalent to (2). Also Theorem 1 is just a classical central limit theorem, not a functional central limit theorem, i.e., it only shows the weak convergence of a sequence of random variables, not the weak convergence of a sequence of random processes.

Of course, right away we can think of two unsolved problems related to Theorem 1.

Problem 1: What would happen if we drop the condition $f \in L^2(-\pi,\pi)$ or equivalently if

$$\sum_{n=0}^{\infty} r_n^2 = \infty?$$

<u>Problem 2:</u> What could happen if we allow the function H to be non-instantaneous, e.g., $Y_t = H(X_t, X_{t+1})$ or $Y_t = H(X_t, X_{t+1}, X_{t+2}, \ldots)$?

It is easy to see that the slower the rate of convergence of r_n to zero is as $n \to \infty$ the higher the dependence of X_n to X_0 is for large n. Hence we should expect non-Gaussian limits as answers to Problem 1. This was shown to be true in the following example given by M. Rosenblatt in [13].

<u>Example 1:</u> Let $r_n = EX_t X_{t+n} = (1+n^2)^{-\alpha/2}$, $0 < \alpha < \frac{1}{2}$, and $Y_t = X_t^2 - 1$. Then the sequence $N^{-1+\alpha} \sum_{t=1}^{N} X_t$ has a non-normal limiting distribution as $N \to \infty$. Moreover the characteristic function of the limiting distribution was shown to have the form

$$\exp \left\langle \frac{1}{2} \sum_{k=2}^{\infty} (2it)^k \frac{C_k}{k} \right\rangle$$

where

$$C_k = \int_0^1 \!\!\cdots\!\! \int |x_1 - x_2|^{-\alpha} |x_2 - x_3|^{-\alpha} \cdots |x_k - x_1|^{-\alpha} dx_1 \ldots dx_k > 0 \text{ for all}$$

$k \geq 2$.

It is this example which opened the way to the later studies of the convergence of Z_n^N to non-normal limits. Other answers to Problem 1 and Problem 2 were not known until 1975.

3. Instantaneous Filters

In 1975, M. Taqqu ([19]) broke through with new ideas and significant results as answers to Problem 1. Taqqu considered the case

(4) $Y_t = H(X_t)$, $EY_t = 0$ and $EY_t^2 < \infty$ for all t

with

(5) $$r_n = n^{-\alpha}L(n)$$

where $0 < \alpha < 1$ and $L(n)$ is a slowly varying function. The condition (4) implies that H has a Hermite expansion

$$H(x) = \sum_{n=1}^{\infty} C_n H_n(x) \text{ with } \sum_{n=1}^{\infty} C_n^2 n! < \infty,$$

where $H_n(x)$ are the Hermite polynomials with leading coefficient equal to one. Taqqu called the number

$$k = \inf \{n | C_n > 0\}$$

the <u>Hermite rank of H</u>.

It turns out this rank will play a very important role in all the studies presented later in this paper. (Note that the importance of the so-called Hermite rank was not found by Hermite, it seems to us, by giving credit to the right person, it is more proper to call this number the Taqqu rank of H).

Let k be the Hermite rank of H. We have, now,

$$Y_t = \sum_{n=k}^{\infty} C_n H_n(X_t).$$

write

$$Y_t^* = \sum_{n=k+1}^{\infty} C_n H_n(X_t)$$

$$Z_n^{N*} = A_N^{-1} \sum_{t=nN}^{(n+1)N-1} Y_t^*$$

and

$$Z_{n,k}^{N} = A_N^{-1} \sum_{t=nN}^{(n+1)N-1} C_k H_k(X_t).$$

By computing the orders in N of the variances of Z_n^N and Z_n^{N*} as $N\to\infty$, he was able to show that , when $k\alpha < 1$, for each n, $Var(Z_n^{N*}) = O(Var\ Z_n^N)$ as $N\to\infty$

and hence

$$Var(Z_n^N) \sim Var(Z_{n,k}^N) \text{ as } N\to\infty.$$

As a result, Z_n^N and $Z_{n,k}^N$ should have the same limiting distribution, if one of them has a limiting distribution. Using this fact, he arrived at the following important theorem.

Theorem 2: (Taqqu's Reduction Theorem)

Suppose $\alpha k < 1$. If the finite dimensional distributions of $\langle Z_{n,k}^N\rangle_{n=-\infty}^{\infty}$ coverges to the corresponding finite dimensional distributions of a random process $\langle Z_{n,k}\rangle_{n=-\infty}^{\infty}$ as $N\to\infty$, then $\langle Z_n^N\rangle$ coverges in distribution to the same process $\langle Z_{n,k}\rangle$ as $N\to\infty$. Moreover, the limiting process $\langle Z_{n,k}\rangle$ must be self-similar.

Actually Taqqu proved the above theorem in [19] for a process with a continuous parameter set, hence a tightness argument was also needed for the proof. Clearly this is an invariance principle type theoem which reduces the problem of finding the weak limit of $\langle Z_n^N\rangle$ for any $Y_t = H(X_t)$ defined in (4) and (5) to the problem of finding the weak limit of $\langle Z_n^N\rangle$ for $Y_t = H_k(X_t)$ with the condition (5) where H_k are the Hermite polynomials k=1,2...

In the same paper, Taqqu also found the limits $\langle Z_{n,k}\rangle$ for k=1,2. When k=1, the limiting process is a fractional Brownian motion and hence is Gaussian. When k=2, the limiting process is non-Gaussian. He called it the Rosenblatt process because in this case the marginal distribution of $Z_{n,k}$, for each n, is exactly the non-Gaussian distribution given by Rosenblatt in the Example 1.

In 1979, both Dobrushin and Major ([4]) and Taqqu([20]) found the limiting processes $\langle Z_{n,k}\rangle$ for all $k \geq 1$ in the form of Wiener-Ito integrals. However, still, not much is known about the property of the distributions of the limiting processes (see, e.g., [12]). We shall present their result here following Dobrushin and Major.

<u>Theorem 3</u>

Consider the process $\langle Y_t\rangle$ defined in (4) and (5) with $\alpha k < 1$ where k is the Hermite rank of H. Choose

$$A_N = N^{1-k\alpha/2}\ L(N)^{k/2}.$$

Then as $N\to\infty$ the finite dimensional distributions of the sequence of processes $\langle Z_n^N\rangle_{n=-\infty}^{\infty}$ defined in (1) converges to those of the process $\langle Z_n^*\rangle$ given by the formula

$$(6)\qquad Z_n^* = D^{-k/2}\, C_k \int e^{in(x_1+\dots+x_k)}\, \frac{e^{i(x_1+\dots+x_k)}-1}{i(x_1+\dots+x_k)}$$

$$|x_1|^{(\alpha-1)/2}\dots|x_k|^{(\alpha-1)/2}dW(x_1)\dots dW(x_k)$$

where W is the random spectral measure of the white noise process and

$$D = \int_{-\infty}^{\infty} \exp(ix)|x|^{\alpha-1}dx = 2\Gamma(\alpha)\cos\left(\frac{\alpha\pi}{2}\right).$$

Actually, Dobrushin and Major ([14]) proved the above theorem for the case when the parameter set in multidimensional, while Taqqu ([20]) proved the theorem for the case when the parameter set is continuous.

So far in this section, we have only talked about non-Gaussian limits. About results on Gaussian limits, in 1983 Breuer and Major ([2]) generalized the part (ii) of

Theorem 1 to the following form.

Theorem 4

Consider the process $\langle Y_t \rangle$ defined in (4) and (5) where H has the Hermite rank k. Suppose

$$(7) \qquad \sum_{n=0}^{\infty} |r_n|^k < \infty.$$

Choose $A_N = N^{1/2}$ and write

$$Z_n^N(H_\ell) = A_N^{-1} \sum_{t=nN}^{(n+1)N-1} H_\ell(X_t).$$

Then

(i) The limits

$$\lim_{n\to\infty} E(Z_0^N(H_\ell))^2 = \sigma_\ell^2 \cdot \ell\,!$$

exists for all $\ell \geqslant k$ and the infinite sum

$$\sigma^2 = \sum_{\ell=k}^{\infty} C_\ell^2 \sigma_\ell^2\, \ell\,! < \infty.$$

(ii) The finite dimensional disbributions of the sequence $\langle Z_n^N \rangle$ defined in (1) converge to those of the sequence $\langle \sigma Z_n^* \rangle$ as $N\to\infty$, where Z_n^*, $n=0,\pm 1,\ldots$ are independent standard normal random variables.

In their paper Breuer and Major actually proved the above theorem for stationary Gaussian fields, i.e., for the case of multidimensional parameter sets. In the same paper, they also gave a theorem with conditions slightly weaker than the condition (7). Their results are clearly a sharp improvement of the part (ii) of Theorem 1. Similar results can be found in [6].

The above theorem can be generalized, in a rather straightforward way, to the case

$Y_t = H(X_{t+t_1},\ldots,X_{t+t_d})$ for some fixed $t_1,\ldots,t_d$,

under the same condition in (7) ([7]).

4. Non-Instantaneous Filters

In this section we shall consider limiting theorems for non-instantaneous filters. They are generalizations of the part (i) of Theorem 1 and they give some answers to Problem 2. Let

$$X_t = \int_{-\pi}^{\pi} e^{itx}\, Z_G(dx) \quad t=0,\pm 1,\pm 2,\ldots,$$

be the spectral respresentation of the stationary process $\langle X_t \rangle$ with

$$r_n = \int_{-\pi}^{\pi} e^{inx}\, G(dx)$$

and

$$EZ_G(dx)\overline{Z_G(dy)} = \delta_{x,y}\, G(dx).$$

Then any process $\langle Y_t \rangle$, $EY_t = 0$ and $EY_t^2 < \infty$ for all t and subordinate to $\langle X_t \rangle$, can be written ([9]) as

$$(8) \qquad Y_t = \sum_{j=1}^{\infty} \frac{1}{j!} \int e^{it(x_1+\ldots+x_j)} \alpha_j\,(x_1,\ldots,x_j)\; Z_G(dx_1)\ldots Z_G(dx_j)$$

with

(i) $\alpha_j\,(x_1,\ldots,x_j) = \overline{\alpha_j(-x_1,\ldots,-x_j)}$ for all $x_1,\ldots,x_j$

(ii) $\sum_{j=1}^{\infty} \frac{1}{j!} \int |\alpha_j(x_1,\ldots,x_j)|^2\, G(dx_1)\ldots G(dx_j) < \infty.$

Again in this section, we shall assume the covariance function of $\langle X_t \rangle$

$$(9) \qquad r_n = n^{-\alpha} L(n).$$

Define

$G_N(A) = \frac{N^{\alpha}}{L(N)}\, G(N^{-1}A)$ for every Borel set A and

N=1,2,... ,

and

$$K_N(x_1,\dots x_j) = \frac{\exp\{i(x_1+\dots+x_j)\}-1}{[\exp[\frac{i}{N}(x_1+\dots+x_j)\}-1]N}.$$

In their 1979 paper ([4]) Dobrushin and Major also proved the following theorem for non-instantaneous filters.

<u>Theorem 5</u>

Consider a process $\{Y_t\}$ defined as in (8), with a Hermite rank k, i.e., k is the largest integer such that $\alpha_j = 0$ in $(L_G^2)^j$ for all $j < k$. Suppose $\alpha_k(0,0,\dots,0) \neq 0$. Suppose r_n satisfies (9) and $\alpha k < 1$. Also assume the following relation

$$(10)\quad \sum_{j=k+1}^{\infty} \frac{1}{j!} \frac{N^{-(j-k)\alpha}}{L(N)^{j-k}} \int |\alpha_j(\frac{x_1}{N},\dots,\frac{x_j}{N})|^2 K_N(x_1,\dots,x_j) G_N(dx_1)\dots G_N(dx_j) \to 0 \text{ as } N\to\infty$$

be satisfied. Choose

$$A_N = N^{1-k\alpha/2} L(N)^{k/2}.$$

Then the finite dimensional distributions of the processes $\{Z_n^N\}$ as defined in (1) converge to those of the process $\{\frac{1}{k!}\alpha_k(0,\dots,0)Z_n^*\}$ as $N\to\infty$ where Z_n^* is defined in (6).

In their paper, this theorem was again proved for the case of a multidimensional parameter set. The condition (10) simply says that the Hermite rank term dominates all the other terms. Dobrushin and Major also showed that if Y_t in (8) is actually a function of finitely many X_t's i.e.

$$(11)\quad Y_t = H(X_{t+t_1},\dots,X_{t+t_d}) \quad \text{for some fixed } t_1,\dots,t_d,$$

then the condition (10) is automatically satisfied. This gives a partial answer to Problem 2 for the non-Gaussian limit case. An open problem here is to consider the case

(12) $Y_t = H(X_t)$ with Y_t a vector function of X_t

where $X_t = (X_{t,1}, X_{t,2}, \ldots X_{t,d})$ is a stationary vector-valued Gaussian process, for both Gaussian and non-Gaussian limits. Clearly (12) includes (11) but not vice versa.

The condition $\alpha_k(0,0,\ldots,0) \neq 0$ in Theorem 5 is crucial. If $\alpha_k(0,\ldots,0) = 0$, the limiting process is degenerate. Therefore, in order to obtain a non-degenerate limit, a different A_N with a lower order in N must be used. This could change the fact that the Hermite rank term is dominating and it could create much complication.

The follwowing example by Rosenblatt ([14]) shows that with the condition $\alpha_k(0,\ldots,0) = 0$ we could obtain a non-Gaussian limit with $A_N = N^\alpha$ where $0 < \alpha < \frac{1}{2}$ in contrast to the above cases where the order α is always bigger than $\frac{1}{2}$.

Example 2: Let

$$Y_t = H_2(X_t) = X_t^2 - 1$$

with $r_n \sim n^{-\gamma}$, $0 < \gamma < \frac{1}{2}$. Define

$$U_t = \sum_{m=-\infty}^{\infty} a_m Y_{t-m}$$

with

$$a_m = \begin{cases} 0 & \text{when } m = 0, \\ m^{-\beta-1} & m > 0, \\ -|m|^{-\beta-1} & m < 0 \end{cases}$$

where $\beta > 0$. Choose $A_N = N^{1-\beta-\gamma}$ with conditions $2(\beta + \gamma) > 1 > \beta + \gamma$. Then $A_N^{-1} \sum_{t=1}^{N} U_t$ can be shown to have a non-Gaussian limiting distribution as $N\to\infty$. The order $\alpha = 1-\beta-\gamma$ here satisfies $0 < \alpha < \frac{1}{2}$. Note in this example the Hermite rank of $\langle U_t\rangle$ is 2 and it can easily be computed

$$\alpha_2(x_1,x_2) = 2\sum_{m=1}^{\infty} m^{-\beta-1} \sin m(x_1+x_2).$$

Hence $\alpha_2(0,0) = 0$. This example was generalized by Major in [10].

For more complicated cases, let us consider Y_t defined in (8) with $r_n = n^{-\alpha}L(n)$ and a Hermite rank k. Write

$$I_j(t) = \frac{1}{j!}\int e^{it(x_1+\ldots+x_j)}\,\alpha_j(x_1,\ldots,x_j)\, Z_G(dx_1)\ldots Z_G(dx_j)$$

Then, Y_t can be written

$$Y_t = \sum_{j=k}^{s} I_j(t) + \sum_{j=s+1}^{\infty} I_j(t) \tag{13}$$

where $s = \max\langle j: j\alpha < 1\rangle$. We shall call the first term on the right side of (13) the non-Gaussian part and the second term the Gaussian part. Suppose $\alpha_k(0,\ldots,0) = 0$ happens. After we reduce the order in N of A_N to obtain a non-degenerate limit, the dominating term may not be the Hermite rank term $I_k(t)$ any more. In fact, there may be more than one term which are dominating. Intuitively, we would expect the following results under suitable conditions.

<u>Case i</u>: If the dominating terms all belong to the non-Gaussian part, we should obtain a non-Gaussian limit.

<u>Case ii</u>: If the dominating terms all belong to the Gaussian part, we should obtain a Gaussian limit.

<u>Case iii</u>: If the dominating terms appear in both the non-Gaussian part and the Gaussian part, the limit should be a mixture of the two cases.

Some results along this direction were studied in [8].

5. Other Results

A. In [15] Rosenblatt obtained a non-central limit theorem for the finite Fourier transforms

$$Y_n^N(\beta) = A_N^{-1} \sum_{t=(n-1)N}^{nN-1} H(X_t)e^{it\beta}$$

as $N\to\infty$ with the assumptions that

(i) $EH(X_t) = 0$ and $EH(X_t)^2 < \infty$ for all t

(ii) the covariance function of $\langle X_t \rangle$ has the form

$$r_n = |n|^{-\alpha} L(|n|) \sum_{j=0}^{m} S_{\lambda_j} \cos n\lambda_j, \; S_{\lambda_j} > 0,$$

for some m and $0 = \lambda_0 < \lambda_1 < \ldots < \lambda_m$, where $L(t)$ is a slowly varying function.

(iii) $\alpha k < 1$ where k is the Hermite rank of H

(iv) $A_N = N^{1-k\alpha/2} L(N)^{k/2}$.

His result was generalized by Giraitis in [5] to include the case of having different rates α_j's for different frequencies λ_j's.

B. In [1] and [16] Avram and Taqqu, and Surgaillis considered the case that the process X_t is a moving average process (not necessarily Gaussian).

$$X_t = \sum_{i=-\infty}^{t} C_{t-i}\, \xi_i$$

where ξ_i's are i.i.d. with $E\xi_i = 0$ and $E\xi_i^{2n} < \infty$ for all n and $C_k \sim k^{-\beta} L(k)$, $\frac{1}{2} < \beta < \frac{1}{2}(1 + \frac{1}{n})$. In this case they used the so-called "generalized powers" to replace the Hermite polynomials. They proved a non-central limit theorem (see [1]) for the sequence

$$\frac{1}{N^H L(N)} \sum_{k=1}^{[Nt]} P_n (X_k)$$

where (i) $H = 1 - n(\beta - \frac{1}{2})$

(ii) P_n is the nth generalized power associated with the distribution of X_0.

C. In 1975 Maruyama proved in [11] several central limit theorems for non-linear functions of a stationary Gaussian process with conditions on the spectral density function $f(\lambda)$ of X_t (not conditions on the covariance functions for the results we mentioned above). Further results following this line can be found in [6] and [3].

REFERENCES

[1] F. Avram and M. Taqqu, Non-central limit theorems and generalized power. (preprints)

[2] P. Breuer and P. Major, Central limit theorems for non-linear functionals of Gaussian Fields. J. Multivariate Analysis 13 (1983) pp. 425-441.

[3] D. Chambers and E. Slud, Central limit theorems for non-linear functionals of stationary Gaussian processes. (preprints)

[4] R.L. Dobrushin and P. Major, Non-central limit theorems for non-linear functionals of Gaussian fields. Z.W. verb. Geb. 50(1979) pp.27-52.

[5] L. Giraitis, Convergence of certain non-linear transformations of a Gaussian sequence to self-similar processes. Translation of Lithuanian Math. J. 23(1983) pp. 31-39.

[6] L. Giraitis and D. Surgailis, CLT and other limits theorems for functionals of Gaussian processes. (preprint)

[7] H.C. Ho and T.C. Sun, A central limit theorem for non-instantaneous filters of a stationary Gaussian process. (preprints)

[8] H.C. Ho and T.C. Sun, Limit theorems for non-linear functionals of a stationary Gaussian process. (preprints)

[9] P. Major, Multiple Wiener-Ito Integrals. Lecture Notes, Math No. 849, Springer-Verlag (1981).

[10] P. Major, Limit theorems for non-linear functions of Gaussian sequences. Z.W. verb. Geb. 57(1981) pp. 129-158.

[11] G. Maruyama, Nonlinear functionals of Gaussian stationary processes and their applications. Proc. 3rd Japan-USSR Symposium on Probability Th. (Tashkent 1975). Lecture Notes, Math No. 550, Springer-Verlag (1976) pp.375-378.

[12] G. O'Brien and W. Vervaat, Marginal distribution of self-similar processes with stationary increments. Z.W. verb. Geb 64(1983) pp.129-138.

[13] M. Rosenblatt, Independence and dependence. Proc. 4th Berkeley Symposium on Math. Stat. and Prob., Univ. of Calif. (1961) pp.431-443.

[14] M. Rosenblatt, Some limit theorems for partial sums of quadratic forms in stationary Gaussian variables. Z.W. veb. Geb. 49(1979) pp. 125-132.

[15] M. Rosenblatt, Limit theorems for Fourier transforms of functionals of Gaussian sequences. Z.W. verb. Geb. 55(1981) pp. 123-132.

[16] D. Surgailis, Zones of attraction of self-similar multiple integrals. Translation of Lithuanian Math. J. 22(1982) pp. 327-340.

[17] T.C. Sun, A central limit theorem for non-linear functions of a normal stationary process. J. Math. Mech. 12(1963) pp. 945-978.

[18] T.C. Sun, Some further results on central limit theorems for non-linear functions of a normal stationary process. J. Math. Mech. 14(1965) pp. 71-85.

[19] M. Taqqu, Weak convergence to fractional Brownian motion and to the rosenblatt process. Z.W. verb. Geb. 31(1975) pp. 287-302.

[20] M. Taqqu, Convergence of integrated processes of arbitrary Hermite rank. Z.W. verb. Geb. 50(1979) pp. 53-83.

Department of Mathematics
Wayne State University
Detroit, Michigan 48202

Institute of Statistics
Academic Sinica
Nankang, Taipei
Taiwan

Multivariate Appell polynomials and the central limit theorem

L.Giraitis, D.Surgailis

Summary. A CLT for processes of the form $f(X_t)$ is proved, where $f(x)$ is a polynomial and X_t, $t \in \mathbb{Z}$ is a process with long range dependence. Conditions on X_t are formulated in terms of semi-invariants; they are specified for linear processes X_t. The notion of the Appell rank of $f(x)$ plays a basic role in the CLT. Various topics related to Appell polynomials (e.g. expansions, diagram formalism for semi-invariants) are discussed.

1. Introduction

Let X_t, $t \in \mathbb{Z} = \{\cdots, -1, 0, 1, \ldots\}$ be a stationary Gaussian process, with zero mean and the covariance $r(t) = EX_tX_0$. Let be given a (real-valued) function $f(x)$ such that $Ef(X_t) = 0$ and $Ef^2(X_t) < \infty$. If $r(t)$ decreases slowly as $|t| \to \infty$ (e.g., if $\sum |r(t)| = \infty$), the question about the limiting distribution of the normalized sum

$$S_N = \sum_1^N f(X_t) \tag{1.1}$$

is difficult and the answer is known only under some additional conditions on $r(t)$, such as the requirement that $r(t)$ varies regulary at infinity; see Dobrushin and Major [7], Taqqu [29] or Sun [24]. However, it is relatively easy to show that the limiting distribution of S_N coincides (under broad conditions on $r(t)$) with that of

$$S_N^{(k_*)} = \Sigma_1^N H_{k_*}(X_t) \,, \tag{1.2}$$

where $k_* = k_*(f, H)$ is the Hermite rank of $f(x)$, i.e. $f(x)$ has the Hermite expansion

$$f(x) = \Sigma_0^\infty c_n H_n(x)/n! \tag{1.3}$$

with $c_0 = \ldots = c_{k_*-1} = 0$ and $c_{k_*} \neq 0$ (see [28, 2]). In particular, S_N is asymptotically Gaussian if such is $S_N^{(k_*)}$.

A simple explanation of the last fact can be given as follows. Denote by $r_f(t)$ the covariance function of the process $f(X_t)$. By the well-known properties of Hermite polynomials,

$$r_f(t) = \Sigma_0^\infty c_n^2 \, r_{H_n}(t) \,/(n!)^2 \tag{1.4}$$

and

$$r_{H_n}(t) = n! \, (r(t))^n \,, \tag{1.5}$$

which imply easily that the conditions

$$\Sigma \, |r_f(t)| < \infty \tag{1.6}$$

and

$$\Sigma \, |r_{H_{k_*}}(t)| < \infty \tag{1.7}$$

are equivalent ([10], Lemma 5). If, in addition, $\Sigma \, r_f(t) \neq 0$ and $\Sigma \, r_{H_{k_*}}(t) \neq 0$ (i.e. we exclude the possibility of negative dependence), then (1.6) and (1.7) imply the linear growth of the variances of S_N and $S_N^{(k_*)}$ respectively. Therefore, the equivalence of (1.6) and (1.7) indicates that $f(X_t)$ and $H_{k_*}(X_t)$ simultaneously are either weakly dependent or strongly dependent. Moreover, from (1.5) it follows that the condition of weak dependence of $f(X_t)$ is

$$\sum |r(t)|^{k_*} < \infty . \tag{1.8}$$

In fact, Breuer and Major [5] (see also Giraitis and Surgailis [10]) showed that (1.8) implies the CLT for $f(X_t)$, provided $\liminf (\mathrm{Var}\, S_N)/N > 0$.

A natural question in this context is what happens if the underlying Gaussian process X_t is replaced by some (simple?) non-Gaussian process with a long range dependence. In particular, if there exists in such a case an analog of Hermite expansions and the Hermite rank which would be decisive in the CLT for $f(X_t)$. Partial answers to these questions were obtained in our papers [25,26,9] , where we considered linear underlying processes X_t , i.e.

$$X_t = \sum_{-\infty}^{\infty} a(t-s)\xi_s , \tag{1.9}$$

where ξ_s, $s \in \mathbb{Z}$ are i.i.d. random variables with zero mean and variance 1, and $a(t)$, $t \in \mathbb{Z}$ are (real, non-random) weights satisfying the condition

$$\sum a^2(t) < \infty , \tag{1.10}$$

(If $a(t)$ decrease sufficiently slow so that $\sum |a(t)| = \infty$ the process X_t (1.9) exibits a long range dependence, although the limiting distribution of $X_1 + \ldots + X_N$ is almost always Gaussian, see Ibragimov and Linnik [12]

Th.18.6.5). It turned out that the analog of the Hermite rank (of a function $f(x)$) in the case of a linear process X_t is the so-called <u>Appell rank</u> $k_* = k_*(f, A)$, which is by definition the index of the lowest non-zero coefficient in the expansion

$$f(x) = \sum_0^\infty c_n A_n(x)/n! \tag{1.11}$$

in Appell polynomials $A_n(x)$, $n = 0, 1, \ldots$. These polynomials are related to the (marginal) distribution of X_t by the relation

$$\sum_0^\infty z^n A_n(x)/n! = e^{zx}/E\exp(zX_t), \quad z \in \mathbb{C} . \tag{1.12}$$

If $X_t \in \mathcal{N}(0,1)$, then $A_n(x) = H_n(x)$ and $k_*(\cdot, A) = k_*(\cdot, H)$. According to the main result of Giraitis [9], the condition

$$\sum |r(t)|^{k_*} < \infty \quad (k_* = k_*(f, A)) \tag{1.13}$$

implies the CLT for $f(X_t)$ in the case of a linear process X_t, provided $\liminf (\mathrm{Var}\, S_N)/N > 0$ and the coefficients in the expansion (1.11) decrease sufficiently rapidly.

Very recently, we have found that the notion of the Appell rank is more universal and is not restricted to linear processes. Below we formulate rather general conditions on the semiinvariants of a stationary process X_t which involve the Appell rank $k_*(f, A)$ of a function $f(x)$ and which ensure the CLT for $f(X_t)$.

Let X_t, $t \in \mathbb{Z}$ be a strictly stationary process, non necessarily linear, having finite moments of any order. Let $A_n(x)$, $n = 0, 1, \ldots$ be Appell polynomials, related to the (marginal) distribution of X_t by means of (1.12). For any integers $\ell_1 \geq 1, \ldots, \ell_k \geq 1$ we denote by

$$\varkappa_{\ell_1,\ldots,\ell_k}(t_1,\ldots,t_k) \quad , \quad t_1,\ldots,t_k \in \mathbb{Z}$$

the joint semi-invariant of random variables $X_{t_1},\ldots,X_{t_1}$ (ℓ_1 times), ..., $X_{t_k},\ldots,X_{t_k}$ (ℓ_k times). We shall assume that there exist functions $q_{\ell_1,\ldots,\ell_k}(t_1,\ldots,t_k)$, $t_1,\ldots,t_k \in \mathbb{Z}$ such that

$$|\varkappa_{\ell_1,\ldots,\ell_k}(t_1,\ldots,t_k)| \leq q_{\ell_1,\ldots,\ell_k}(t_1,\ldots,t_k),$$

with the following properties:

(q1) For all $N \geq 1$, $k \geq 3$, $t_1,\ldots,t_k = 1,\ldots,N$, $\ell_1,\ldots,\ell_k \in \mathbb{Z}_+$, $1 \leq i \leq k$ and $p \geq 1 + [1/\ell_i]$,

$$\Big(\sum_{t_i=1}^{N} q^p_{\ell_1,\ldots,\ell_i,\ldots\ell_k} \Big)^{1/p}$$

$$\leq C\, q_{\ell_1,\ldots,\widehat{\ell_i},\ldots,\ell_k} , \tag{1.14}$$

where the constant C does not depend on N and $t_1,\ldots,t_k$; $[\cdot]$ is the entire part and $\widehat{\ell_i}$ means that ℓ_i is absent.

(q2) In the case $k=2$, the functions $q_{\ell_1,\ell_2}(t_1,t_2) = q_{\ell_1,\ell_2}(t_1-t_2)$ satisfy

(1.14) for $\ell_1 + \ell_2 \geq 3$ and $p \geq 1 + [1/\min(\ell_1, \ell_2)]$.

Theorem 1. Let $f(x)$ be a polynomial with the Appell rank $k_* = k_*(f, A) \geq 2$; $S_N = \sum_1^N f(X_t)$.

Assume that the conditions (q1), (q2) hold as well as

$$(q3) \qquad \sum q_{1,1}^{k_*}(t) < \infty$$

and

$$(q4) \qquad \sum_{t=1}^{N} \Big(\sum_{s=1,\, |t-s| \geq K}^{N} q_{n,1}(t-s)\Big)^k \leqslant C N^{(k+1)/2}$$

for any integers $n, k \geqslant k_*$, where $C = C(K) \to 0$ $(K \to \infty)$.

If, moreover, $\liminf (\mathrm{Var}\, S_N)/N > 0$, then

$$S_N / (\mathrm{Var}\, S_N)^{1/2} \longrightarrow \mathcal{N}(0, 1).$$

Let us note that the conditions (q1), (q2) do not depend on the Appell rank k_* and are less restrictive than (q3) and (q4). In the case of linear process, (q1) and (q2) are always satisfied and their verication is very simple.

Theorem 2. Let X_t be a linear process (1.9) having finite moments of any order. Then (q1) and (q2) hold. If, moreover,

$$\sum_{t=-\infty}^{\infty} \Big(\sum_{s=-\infty}^{\infty} |a(t-s)\, a(s)|\Big)^{k_*} < \infty, \qquad (1.15)$$

then (q3) and (q4) hold as well.

Note that (1.15) is only slightly stronger than (1.13), as the covariance of linear process (1.9) is

$$r(t) = \sum_{-\infty}^{\infty} a(t-s)\, a(s).$$

In some cases the conditions (1.13) or (1.15) are exact, in the following sense. Let $0 \le a(t) \sim \text{const}\, |t|^{-\beta}$ $(|t| \to \infty)$, with $\beta \in (1/2, 1)$. Then $r(t) \sim$ $\sim \text{const}\, |t|^{1-2\beta}$ and (1.15) is equivalent to

$$(2\beta - 1) k_* > 1 . \tag{1.16}$$

On the other hand, if $(2\beta - 1) k_* < 1$, then $\text{Var}\, S_N$ $\sim \text{const}\, N^{\alpha}$, $\alpha = 1 - ((2\beta - 1) k_* / 2) > 1/2$ and the limiting distribution of $S_N / (\text{Var}\, S_N)^{1/2}$ is non-Gaussian [26].

Roughly speaking, Theorem 1 gives sufficient conditions for the CLT for simple functions (polynomials) of a 'simple' process X_t with a long range dependence. We feel that the 'simplicity' of X_t is somehow related to the condition (q1) and (q2). It would be interesting to verify these conditions for other classes of processes X_t than linear. As for simplicity of $f(x)$, we must say than expanding in Appell polynomials (which are not orthogonal in general) of non-analytic functions is an open problem, see Sect. 6 below.

The case $k_* (f, A) = 1$ is also left open by Theorem 1. Since $A_1(x)$ is linear, the conditions on the semiinvariants of X_t in the case $k_* = 1$ should guarantee the CLT for X_t itself (while the conditions of Theorem 1 do not imply this, in our opinion). Conditions on semi-invariants for the CLT for X_t are well-known (see e.g. Brillinger [4], Th. 4.4.1); in our notation they correspond to (1.14) for $p = \ell_1 = \ldots = \ell_k = 1$ and all $k \geqslant 2$.

The contents of the remaining parts of this paper are the following. In Sect. 2 we present a variance based discussion of the weak dependence of the process $f(X_t)$ similar to the discussion above in the case of a Gaussian process. In Sections 3 and 4 we discuss the basic properties of the univariate and the multivariate Appell polynomials, respectively. Diagram formalism for semi-invariants of Appell polynomials plays an important role in the proof of Theorem 1 based on the semi-invariant method and is treated in Sect.5. In Sect. 6 we discuss the problem of expansions in Appell polynomials. Finally, in Sect. 7 and 8 we prove Theorems 1 and 2, respectively.

2. The variance of S_N.

Let us start with the notation and some basic fact about semi-invariants. By the definition, the (joint) semi-invariant of (real) random variables $Y_1, \dots, Y_k$ is equal to

$$\chi(Y_1, \dots, Y_k)$$
$$= \partial^k \log E \exp(\textstyle\sum_1^k a_j Y_j)/\partial a_1 \cdots \partial a_k \big|_{a_1 = \dots = a_k = 0}.$$

Sometimes we shall use the notation $\chi(Y_1, \dots, Y_k) = \chi(Y_j,\ j = 1, \dots, k)$, $\chi(\underbrace{Y, \dots, Y}_{k}) = \chi_k(Y)$. Without mentioning it explicitly, we shall assume that all expectations and semi-invariants introduced in this text exist.

The following properties of semi-invariants are well-known (see e.g. [15, 17]).

(χ1) $\quad \chi(Y_1, \dots, Y_k) = \sum (-1)^r (r-1)!\, E Y^{V_1} \cdots E Y^{V_r}$,

where the sum is taken over all partitions $\{V_1, \dots, V_r\}$, $r \geq 1$ of $\{1, \dots, k\}$ and

$$Y^V = \prod_{i \in V} Y_i, \qquad V \subseteq \{1, \dots, k\}.$$

(χ2) If $Y_1, \dots, Y_k$ are (jointly) Gaussian and $k > 2$, then $\chi(Y_1, \dots, Y_k) = 0$.

(χ3) If $\{1, \dots, k\} = V_1 \cup V_2$ and the systems $(Y_i,\ i \in V_1)$, $(Y_i,\ i \in V_2)$ are independent, then $\chi(Y_1, \dots, Y_k) = 0$.

Let X_t, $t \in \mathbb{Z}$ be a strictly stationary process and

$$\varkappa_{k,\ell}(t) = \varkappa(\underbrace{X_0, \dots, X_0}_{k}, \underbrace{X_t, \dots, X_t}_{\ell})$$

$k, \ell \geq 1, t \in \mathbb{Z}$ the 'two-point' semi-invariant. We shall assume that $E X_t \equiv 0$ and denote by

$$r(t) = \varkappa_{1,1}(t) = E X_0 X_t$$

the covariance function. Condition (q2′) below is nothing else as (q2), in a slightly different formulation.

(q2′) For any $K < \infty$ and all (integers) $1 \leq k, \ell \leq K$, $k + \ell \geq 3$ there exist $p(t) = p(t, K) \geq 0$ and $q(t) = q(t, K) \geq 0$ satisfying the conditions $\sum p^2(t) < \infty$, $\sum q(t) < \infty$, such that for all $t \in \mathbb{Z}$,

$$|\varkappa_{k,\ell}(t)| \leq \begin{cases} p(t) & \text{if } k = 1 \text{ or } \ell = 1, \\ q(t) & \text{if } k \geq 2 \text{ and } \ell \geq 2. \end{cases} \tag{2.1}$$

Let $A_n(x)$, $n = 0, 1, \dots$ be Appell polynomials related to the marginal distribution of X_t. Note that $A_0(x) = 1$ and $A_1(x) = x$.

Theorem 3. Assume that a (stationary) process X_t satisfies the condition (q2′). Let $f(x)$ be a polynomials with the Appell rank k_*, and denote

$$r_f(t) = E f(X_0) f(X_t),$$

$$S_N = \sum_1^N f(X_t).$$

Then:

(i) If $k_* \geq 2$, conditions

$$\sum |r(t)|^{k_*} < \infty \tag{2.2}$$

and

$$\sum |r_f(t)| < \infty \tag{2.3}$$

are equivalent.

(ii) $k_* = 1$ and (2.3) imply

$$\mathrm{Var}\,(\Sigma_1^N X_t) \leq CN . \tag{2.4}$$

(iii) $k_* = 1$ and (2.2) imply

$$\mathrm{Var}\,(S_N) \leq CN . \tag{2.5}$$

Although somewhat weaker than the corresponding statement in the Gaussian case (Lemma 5, [10]), Theorem 3 says that the process $f(X_t)$ is weakly dependent in the wide sense if the covariance of X_t is summable in power $k_*(f, A)$. For linear processes X_t, Theorem 1 was proved by Giraitis [9]. The proof given below goes along the same lines. It uses essentially a combinatorial formula for the covariance of Appell polynomials which is a particular case of the 'diagram formula' of Sect. 5.

Proof of Theorem 3. As $f(X)$ is a polynomial, its Appell expansion (1.11) is finite, i.e. $c_n = 0$ for all $n > K$ and some $K < \infty$. Therefore

$$r_f(t) = \Sigma_{k_*}^K \Sigma_{k_*}^K c_k c_l E A_k(X_0) A_l(X_t).$$

According to Theorem 4 (iii),

$$E A_k(X_0) A_l(X_t) = \sum d(k_1, \dots, k_r; l_1, \dots, l_r) \tag{2.6}$$

$$\cdot \prod_{i=1}^r \varkappa_{k(i), l(i)}(t) ,$$

where the sum is taken over all $r = 1, 2, \ldots$ and all integers $k(1) \geq 1, \ldots, k(r) \geq 1, \ell(1) \geq 1, \ldots, \ell(r) \geq 1$ such that $k(1) + \ldots + k(r) = k$, $\ell(1) + \ldots + \ell(r) = \ell$; $d(k(1), \ldots, k(r); \ell(1), \ldots, \ell(r))$ is $\binom{k}{k(1), \ldots, k(r)} \binom{\ell}{\ell(1), \ldots, \ell(r)} \frac{1}{r!}$.

Assume that $k \geq \ell$ and consider the following cases:

(c1) among the integers $k(1), \ldots, k(r), \ell(1), \ldots, \ell(r)$ there are at least two which are ≥ 2;

(c2) $k(1) = \ldots = k(r) = \ell(1) = \ldots = \ell(r) = 1$;

(c3) $k(1) \geq 2$, $k(2) = \ldots = k(r) = \ell(1) = \ldots = \ell(r) = 1$ and $r = \ell \geq 2$;

(c4) $k(1) \geq 2$, $\ell(1) = \ell = r = 1$.

It is clear that the four cases exhaust all the possibilities in (2.6), up to the permutation of $k(1), \ldots, k(r)$. We claim that

$$|\Pi_1^r \chi \ldots (t)| \leq q_1(t) \quad \text{in (c1)}, \quad (2.7)$$

$$\Pi_1^r \chi \ldots (t) = (r(t))^\ell \quad \text{in (c2)}, \quad (2.8)$$

$$|\Pi_1^r \chi \ldots (t)| \leq \varepsilon |r(t)|^\ell + q_1(t) \quad \text{in (c3)}, \quad (2.9)$$

$$|\Pi_1^r \chi \ldots (t)| \leq p_1(t) \quad \text{in (c4)}. \quad (2.10)$$

Here, $\sum p_1^2(t) < \infty$, $\sum q_1(t) < \infty$ and $\varepsilon > 0$ is arbitrary small.

In fact, (2.8) is obvious, while (2.7) and (2.10) follow immediately from (2.1). In (c3), from (2.1) it follows that

$$|\Pi_1^r \chi \ldots (t)| \leq |r(t)|^{r-1} p(t)$$

$$\leq \varepsilon |r(t)|^{2(r-1)} + (1/\varepsilon) p^2(t).$$

As $r = l$ and $2(l-1) \geq l$ for $l \geq 2$, this implies (2.9).

Let $k_* \geq 2$. According to (2.6) and (2.7) - (2.9), $r_f(t)$ can be represented as

$$r_f(t) = c_{k_*}^2 k_*! (r(t))^{k_*} + \sum_{k_*+1}^{K} c_n^2 n! (r(t))^n$$

$$+ \theta_\varepsilon(t) + q_2(t), \tag{2.11}$$

where $\sum q_2(t) < \infty$, $|\theta_\varepsilon(t)| \leq \varepsilon |r(t)|^{k_*}$ and ε is arbitrary small (e.g. less than $c_{k_*}^2 k_*! / 4$). As $r(t) \to 0$ $(|t| \to \infty)$, so $|r(t)|^k \leq \varepsilon |r(t)|^{k_*}$ if $k > k_*$ and t is sufficiently large, with ε arbitrary small. This proves Theorem 3 (i).

(ii). If $k_* = 1$, then (2.7) - (2.1o) imply again that

$$r_f(t) = r(t) + p_3(t) + q_3(t), \tag{2.12}$$

where $\sum p_3^2(t) < \infty$, $\sum |q_3(t)| < \infty$. Consequently, (2.3) implies

$$\sum r^2(t) < \infty \tag{2.13}$$

Denote $f_1(x) = f(x) - c_1 x = f(x) - c_0 A_0(x) - c_1 A_1(x)$. Then $k_*(f_1, A) \geq 2$ which gives by (2.13)

$$\sum |r(t)|^{k_*(f_1)} < \infty \tag{2.14}$$

According to Theorem 3(i), (2.14) implies $\sum |r_{f_1}(t)| < \infty$. From this it follows that

$$Var\left(\sum_1^N f_1(X_t)\right) = Var\left(S_N - c_1 \sum_1^N X_t\right) \leq CN. \tag{2.15}$$

Similarly, from (2.3) it follows that $\mathrm{Var}\ S_N \leq CN$;
i.e. one has (2.4) as $c_1 \neq 0$.

(iii). Again, $\sum |r(t)| < \infty$ implies $\mathrm{Var}\ (\sum_1^N X_t) \leq CN$ and (2.14), where $f_1(x)$ is the same as above. As $k_*(f_1) \geq 2$, by Theorem 3 (i) we have again that $\sum |r_{f_1}(t)| < \infty$ and consequently (2.15), which yields (2.5).

3. Appell polynomials

According to Sheffer [21] (see also Bourbaki [6]), a set of polynomials $A_n(x)$, $n = 0, 1, \ldots$ ($A_n(x)$ being of degree n) is called an Appell set if it satisfies one of the two equivalent conditions:

$$A_n'(x) = n A_{n-1}(x) , \quad n = 1, 2, \ldots \tag{3.1}$$

or

$$\sum_0^\infty z^n A_n(x)/n! = e^{zx} A(z) , \tag{3.2}$$

where the last equality is the equality of formal power series (in z). Examples of Appell polynomials are the polynomials of Hermite, Bernoulli, Euler, and many others. Appell polynomials were studied in connection with differential equations, expansions of analytic functions [3] , operator calculus [19]. Below we shall consider a special class of Appell polynomials with generating function

$$\sum_0^\infty z^n A_n(x)/n! = e^{zx} \left(\int e^{zx} d\mu \right)^{-1} , \tag{3.3}$$

where μ is a probability measure on the real line $\mathbb{R}$ and $\int \ldots = \int_{\mathbb{R}} \ldots$. Sometimes we shall write $A_n(x) = A_n^{(\mu)}(x)$, in order to indicate the connection with the measure μ. In the physical literature such polynomials are usually called <u>Wick polynomials</u> or <u>Wick powers</u> [23] , or <u>generalized powers</u> [8] .However, we prefer to call them Appell polynomials and in the sequel this term will refer to the polynomials with the generating function (3.3) only.

Let us present some useful formulae involving Appell polynomials:

$$A_n(x) = \sum_0^n \binom{n}{k} (m^{-1})_k \, x^{n-k} , \tag{3.4}$$

$$x^n = \sum_0^n \binom{n}{k} m_k \, A_{n-k}(x) , \tag{3.5}$$

$$A_{n+1}(x) = x \, A_n(x) - \sum_0^n \binom{n}{k} \varkappa_{n-k+1} A_k(x) , \tag{3.6}$$

$$\int A_n(x) d\mu = \delta_{0n} . \tag{3.7}$$

Here δ_{0n} is Kronecker's symbol, $\varkappa_k$ is the k-th semi-invariant and $m_k = \int x^k d\mu$ is the k-th moment of μ, respectively;

$$(m^{-1})_n = \sum (-1)^r \varkappa_{k_1} \cdots \varkappa_{k_r} \, b(k_1, \ldots, k_r)$$

where the sum is taken over all $r \geq 1$ and all integers $k_1 \geq 1, \ldots, k_r \geq 1$ such that $k_1 + \ldots + k_r = n$, and $b(k_1, \ldots, k_r) = \binom{n}{k_1, \ldots, k_r} \frac{1}{r!}$.

The notation $(m^{-1})_n$ is suggested by the relation

$$\sum_0^n \binom{n}{k} m_k (m^{-1})_{n-k} = \delta_{0n} .$$

Formulae (3.4) - (3.7) are well-known [8, 19, 27]. In particular, (3.7) follows by integration of both sides of (3.3) with respect to μ, which yields $\sum_0^\infty z^n \cdot \int A_n \, d\mu / n! = 1$. The relations (3.4) - (3.4) easily generalize to the multivariate case, see Sect. 4.

Let us present some lower Appell polynomials:

$$A_0(x) = 1,$$
$$A_1(x) = x - m_1,$$
$$A_2(x) = x^2 - 2m_1 x + 2m_1^2 - m_2,$$
$$A_3(x) = x^3 - 3m_1 x^2 + 3x(2m_1^2 - m_2) + 6m_1 m_2 - m_3 - 6m_1^3.$$

Among Appell polynomials, only Hermite ones are orthogonal [22]. In particular, the relations $m_1 = 0$, $m_2 = 1$ and $\int x A_n^{(M)}(x)\, d\mu = 0 \quad \forall n \geq 2$ together with (3.6), (3.7) imply that $\varkappa_{n+1} = 0 \quad \forall n \geqslant 2$, i.e. μ is Gaussian measure and $A_n^{(M)}(x)$ coincide with the Hermite polynomials $H_n(x)$.

When dealing with Appell polynomials over sum of independent random variables, the so-called <u>binomial</u> (or <u>multinomial</u>) formulae are useful [1, 8], but we shall not need them below.

4. Multivariate Appell polynomials

There are at least two reasons to introduce multivariate Appell polynomials in this paper. The first one is that it provides a possibility to extend the above mentioned results to vector-valued processes X_t, or to functionals of the form $f(X_t, X_{t+1}, \dots, X_{t+k})$ of a scalar-valued process X_t. The second reason is that the diagram formula for semi-invariants (which is the main object of the next section) is more easily proved and even written down in the multivariate case.

Multivariate Appell polynomials

$$A_{n_1,\dots,n_k}(x_1,\dots,x_k) = A^{(\mu)}_{n_1,\dots,n_k}(x_1,\dots,x_k)\,, \tag{4.1}$$

$n_1, \dots, n_k = 0, 1, \dots$ can be defined again by the relations

$$\partial A_{n_1,\dots,n_i,\dots,n_k} / \partial x_i = n_i A_{n_1,\dots,n_i-1,\dots,n_k}\,, \tag{4.2}$$

$$\int A_{n_1,\dots,n_k}(x_1,\dots,x_k)\, d\mu = \delta_{0,n_1+\dots+n_k} \tag{4.3}$$

or (equivalently) by formal power series

$$\sum z_1^{n_1} \dots z_k^{n_k} A_{n_1,\dots,n_k}(x_1,\dots,x_k)/n_1! \dots n_k!$$

$$= \exp\left(\textstyle\sum_1^k z_j x_j\right)\left(\int \exp\left(\textstyle\sum_1^k z_j x_j\right) d\mu\right)^{-1},$$

where μ is a probability measure in $\mathbb{R}^k$ and $\int = \int_{\mathbb{R}^k}$. if $\mu = \nu_1 \times \dots \times \nu_k$ is a product measure, the polynomials $A^{(\mu)}_{n_1,\dots,n_k}$ are (tensor) products of univariate Appell polynomials $A^{(\nu_1)}_{n_1}, \dots, A^{(\nu_k)}_{n_k}$. In the multivariate case the formulae (3.4) - (3.7) are valid again, with appropriate modifications.

It is more convenient, however, to consider instead of polynomials (4.1) of degree n_i in variable x_i, $i=1,\dots,k$, polynomials in $n_1+\dots+n_k=n$ real variables which are linear in each of them. Following the tradition in mathematical physics, we shall denote such polynomials by

$$:y_1\cdots y_n: \;=\; :y_1\cdots y_n:^{(\nu)}. \tag{4.4}$$

By the definition,

$$:y_1\cdots y_n: \;=\; \partial^n\Big[\exp\big(\textstyle\sum_1^n z_j x_j\big)\times \tag{4.5}$$

$$\times\Big(\int\exp\big(\textstyle\sum_1^n z_j x_j\big)d\nu\Big)^{-1}\Big]\Big/\partial z_1\cdots\partial z_n\Big|_{z_1=\dots=z_n=0},$$

where ν is a probability measure in $\mathbb{R}^n$. The relations between (4.4) and (4.1) are

$$:y_1\cdots y_n:^{(\nu)} \;=\; A^{(\nu)}_{1,\dots,1}(y_1,\dots,y_n) \tag{4.6}$$

and

$$A^{(\mu)}_{n_1,\dots,n_k}(x_1,\dots,x_k) \;=\; :x_1^{n_1}\dots x_k^{n_k}:^{(\nu)}; \tag{4.7}$$

in the last formula ν is concentrated on the k-dimensional hyperplane $y_1=\dots=y_{n_1},\dots,y_{n_1+\dots+n_{k-1}+1}=\dots=y_n$ and coincides there with μ. In particular, if ν is concentrated on the diagonal $y_1=\dots=y_n$ and its marginal distribution is μ, then

$$:\underbrace{x\ \dots\ x}_{n}:^{(\nu)} \;=\; :x^n: \;=\; A^{(\mu)}_n(x).$$

We shall move one step further in our discussion, by considering polynomial forms in random variables instead of polynomials in real variables. Namely, any polynomial (4.4) can be considered as a random variable on the probability space $(\mathbb{R}^n, \nu)$. Conversely, for any random variables $Y_1, \dots, Y_n$ on some probability space (Ω, P) one can define a new random variable $:Y_1 \cdots Y_n:$ as

$$:Y_1 \dots Y_n: \;=\; :y_1 \dots y_n:^{(\nu)}\Big|_{y_1=Y_1, \dots, y_n=Y_n}, \tag{4.8}$$

where $\nu(dx) = P((Y_1, \dots, Y_n) \in dx)$. The advantage of (4.8) is that we do not need any more to indicate the distribution ν on which (4.4) depend. Note that $:Y_1 \dots Y_n:$ is invariant under the permutations of $Y_1, \dots, Y_n$. We shall call $:Y_1 \cdots Y:$ the <u>Appell product</u> of random variables $Y_1, \dots, Y_n$ (in the literature, the term <u>Wick product</u> is more common). We have

$$:Y_1: \;=\; Y_1,$$

$$:Y_1 Y_2: \;=\; Y_1 Y_2 - Y_1 EY_2 - Y_2 EY_1 + 2EY_1 EY_2 - E(Y_1 Y_2),$$

$$\begin{aligned} :Y_1 Y_2 Y_3: \;=\; & Y_1 Y_2 Y_3 - Y_1 Y_2 EY_3 - Y_1 Y_3 EY_2 - Y_2 Y_3 EY_1 \\ & + 2(EY_1 EY_2 Y_3 + EY_2 EY_1 Y_3 + EY_3 EY_1 Y_2) + 2(Y_1 EY_2 EY_3 + Y_2 EY_1 EY_3 \\ & + Y_3 EY_1 EY_2) - Y_1 EY_2 Y_3 - Y_2 EY_1 Y_3 - Y_3 EY_1 Y_2 - 6EY_1 EY_2 EY_3 - EY_1 Y_2 Y_3 \end{aligned}$$

and

$$:\underbrace{Y \dots Y}_{n}: \;\equiv\; :Y^n: \;=\; A_n^{(\mu)}(Y),$$

where $\mu(dx) = P(Y \in dx)$. If W is a finite set, and Y_i, $i \in W$ is a system of random variables, we shall demote their product by

$$Y^W = \prod_{i \in W} Y_i \ , \tag{4.9}$$

their Appell product by

$$:Y^W: \tag{4.10}$$

and their semiinvariant by

$$\chi(Y^W) = \chi(Y_i \ , i \in W) \ . \tag{4.11}$$

Of course, $:Y^W:$ is not the same as $:(Y^W):$ $= Y^W - E(Y^W)$, as well as $\chi(Y^W)$ is not the first semi-invariant, or the expectation, of the random variable Y^W. Therefore the notation (4.9) - (4.10) requires some care. With the help of it, the relations (3.4) - (3.5) can be generalized as follows ([27], Prop. 1):

$$:Y^W: = \sum_{U \subseteq W} Y^U \sum_{\{V\}} (-1)^r \chi(Y^{V_1}) \cdots \chi(Y^{V_r}) \ , \tag{4.12}$$

$$\begin{aligned} Y^W &= \sum_{U \subseteq W} :Y^U: \sum_{\{V\}} \chi(Y^{V_1}) \cdots \chi(Y^{V_r}) \\ &= \sum_{U \subseteq W} :Y^U: E(Y^{W \setminus U}) \ , \end{aligned} \tag{4.13}$$

where the sum $\sum_{U \subseteq W}$ is taken over all subsets $U \subseteq W$ including $U = \emptyset$ and the sum $\sum_{\{V\}}$ is taken over all partitions $\{V\} = \{V_1, \ldots V_r\}$, $r \geq 1$ of $W \setminus U$; $Y^\emptyset = :Y^\emptyset: = \chi(Y^\emptyset) = 1$ by definition. (4.12) can be most easily proved by observing that the right-hand side of it satisfies the recursive relation (the analog of (3.6)), namely

$$:Y^{W}: = Y_i :Y^{W\setminus\{i\}}: - \sum_{V\ni i} \chi(Y^V) :Y^{W\setminus V}: , \qquad (4.14)$$

where $i \in W$ is any element and the sum $\sum_{V\ni i}$ is taken over all $V \subseteq W$ which contain it. In turn, (4.14) follows easily from the definitions (4.8), (4.5) [27]. Let us remark, too, that the formulae (4.12) - (4.14) become somewhat simpler in the Gaussian case (i.e. when the system Y_i, $i \in W$ is Gaussian), due to the fact that $\chi(Y^V) = 0$ for all $V \subseteq W$ which contain more than two elements.

In the physical literature, Appell (or Wick) products are sometimes called the 'renormalized products', the 'renormalization' consisting in the substraction from $Y_1, \dots, Y_n$ all 'internal interactions' [11]. In the probabilistic language one should speak rather about the 'substraction of all internal correlations betweed $Y_1, \dots Y_n$.' This intuition is confirmed by the comparison of the semi-invariants $\chi(Y^{W_1}, \dots, Y^{W_m})$ and $\chi(:Y^{W_1}:, \dots, :Y^{W_m}:)$ (which are some measure of dependence between the random variables). Namely, the second semi-invariant is as a rule much smaller than the first, due to the fact that only 'interactions' (or correlations) between different groups $(Y_j, j \in W_i)$, $i = 1, \dots, m$ contribute to its value.

Let us note, finally, that an interesting attempt to introduce 'continuous time Wick product' was made in [18].

5. Diagrams and semi-invariants of Appell polynomials

Diagrams are widely used by physicists. Roughly speaking, they help to reduce Gaussian functional integrals (i.e., expectations with respect to a Gaussian process) to a sum of finite dimensional integrals [11]. Recently also probabilists use diagrams, e.g. in order to obtain exact estimates of semi-invariants in the limit theorems for processes with long-range dependence [5, 10].

A <u>diagram</u> is a partition $\{V\} = \{V_1, \dots, V_r\}$ of a (finite) set W by non-empty subsets $V_1, \dots, V_r$, $r = 1, 2, \dots$ which are called the <u>edges</u> of the diagram $\{V\}$. The number of elements of a finite set V will be denoted by $|V|$.

Assume that W is a union of disjoint subsets $W_1, \dots, W_k$. It is convenient to imagine W as a table which rows are $W_1, \dots, W_k$. An edge V of a diagram $\{V\}$ is called <u>free</u> if it consists of a single point (i.e. $|V| = 1$), and <u>flat</u>, if V belongs to some W_i, $i = 1, \dots, k$. Sometimes flat edges are also called <u>loops</u> [17, 11].

A diagram $\{V\} = \{V_1, \dots, V_r\}$ will be called <u>Gaussian</u> if $|V_1| = \dots = |V_r| = 2$, and <u>connected</u> if the rows $W_1, \dots, W_r$ of the table W cannot be divided into two groups, each of which is partitioned by the diagram separately. For example, the diagram in Figure 1(a) is connected and in Figure 1(b) is not connected.

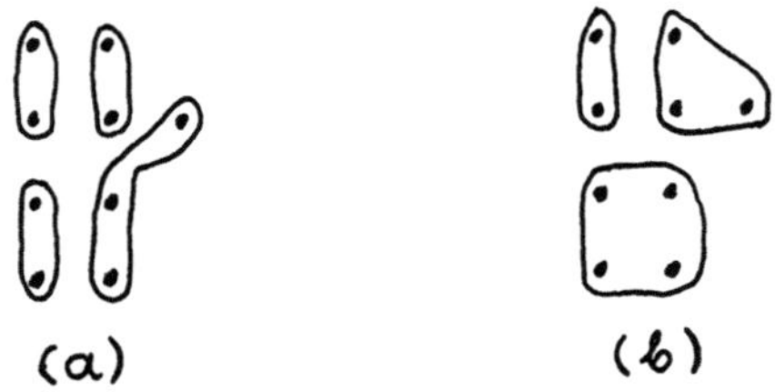

Figure 1.

Let be given a system Y_i , $i \in W$ of random variables. For each diagram $\{V\} = \{V_1, \ldots, V_r\}$ we define its 'contribution'

$$I_{\{V\}} = \Pi_1^r \, \chi(Y^{V_i}) .$$

<u>Theorem 4</u>. Each of the quantities

(i) $E(Y^{W_1} \cdots Y^{W_k}) \equiv E(Y^W)$,

(ii) $E(:Y^{W_1}: \ldots :Y^{W_k}:)$,

(iii) $\chi(Y^{W_1}, \ldots, Y^{W_k})$,

(iv) $\chi(:Y^{W_1}:, \ldots, :Y^{W_k}:)$

is equal to

$$\Sigma \, I_{\{V\}} \quad , \tag{5.1}$$

where the sum is taken over

(i) all diagrams,

(ii) all diagrams without flat edges,

(iii) all connected diagrams,

(iv) all connected diagrams without flat edges, respectively.

Remark. In the Gaussian case Theorem 4 is known as Wick theorem [23, 11, 17] and the diagrams in (i) - (iv) are Gaussian. In the case when $EY_i = 0 \quad \forall i \in W$ the diagrams in (i) - (iv) are without free edges.

Proof of Theorem 4. (i) and (iii) are well-known

(ii) and (iv) can be proved by induction, in a similar way. For example, let us prove (iv). It is convenient to use the induction to prove a more general statement which combines both (iii) and (iv). Namely, for any $0 \le s \le k$,

$$\chi(:Y^{W_1}:, \dots, :Y^{W_s}:, Y^{W_{s+1}}, \dots, Y^{W_k}) = \sum I_{\{V\}} \qquad (5.2)$$

where the sum is taken over all connected diagrams without flat edges in $W_1, \dots, W_s$.

Note that (5.2) holds for any $s \le k$ provided $|W_1| = \dots = |W_s| = 1$, as $:Y_i:$ and Y_i differ only by constant which has no effect on the semi-invariant. Therefore it suffices to prove the induction in the number of elements of W_s.

Suppose that (5.2) holds if either $s < u$ or $s = u$ and $|W_s| < w$. We shall prove it for $s = u$ and $|W_s| = w$. By (4.13),

$$:Y^{W_s}: = Y^{W_s} - \sum_{U \subset W_s} :Y^U: \sum_{\{G\}} I_{\{G\}}, \qquad (5.3)$$

where $|U| < |W_s| = w$ and the sum $\sum_{\{G\}}$ is taken over all diagrams $\{G\}$ over $W_s \setminus U$. Denote by χ the left hand side of (5.2). Substituting (5.3) into χ and using the inductive assumption, we obtain

$$\chi = \sum' I_{\{V\}} - \sum_{U \subset W_s} \sum'' I_{\{V\}}, \qquad (5.4)$$

.

where Σ' is the sum over all connected diagrams $\{V\}$ without flat edges in $W_1, \ldots, W_{s-1}$, and Σ'' is the sum over all connected diagrams $\{V\}$ without flat edges in $W_1, \ldots, W_{s-1}$ and U, such that $W_s \setminus U$ is a union of (flat) edges of $\{V\}$. Clearly (5.4) implies (5.2).

If $k = 1$, all diagrams consist of flat edges. This implies

Corollary 1. $E :Y^W: = 0 \quad (W \neq \emptyset)$.

6. Expansions in Appell polynomials

The notion of the Appell rank is closely related to the problem of Appell expansions in the space $L^2(\mu)$ which seems to be difficult and little studied. In the literature we have found only results concerning Appell expansions of analytic functions [3, 13]. As a consequence, the definition of the Appell rank given below is restricted to analytic functions and is not very satisfactory.

Let $A_n(x) = A_n^{(\mu)}(x)$, $n \geqslant 0$ be the Appell set of polynomials, corresponding to a probability distribution μ on $\mathbb{R}$. Let $L^2(\mu)$ denote the Hilbert space of all complex functions $f(x)$ such that $\int |f|^2 d\mu < \infty$. We shall denote by $n(\mu)$ the <u>dimension</u> of $L^2(\mu)$, which is the maximal number of linearly independent vectors. Here, $n(\mu) < \infty$ if and only if μ is purely atomic with $n(\mu)$ atoms. As $A_n(x)$ is polynomial of degree n exactly, it follows that $A_n(x)$, $0 \leqslant n < n(\mu)$ are linearly independent in $L^2(\mu)$ [20].

Let us consider first Appell expansions of polynomials. Any polynomial $f(x)$, $x \in \mathbb{R}$ of degree $n < \infty$ can be represented uniquely as

$$f(x) = \sum_{k \leq n} c_k A_k(x)/k! , \tag{6.1}$$

where the equality (6.1) holds for each $x \in \mathbb{R}$. As a consequence, (6.1) holds μ - a.e. in $\mathbb{R}$ and in the $L^2(\mu)$-sense as well but it is not a unique representation in $L^2(\mu)$ if $n \geqslant n(\mu)$. If, however, $n < n(\mu)$, then the representation (6.1) is unique in $L^2(\mu)$ among all elements of $L^2(\mu)$ which are finite linear combinations of A_j, $0 \leq j < n(\mu)$. (This fact is a consequence of the linear independence of A_n, $n < n(\mu)$ in $L^2(\mu)$.) These considerations lead to

<u>Definition 1</u>. Let $f(x)$ be a polynomial of degree $n < n(\mu)$ having the representation (6.1). Then its Appell rank $k_*(f, A)$ is $\min(k \geq 0 : c_k \neq 0)$.

Of course it would be much better if the pepresentation (6.1) of polynomials were unique in the class of all elements of $L^2(\mu)$ given by infinite Appell series convergent in $L^2(\mu)$, but that might be not true in general. Note that the condition $n \geq n(\mu)$ implies $n(\mu) < \infty$, which means that $L^2(\mu)$ is finite dimensional and any element $f \in L^2(\mu)$ can be written uniquely as a linear combination of $A_0, \dots, A_{n(\mu)-1}$.

Let us consider now Appell expansions of analytic function Denote by $E[1, \tau]$ the class of all entire functions of the growth order ≤ 1 and the normal type $\tau \in (0, \infty)$. In other words, $f \in E[1, \tau]$ if it satisfies the inequality

$$\max\{|f(z)| : |z| = r\} \leq \exp(Kr)$$

for any $K > \tau$ and all $r \geq r_0(K) < \infty$. Set also $E[1, \tau) = \cup_{\tau' < \tau} E[1, \tau']$.

<u>Theorem 5</u> (Kazmin [13]). Let $A(z) = (\int e^{zx} d\mu)^{-1}$ be analytic in $\{|z| < \sigma\}$ and do not vanish in $\{|z| < \tau\}$, where $\tau < \sigma$. Then

(i) Any function $f \in E[1, \tau)$ can be expanded as

$$f(z) = \sum_0^\infty c_n A_n(z) / n! \qquad (6.2)$$

where

$$\limsup_{n\to\infty} |c_n|^{1/n} < \tau \tag{6.3}$$

and the series (6.2) converge uniformly on each compact set of the complex plane. Moreover, the expansion (6.2) is unique provided the coefficients satisfy (6.3).

(ii) If $c_n \in \mathbb{C}$ satisfy (6.3), then the series (6.2) converge uniformly on compact sets and their sum $f \in E[1,\tau)$.

In order to consider the uniqueness in $L^2(\mu)$ of the expansions (6.2), we shall introduce

Condition (A). For any entire function $f(z)$ the relation $f(x)=0$ μ-a.e. in $\mathbb{R}$ implies $f(z) \equiv 0$.

From Theorem 5 and the relations (5.1), (3.7) it follows.

<u>Corollary 2</u>. Assume that condition (A) holds as well as the conditions of Theorem 5, with τ replaced by 2τ. Then, for each $f \in E[1,\tau)$ the series (6.2) converge in $L^2(\mu)$ and

$$c_n = \int_{\mathbb{R}} f^{(n)}(x)\, d\mu, \quad n=0,1,\dots . \tag{6.4}$$

The expansion (6.2) is unique in the class of all $L^2(\mu)$-convergent Appell expansions, with the coefficients satisfying (6.3).

Motivated by this Corollary, we introduce

<u>Definition 2</u>. Let μ satisfy the conditions of Corollary 2. The Appell rank of $f \in E[1,\tau)$ is $\min\{k \geq 0 : c_k \neq 0)$, where c_n are given by (6.4).

An interesting open problem is to find out under which conditions on μ the system $\{A_n\}_0^\infty$ is a <u>basis</u> in $L^2(\mu)$ (not necessarily orthogonal), which means that <u>any</u> function $f \in L^2(\mu)$ can be expanded uniquely in the Appell series (1.11), convergent in $L^2(\mu)$. Any basis in a Banach space is a <u>minimal system</u> [14]. A necessary and sufficient condition for a system $\{f_n\}_0^\infty \subset L^2(\mu)$ to be minimal in $L^2(\mu)$ is the existence of a <u>biorthogonal system</u> $\{g_n\}_0^\infty \subset L^2(\mu)$, i.e. such that $\int f_n \bar{g}_m \, d\mu = \delta_{nm}$, $n,m = 0,1,\ldots$ [14].

<u>Theorem 6</u>. Assume that for some $r > 0$,

$$\int e^{r|x|} \, d\mu < \infty . \qquad (6.5)$$

Then the system of Appell polynomials $\{A_n\}_0^\infty$ is minimal in $L^2(\mu)$ if and only if there exists the density $p(x) = d\mu/dx \in C^\infty(\mathbb{R})$ such that

$$Q_n \equiv p^{-1} d^n p / dx^n \in L^2(\mu), \quad n = 0,1,\ldots . \qquad (6.6)$$

In the latter case, the system $\{Q_n\}_0^\infty$ is biorthogonal with $\{A_n\}_0^\infty$.

Below we give a short proof of the necessity of conditions (6.6) for the existence of the biorthogonal system $\{Q_n\}_0^\infty$ which is the more interesting part of Theorem 6.

Let $\{Q_n\}_0^\infty$ be biorthogonal with $\{A_n\}_0^\infty$. By the Cauchy formula and (6.5), one can check that the generating series (3.2) converge in $L^2(\mu)$ for all $z \in \mathbb{C}$, $|z| < r'$ and some $r' > 0$. Therefore, multiplying both sides of (3.2) by Q_n and integrating with respect to $d\mu$, one has

$$\int e^{zx} Q_n(x)\, d\mu = z^n \varphi(z), \tag{6.7}$$

where

$$\varphi(z) = \int e^{zx}\, d\mu \quad . \tag{6.8}$$

If $r' < r/2$, then $e^{zx} Q_n \in L^1(\mu)$ by (6.5). Therefore both sides of (6.7)can be extended analytically to $\{z \in \mathbb{C} : |\mathrm{Re}\, z| < r''\}$ for some $r'' > 0$ ([16], Th.7.1.1.). In particular,

$$\int e^{itx} Q_n(x)\, d\mu = (it)^n \varphi(it) \quad \forall t \in \mathbb{R} . \tag{6.9}$$

Note that the left hand side of (6.9) is bounded in t for any $n \geq 0$. In the case $n = 1$ this yields $|\varphi(it)| \leq \mathrm{const}/|t|$ and consequently the existence of the density $p = d\mu/dx \in L^2(\mathbb{R}, dx)$. Similarly it follows from (6.9) that p belongs to the Sobolev spaces

$$W^{n,2} = \{f \in L^2(\mathbb{R}, dx) : \int |\hat{f}|^2 (1 + |t|^{2n})\, dt < \infty\}, \ n = 0, 1, \ldots$$

and its n-th Sobolev derivative coincides with $Q_n(x)p(x)$. By the Sobolev embedding theorem ([30],Ch. VI.6), $p \in C^\infty(\mathbb{R})$ and $p^{(n)}(x) = Q_n(x)\, p(x)$. As $Q_n \in L^2(\mu)$, this implies (6.6).

7. Proof of Theorem 1

According to the conditions of this theorem it suffices to prove that for any $k \geqslant 3$,

$$\chi_k(S_N) = o(N^{k/2}), \tag{7.1}$$

where

$$S_N = \sum_{n \geq k_*} c_n \sum_1^N A_n(X_t) \equiv \sum_{n \geq k_*} c_n S_N^{(n)}.$$

By Theorem 4 (iv),

$$\chi(S_N^{(n(1))}, \ldots, S_N^{(n(k))})$$

$$= \sum_{\{V\}} \sum_{t_1, \ldots, t_k = 1}^{N} \chi(X^{V(1)}) \ldots \chi(X^{V(r)})$$

$$\equiv \sum_{\{V\}} R_{\{V\}}(\chi). \tag{7.2}$$

Here $\sum_{\{V\}}$ is the sum over all connected diagrams $\{V\} = \{V(1), \ldots, V(r)\}, r \geq 1$ without flat edges, over the table

$$T = \begin{pmatrix} t_1 & \ldots & t_1 \\ t_2 & \ldots & t_2 \\ \ldots & & \\ t_k & \ldots & t_k \end{pmatrix} \quad \begin{matrix} \text{(n(1) elements)} \\ \text{(n(2) elements)} \\ \ldots \\ \text{(n(}k\text{) elements).} \end{matrix} \tag{7.3}$$

This table consists of k rows $T(1), \ldots, T(k)$ with $|T(1)| = n(1), \ldots, |T(k)| = n(k)$ elements in each. To each elemen $s \in T$ there corresponds random variable X_s, and $\chi(X^V)$ $(V \subseteq T)$ is the semi-invariant of X_s, $s \in V$. (See also Sect. 5 for terminology and notation.) We shall prove that for any $n(1) \ge k_*, \ldots, n(k) \ge k_*$ and any diagram $\{V\}$ in (7.2),

$$R_{\{V\}}(\chi) = o(N^{k/2}) \qquad (k \ge 3). \tag{7.4}$$

The proof is based on induction in k and divided into several parts (lemmas). In order to use effectively the inductive assumption and the conditions (q1) - (q4), we shall introduce a slightly more general set of conditions of this type, which are formulated below.

Let us denote by $\mathcal{V}_1(i)$ (resp., $\mathcal{V}_2(i,j)$) the class of all subsets $V \subset T$ such that $V \subset T(i)$ (resp., $V \subset T(i) \cup T(j)$), $i,j = 1, \ldots, k, i \ne j$. Moreover, set

$$\mathcal{V}_1 = \bigcup_{i=1,\ldots,k} \mathcal{V}_1(i),$$

$$\mathcal{V}_2 = \bigcup_{i,j=1,\ldots,k,\, i\ne j} \mathcal{V}_2(i,j),$$

$$\mathcal{V}_3 = \{ V \subseteq T : V \notin \mathcal{V}_1 \cup \mathcal{V}_2 \}.$$

For any non-empty subset $V \subseteq T$, let be given a function $q_V(t_1, \ldots, t_k) \ge 0, t_1, \ldots, t_k = 1, \ldots, N$ depending on the variables $t_i, i \in V$ only, uniformly bounded in $t_i, i \in V$ and N. Set also

$$l(i) = l(i; V) = |V \cap T(i)|,$$

$V \subseteq T, i = 1, \ldots, k$. We shall assume that the functions q_V, $V \subseteq T$ satisfy the following conditions.

(Q1) If $V \in \mathcal{V}_1(i)$ $(i = 1, \ldots, k)$ and $p = n(i)/l(i)$, then

$$\Big(\sum_{t_i=1}^{N} q_V^p \Big)^{1/p} \le C_V N^{1/2p}. \tag{7.5}$$

(Q2) If $V \in \mathcal{V}_2(i,j)$, $i,j = 1,\ldots,k$, $i \neq j$, $K \geq 0$, then $q_V(t_i,t_j) = q_V(t_j,t_i)$ and

$$(\sum_{t_i=1,\ldots,N,\ |t_i-t_j|\geq K} q_V^p(t_i,t_j))^{1/p}$$

$$\leq C_V(K)\, q_{V\cap T(j)}(t_j)\, N^{\gamma} \qquad (7.6)$$

for all $\ell(i), \ell(j), p, \gamma$ which satisfy one of the following conditions:

(Q2a) $p \geq 1$, $\ell(i) \geq 2$, $\ell(j) \geq 2$, $\gamma = \ell(j)/2n(j)$;

(Q2b) $p \geq 2$, $\ell(i) \geq 2$, $\ell(j) = 1$, $\gamma = 1/2n(j)$;

(Q2c) $p \geq k_*$, $\ell(i) = \ell(j) = 1$, $\gamma = 1/2n(j)$;

(Q2d) $p = 1$, $\ell(i) \geq k_*$, $\ell(j) = 1$, $\gamma = 1/2$.

(Q3) If $V \in \mathcal{V}_3$, then

$$(\sum_{t_i=1}^{N} q_V^p)^{1/p} \leq C_V\, q_{V\setminus T(i)} \qquad (7.7)$$

for any $i = 1,\ldots,k$ such that one of the two conditions hold:

(Q3a) $p \geq 2$, $\ell(i) \geq 1$;

(Q3b) $p \geq 1$, $\ell(i) \geq 2$.

The constants C_V, $C_V(K)$ in (7.5) - (7.7) do not depend on N and $t_1,\ldots,t_k$. Moreover, for each $V \in \mathcal{V}_2$,

$$C_V(K) \to 0 \quad (K \to \infty).$$

Below $C_V(0)$ will denoted by c_V

We leave to the reader the verification of the fact that the functions $q_{\ell(1),\ldots,\ell(k)}(t_1,\ldots,t_k) \equiv q_V(t_1,\ldots,t_k)$, $V \in \mathcal{V}_2 \cup \mathcal{V}_3$ in (q1) - (q4) satisfy the conditions (Q1) - (Q3), with appropriate q_V, $V \in \mathcal{V}_1$.

The induction step $k-1 \longrightarrow k$ in the proof of (7.4) is based on Lemma 1 below. To formulate it, we introduce the following notation.

Let $\{V\} = \{V(1), \ldots, V(r)\}$ be a diagram over the table T (7.3). Set

$$T' = T \setminus T(1)$$

and

$$V'(i) = V(i) \cap T'.$$

We shall assume that $V'(1), \ldots, V'(r')$, $r' \leq r$ are not empty, while $V(r'+1), \ldots, V(r)$ belong to $T(1)$. Then

$$\{V'\} = \{V'(1), \ldots, V'(r')\}$$

is a diagram over the table T' which has $k-1$ rows. Similarly as in (7.2), denote

$$R_{\{V\}} \equiv R_{\{V\}}(q) = \sum_{t_1,\ldots,t_k=1}^{N} q_{V(1)} \cdots q_{V(r)} \tag{7.8}$$

and

$$R_{\{V'\}} \equiv \sum_{t_2,\ldots,t_k=1}^{N} q_{V'(1)} \cdots q_{V'(r')} .$$

<u>Lemma 1</u>. Assume that q_V, $V \subseteq T$ satisfy the conditions (Q1) - (Q3). Then

$$R_{\{V\}} \leq C R_{\{V'\}} N^{\gamma(1)}, \tag{7.9}$$

where C depends only on the constant $C_{V(i)}, V(i) \cap T(1) \neq \emptyset$ in (7.5) - (7.6) and vanishes if one of them vanishes. The nuhber $\gamma(1) \geq 0$ satisfies one of the two conditions:

(Γ_1) $\gamma(1) \leq 1/2$, and the edges $V(i), i = 1, \ldots, r$, $V(i) \cap T(1) \neq \emptyset$ belong to $\mathcal{V}_1 \cup \mathcal{V}_2$;

(Γ_2) $\gamma(1) \leq 1/2 \sum_{V(i) \in \mathcal{V}_1(1)} |V(i)|/n(1) +$

$$+ 1/2 \sum_{j=2}^{k} \sum_{V(i) \in \mathcal{V}_2(1,j)} |V(i) \cap T(j)|/n(j).$$

Proof. We shall assume that the first s $(\leq r')$ edges $V(1), \ldots, V(s)$ intersect with $T(1)$ (but do not belong to it), the edges $V(s+1), \ldots, V(r')$ belong to $T' = T \setminus T(1)$ and $V(r'+1), \ldots, V(r)$ belong to $T(1)$.

Recall that $V \in \mathcal{V}_2$ is called Gaussian if $|V| = 2$. A set $V \in \mathcal{V}_2$ will be called a triangle if it is not Gaussian and its intersection with some row $T(i)$, $i = 1, \ldots, k$ consists of a single element. As it follows from (Q2b) - (Q2d), Gaussian edges and 'large' triangles $V \in \mathcal{V}_2(1,j)$ with $|V \cap T(1)| > n(1)/2$ will play a particular role in the proof. We shall consider different cases, the main among which are:

Case 1. Among $V(1), \ldots, V(s)$ there is a triangle (say, $V(1)$), such that $|V(1) \cap T(1)| > n(1)/2$.

Case 2 is the alternative of Case 1.

We shall start from Case 2 as it is more simple. By Hölder's inequality

$$|\sum h_1 \cdots h_r| \le \prod_1^r (\sum |h_i|^{p(i)})^{1/p(i)},$$

$1/p(1) + \ldots + 1/p(r) = 1$, we have

$$\sum_{t_1} q_{V(1)} \cdots q_{V(r)}$$

$$\le \prod_1^s q'_{V(i)} \prod_{s+1}^{r'} q_{V'(i)} \prod_{r'+1}^{r} q'_{V(i)}, \tag{7.10}$$

where

$$q'_{V(i)} = (\sum_{t_1} q_{V(i)}^{p(i)})^{1/p(i)}, \tag{7.11}$$

$$p(i) = n(1)/\ell(i,1),$$

$$\ell(i,j) = |V(i) \cap T(j)|.$$

By (Q1),

$$\prod_{r'+1}^{r} q'_{V(i)} \le CN^{\gamma'}, \tag{7.12}$$

where

$$\gamma' = 1/2 \sum_{r'+1}^{r} |V(i)|/n(1) . \tag{7.13}$$

If $V(i) \in \mathcal{V}_3$ $(1 \le i \le s)$, then either $\ell(i,1) \ge 2$ and (Q3a) applies with $p = p(i) \ge 1$, or $\ell(i,1) = 1$ and (Q3b) applies with $p = p(i) = n(1) \ge k_* \ge 2$. Therefore, in both cases,

$$q'_{V(i)} \le C q_{V'(i)} \tag{7.14}$$

If, on the other hand, $V(i) \in \mathcal{V}_2(1,j)$ $(1 \le i \le s,\ 2 \le j \le k)$ then either it is not a triangle and satisfies (Q2a) as $p(i) \ge 1$, or $V(i)$ is a triangle but then $p(i) \ge 2$

according to the definition of Case 2, and $V(i)$ satisfies (Q2b). Consequently,

$$q'_{V(i)} \leq C q_{V'(i)} N^{|V'(i)|/2n(j)} . \tag{7.15}$$

From (7.10) - (7.15) it follows the inequality (7.9) in Case 2, with $\gamma(1)$ satisfying (Γ_2) .

In Case 1, there are several subcases:

1a). Let $\ell(1,1) = |V(1) \cap T(1)| \geq k_*$ and $s = 1$ (i.e. the only edges among $V(2), \ldots, V(r)$ which intersect with $T(1)$ are flat and belong to $T(1)$). According to (Q2d),

$$\sum_{t_1} q_{V(1)} \cdots q_{V(r)} \leq C \prod_2^{r'} q_{V'(i)} \sum_{t_1} q_{V(1)}$$

$$\leq C \prod_1^{r'} q_{V'(i)} N^{1/2}$$

and (7.9) holds with $\gamma(1)$ satisfying (Γ_1) .

1b). Let $\ell(1,1) < k_*$ and $s = 1$. By (7.10) and (7.12),

$$\sum_{t_1} q_{V(1)} \cdots q_{V(r)} \leq C q'_{V(1)} \prod_2^{r'} q_{V'(i)} N^{\gamma'} .$$

Now,

$$q'_{V(1)} = \left(\sum_{t_1} q^p_{V(1)}\right)^{1/p}$$

$$\leq \left(\sum q^2_{V(1)}\right)^{1/2} \left(\sum_{t_1=1}^{N} 1\right)^{(2-p)/2p}$$

$$\leq C q_{V'(1)} N^{\gamma''} , \tag{7.16}$$

where $p = p(1) = n(1)/\ell(1,1) < 2$ and

$$\gamma'' = 1/2n(j) + (2-p)/2p$$

by (Q2b), provided $V(1) \in \mathcal{V}_2(1,j)$. From the relations $n(1) \geq k_*$, $n(j) \geq k_*$, $(2-p)/2p = (2\ell(1,1)-n(1))/2n(1)$, $\gamma' = (n(1)-\ell(1,1))/2n(1)$ and $\ell(1,1) \leq k_* - 1$ it follows easily that $\gamma' + \gamma'' \leq 1/2$. Therefore, (7.9) is true with $\gamma(1)$ satisfying (Γ_1).

1c) Assume that $\ell(1,1) \leq n(1)-1$, $s \geq 2$ and there exists an edge among $V(2), \ldots, V(s)$ (say, $V(2)$) such that $|V(2)| \geq 3$. Again, by Hölder's inequality,

$$\sum_{t_1} q_{V(1)} \cdots q_{V(r)}$$

$$\leq \tilde{q} \prod_3^s q'_{V(i)} \prod_{s+1}^{r'} q_{V'(i)} \prod_{r'+1}^{r} q'_{V(i)}, \qquad (7.17)$$

where r' and $q'_{V(i)}$, $i \geq 3$ are the same as in (7.10), and $\tilde{q} = \tilde{q}(t_2, \ldots, t_k)$ is equal to

$$\tilde{q} = (\sum_{t_1} (q_{V(1)} q_{V(2)})^p)^{1/p}$$

$$\leq (\sum_{t_1} q_{V(1)}^{2p})^{1/2p} (\sum_{t_1} q_{V(2)}^{2p})^{1/2p} \equiv \tilde{q}_1 \cdot \tilde{q}_2, \qquad (7.18)$$

where $p = n(1)/(\ell(1,1)+\ell(2,1)) \geq 1$. As $2p \geq 2$ and $|V_1| \geq 3$, $|V_2| \geq 3$, so

$$\tilde{q}_1 \leq C q_{V'(1)} N^{1/2n(j)} \qquad (7.19)$$

by (Q2b), provided $V(1) \in \mathcal{V}_2(1,j)$, and

$$\tilde{q}_2 \le \begin{cases} C\, q_{V'(2)} & \text{if } V(2) \in \mathcal{V}_3 , \\ C\, q_{V'(2)} N^{1/2n(j)} & \text{if } V(2) \in \mathcal{V}_2(1,j) \end{cases} \qquad (7.20)$$

$j = 2, \ldots, k$. From (7.13) and (7.17) - (7.20) we have again (7.9), with $\gamma(1)$ satisfying (Γ_2') .

1d). Let $\ell(1,1) \le n(1)-1$, $V(2), \ldots, V(s)$ be Gaussian and $(2 \le)\, s < k_*$. Then

$$\sum_{t_1} q_{V(1)} \cdots q_{V(r)} \le C \prod_1^s \tilde{q}_i \prod_{s+1}^{r'} q_{V'(i)} , \qquad (7.21)$$

where

$$\tilde{q}_i = \Big(\sum_{t_1} q_{V(i)}^{k_*} \Big)^{1/k_*}$$
$$\le C N^{1/2k_*} q_{V'(i)} , \quad i = 2, \ldots, s$$

according to (Q2c), while

$$\tilde{q}_1 = \Big(\sum_{t_1} q_{V(1)}^{p} \Big)^{1/p}$$

with $p = k_* / (k_* - s + 1)$. If $p \ge 2$, then $\tilde{q}_1 \le C q_{V(1)} N^{1/2k_*}$ by (Q2b) and (7.9) holds with $\gamma(1) = s/2k_* < 1/2$. One the other hand, if $p > 2$, then $\tilde{q}_1 \le C q_{V'(1)} N^{\gamma''}$ (see (7.16)), with $\gamma'' \le 1/2k_* + (2-p)/2p = 1/2k_* + (k_* - 2s + 2)/2k_*$. Again,

$$\gamma(1) = \gamma'' + (s-1)/2k_* \le (k_* - s + 2)/2k_* \le 1/2 .$$

The last case is

1e). Let $\ell(1,1) \le n(1) - 1$, $V(2), \ldots, V(s)$ be Gaussian and $s \ge k_*$. Then

$$\Sigma_{t_1} q_{V(1)} \cdots q_{V(r)} \le C \prod_1^s \tilde{q}_i \prod_{s+1}^{r'} q_{V'(i)} ,$$

where

$$\tilde{q}_i = (\Sigma_{t_1} q^s_{V(i)})^{1/s} , \quad i = 1, \dots, s .$$

As $s \ge k_* \ge 2$, from (Q2b) and (Q2c) it follows that

$$\tilde{q}_i \le C q_{V'(i)} N^{1/2n(j)}$$

provided $V(i) \in \mathcal{V}_2(1, j)$, $2 \le j \le k$, $i = 1, \dots, s$.
This implies (7.9) with $\gamma(1)$ satisfying (Γ_2).

Lemma 1 is proved, as the statement concerning the constant C in (7.9) follows easily from the proof of this inequality which was given above.

Lemma 2. Let the conditions of Lemma 1 hold. Then

$$R_{\{V\}} \le C N^{k/2} \quad (k \ge 1) \tag{7.22}$$

where the constant C satisfies the following conditions:

(i) If the diagram $\{V\}$ has an edge belonging to $\mathcal{V}_3$, then $C = o(1)$.

(ii) If $\{V\}$ has no such edge, then C depends only on the constants C_V , $V \in \{V\}$ in (Q1) - (Q3) and vanishes if one of them vanishes.

Proof. For each $j = 1, \dots, k$, we define

$$\mu(j) = 1/2 \, \Sigma_{V(i) \in \mathcal{V}_1(j)} |V(i)|/n(j) +$$

$$+ \tfrac{1}{2} \Sigma_{m \ne j} \Sigma_{V(i) \in \mathcal{V}_2(j,m)} |V(i) \cap T(m)|/n(m).$$

Note that $\mu(1)$ coincides with the right hand side of the inequality (Γ_2) of Lemma 1 and $\mu(j), j \neq 1$ is the corresponding quantity for the row $T(j)$ instead of $T(1)$. The sum $\sum_1^k \mu(j)$ has the the following meaning. For each $V \subseteq T$, let us define

$$\mu(V) = 1/2 \sum_{j=1}^{k} |V \cap T(j)| / n(j).$$

Then μ is a measure on T and $\mu(T) = k/2$. Clearly,

$$\sum_1^k \mu(j) = \mu\Big(\bigcup_{i=1,\dots,r\,:\,V(i) \in \mathcal{V}_1 \cup \mathcal{V}_2} V(i) \Big)$$

and, consequently,

$$\sum_1^k \mu(j) \leq k/2. \tag{7.23}$$

- Note that (7.22) holds by (Q1) for $k=1$. Assume that Lemma 2 is true if the number of rows of the table does not exceed $k-1$. The proof of the induction step $k-1 \to k$ will be based on Lemma 1. Denote by ${}_jT' = T \setminus T(j)$ $(j=1,\dots,k)$ the (sub)table with $k-1$ rows, and by ${}_jV'(i) = V(i) \cap {}_jT'$ the diagram over ${}_jT'$ induced by the diagram $\{V\}$, similarly as in the proof of Lemma 1. According to Lemma 1,

$$R_{\{V\}} \leq C R_{\{{}_jV'\}} N^{\gamma(j)}, \quad j = 1, \dots, k. \tag{7.24}$$

There are two cases:

1) $\gamma(j) \leq \mu(j)$ for all $j = 1, \dots, k$, or
2) $\gamma(j) > \mu(j)$ for some $j = 1, \dots, k$, say, $j=1$

In the first case, by (7.23) we have that

$$\min(\gamma(j) : j=1,\dots,k) \le 1/2 . \tag{7.25}$$

Hence, (7.22) follows by (7.24) and the inductive assumption. Moreover, if the diagram $\{V\}$ satisfies the condition (i) of Lemma 2, then $\sum_1^k \mu(j) < k/2$ and the minimum in (7.25) is strictly less then 1/2. Together with the inductive assumption, this implies $R_{\{V\}} = o(N^{k/2})$, or $C = o(1)$ in (7.22).

In the second case, $\gamma(1)$ satisfies the condition (Γ_1) of Lemma 1 which means that $\gamma(1) \le 1/2$ and (7.22) holds according to (7.9) and the inductive assumption. Note that (Γ_1) excludes the possibility of $\{V\}$ having and edge from $\mathcal{V}_3$ and $\{_1V'\}$ having no such edge. Again, by the inductive assumption and Lemma 1, this implies that the constant C in (7.22) satisfies (i) and (ii). Lemma 2 is proved.

Lemma 3. Let q_V, $V \subseteq T$ satisfy the conditions (Q1) -(Q3) and $\{V\} = \{V_1,\dots,V_r\}$ be a connected diagram without flat edges. Then

$$R_{\{V\}} = o(N^{k/2}) \quad (k \ge 3) . \tag{7.26}$$

Proof. If $\{V\}$ has an edge from $\mathcal{V}_3$, then (7.26) follows from Lemma 2(i).

Assume that $V_i \in \mathcal{V}_2$, $i = 1,\dots,r$. Define new functions $q_V^{(0)}$, $V \in \mathcal{V}_2(i,j)$ $(i,j=1,\dots,k,\ i\ne j)$ by

$$q_V^{(0)}(t_i,t_j) = \begin{cases} q_V(t_i,t_j) & \text{if } |t_i - t_j| \le K , \\ 0 & \text{if otherwise.} \end{cases}$$

Then $R_{\{V\}} = R_{\{V\}}(q_V)$ (7.8) can be rewritten as

$$R_{\{V\}} = R_{\{V\}}(q_V^{(0)}) + \sum_1^r R_{\{V\}}(q_V^{(j)}),$$

where $R_{\{V\}}(q_V^{(0)})$ is the sum (7.8) with q_V replaced by $q_V^{(0)}$, and $R_{\{V\}}(q_V^{(j)})$ $(j=1,\dots,r)$ is the sum (7.8) corresponding to some functions $q_V^{(j)}$, $V \in \mathcal{V}_2$ which satisfy the inequality

$$q_V^{(j)} \le \begin{cases} q_V & , \text{ if } V \ne V(j), \\ q_V - q_V^{(0)} & , \text{ if } V = V(j). \end{cases} \tag{7.27}$$

By the connectedness of $\{V\}$,

$$R_{\{V\}}(q_V^{(0)}) \le CN, \tag{7.28}$$

where $C = C(K) < \infty$ if $K < \infty$. A short proof of (7.28) is given in [10]. On the other hand, from (7.27) it follows that $q_V^{(j)}$, $V \in \mathcal{V}_2$ satisfy (Q2) with some constants $C_V^{(j)} = C_V^{(j)}(0)$ which are related to the (old) constants $C_V(K)$ in (7.6) by the relations

$$C_V^{(j)} \le \begin{cases} C_V(0) & , \text{ if } V \ne V(j), \\ C_V(K) & , \text{ if } V = V(j). \end{cases} \tag{7.29}$$

As $C_V(K) \to 0$ $(K \to \infty)$, by (7.29) and Lemma 2(ii) we have that

$$R_{\{V\}}(q_V^{(j)}) \le C_j N^{k/2}, \quad j = 1, \dots, k, \tag{7.30}$$

where $C_j = C_j(K) \to 0$ $(K \to \infty)$. Npw, Lemma 3 follows from (7.28) and (7.30).

From Lemma 3 it follows the relation (7.4) and hence Theorem 1.

8. Proof of Theorem 2

Let X_t, $t \in \mathbb{Z}$ be the linear process (1.9) with the weights $a(t)$ satisfying (1.10). According to the property $(\varkappa 3)$ of semi-invariants,

$$\varkappa_{\ell_1,\dots,\ell_k}(t_1,\dots,t_k)$$

$$= \varkappa_{\sum_{i=1}^{k}\ell_i}(\xi_0) \sum_{-\infty}^{\infty} a^{\ell_1}(t_1-s)\cdots a^{\ell_k}(t_k-s) \,. \quad (8.1)$$

Set

$$q_{\ell_1\cdots\ell_k}(t_1,\dots,t_k)$$

$$= |\varkappa_k(\xi_0)| \sum_{-\infty}^{\infty} |a^{\ell_1}(t_1-s)|\cdots|a^{\ell_k}(t_k-s)| \quad (8.2)$$

$k \geq 2$. Then $|\varkappa_{\ell_1,\dots,\ell_k}| \leq q_{\ell_1,\dots,\ell_k}$. Let us prove that $q_{\ell_1,\dots,\ell_k}$ satisfy the conditions (q1) and (q2).

Take $i = 1$ in (q1), for simplicity. Then

$$\sum_{t_1} q^{p}_{\ell_1,\dots,\ell_k}$$

$$\leq C \sum_{t_1} \Big(\sum_s |a^{\ell_1}(t_1-s)\cdots a^{\ell_k}(t_k-s)|\Big)^{[p]}$$

$$\cdot \Big(\sum_s |a^{\ell_2}(t_2-s)\cdots a^{\ell_k}(t_k-s)|\Big)^{p-[p]}$$

$$\leq C q^{p-[p]}_{\ell_2,\dots,\ell_k} \sum_{s_1}\cdots\sum_{s_{[p]}} \prod_{j=2}^{k} |a^{\ell_j}(t_j-s_1)\cdots a^{\ell_j}(t_j-s_{[p]})|$$

$$\cdot\, b(s_1,\dots,s_{[p]}) \,,$$

where

$$b(s_1, \dots, s_{[p]}) = \sum_{t_1} |a^{l_1}(t_1 - s_1) \cdots a^{l_1}(t_1 - s_{[p]})|$$

is bounded by a constant if either $[p] \geq 2$ or $[p] = 1$ and $l_1 \geq 2$. This proves (q1). The condition (q2) can be verified analogously.

To prove the second part of Theorem 2, note that (1.15) is the same as (q3) as

$$q_{1,1}(t) = \sum_{-\infty}^{\infty} |a(s)a(t-s)| .$$

We shall prove that (1.15) implies (q4) as well. It is easy to see that (q4) is equivalent to

$$\sum_{s=1}^{N} \left(\sum_{t=1}^{N} q_{n,1}(t-s)\right)^k = o(N^{(k+1)/2}) \tag{8.3}$$

for $k \geq k_*$. Denote the left hand side of (8.3) by Δ. Then

$$\Delta \leq \sum_{s=-\infty}^{\infty} \left(\sum_{t=1}^{N} q_{n,1}(t-s)\right)^k$$

$$= \sum_{t_1, \dots, t_k = 1}^{N} \psi(t_1 - t_k, \dots, t_{k-1} - t_k),$$

where

$$\psi(t_1, \dots, t_{k-1}) = \sum_{s=-\infty}^{\infty} q_{n,1}(t_1 - t_k - s) \cdots$$

$$\cdots q_{n,1}(t_{k-1} - t_k - s)\, q_{k,1}(-s).$$

Now,

$$\Delta \leq \sum_{t_k=1}^{N} \sum_{|t_1-t_k|\leq K, \ldots, |t_{k-1}-t_k|\leq K} \psi(t_1-t_k, \ldots, t_{k-1}-t_k)$$

$$+ \sum_{j=1}^{k-1} \sum_{t_k=1}^{N} \sum_{t_1, \ldots, t_{k-1} : |t_j - t_k| \geq K} \psi(t_1-t_k, \ldots, t_{k-1}-t_k)$$

$$\equiv \Delta_0 + \sum_{j=1}^{k-1} \Delta_j .$$

Here

$$\Delta_0 \leq CN, \tag{8.4}$$

where $C = C(K) < \infty$ if $K < \infty$. On the other hand, if

$$\sum_{t_1, \ldots, t_{k-1}=-\infty}^{\infty} \psi^2(t_1, \ldots, t_{k-1}) < \infty , \tag{8.5}$$

then

$$\Delta_j \leq \sum_{t_k=1}^{N} \Big(\sum_{t_1, \ldots, t_{k-1}=1}^{N} 1 \Big)^{1/2} \times$$

$$\times \Big(\sum_{t_1, \ldots, t_{k-1} : |t_j| \geq K} \psi^2(t_1, \ldots, t_{k-1}) \Big)^{1/2}$$

$$\leq \varepsilon(K) \, N^{(k+1)/2} ,$$

where $\varepsilon(K) \to 0 \;\; (K \to \infty)$. Together with (8.4) this implies (8.3).

It remains to prove (8.5). By (8.2),

$$\sum_{t_1, \ldots, t_{k-1}} \psi^2(t_1, \ldots, t_{k-1}) =$$

$$= C \sum_{t_1,\dots,t_{k-1}} \Big(\sum_{s_1,\dots,s_k} |a^n(s_1) \cdots a^n(s_k)| \times$$

$$\times \sum_{t=-\infty}^{\infty} |a(t_1-t-s_1) \dots a(t_{k-1}-t-s_{k-1}) a(-t-s_k)| \Big)^2 .$$

By Minkowski's inequality, the right hand side does not exceed

$$C \Big(\sum_{s_1,\dots,s_k} |a^n(s_1) \cdots a^n(s_k)| \, \Phi^{1/2} \Big)^2 ,$$

where

$$\Phi = \sum_{t_1,\dots,t_{k-1}} \Big(\sum_t |a(t_1-t-s_1) \cdots$$

$$\cdots a(t_{k-1}-t-s_{k-1}) a(-t-s_k)| \Big)^2$$

$$= \sum_t q_{1,1}^{k}(t) < \infty .$$

Theorem 2 is proved.

References

[1] Avram, F., Taqqu, M.S. Non-central limit theorems and generalized powers. Cornell University 1985.

[2] Berman, S.M. High level sojourns for strongly dependent Gaussian Processes. Z.Wahrsch.verw.Gebiete 50 (1979), 223-236.

[3] Boas, R.P., Buck, R.C. Polynomials expansions of analytic functions. Springer: New-York, 1964.

[4] Brillinger, D.R. Time series. Date analysis and theory. Holt, Rineart and Winston: New-York 1975.

[5] Breuer, P., Major, P. Central limit theorem for non-linear functionals of Gaussian fields. J.of multivariate analysis 13 (1983), 425-441.

[6] Bourbaki, N. Éléments de mathématique. Livre IV. Fonctions d'une variable réele. Herman: Paris, 1958

[7] Dobrushin, R.L., Major, P. Non-central limit theorem for non-linear functionals of Gaussian fields. Z.Wahrsch. verw. Gebiete 50 (1979), 27-52.

[8] Feinsilver, P.J. Special functions, probability semigroups, and Hamiltonian flows. Lecture Notes in Math., 696, Springer: New-York, 1978.

[9] Giraitis, L. Central limit theorem for functionals of linear processes. (Russian). Litovsk.mat.sb. 25 (1985) No 1, 43-57.

[10] Giraitis, L., Surgailis, D. CLT and other limit theorems for functionals of Gaussian processes. Z.Wahrsch.verw. Gebiete (to appear).

[11]. Glimm, J., Jaffe, A. Quantum physics. A functional integral point of view. Springer: New-York-Berlin 1981.

[12] Ibragimov, I.A., Linnik, J.V. Independent and stationary sequences of random variables. Walters-Noordhoff: Groningen 1971.

[13] Kaz'min, J.A. Appell polynomial series expantions. (Russian) Mat.Zametki 5 (1969), 509-520.

[14] Krein, S.G. (ed.). Functional analysis. (in Russian). Nauka: Moscow 1972.

[15] Leonov, V.P., Shiryaev, A.N. Sur le calcul des semi-invariants. Teor.Verojatn. i Primenen. 4 (1959), 342-355.

[16] Lukacs, E. Characteristic functions (2 nd ed). Griffin: London, 1970

[17] Malyshev, V.A. Cluster expansions in lattice models of statistical physics and quantum field theory.(Russian). Uspeki Mat.Nauk 35 (1980), No 2, 3-53.

[18] Ruiz de Chavez, J. Conpensation multiplicative et "Produits de Wick". In: Seminaire de Probabilités XVIII Proceedings. Lecture Notes in Math., 1123, Springer: Berlin 1985, pp. 242-247.

[19] Rota, G.C. Finite Operators Calculus. Academic Press: New York 1975.

[20] Szegö, G. Orthogonal polynomials. American mathematical society Colloquium publ. vol.XXIII. Rew.ed New-York 1959.

[21] Sheffer, I.M. Note an Appell polynomials. Bull.Amer.Math. Soc. 51 (1945), 739-744.

[22] Shohat, J.The relation of the classical orthogonal polynomials to the polynomials of Appell. Amer.J.Math. 58 (1936), 453-464.

[23] Simon, B. The $P(\psi)_2$ Euclidean (Quantum) Field Theory. Princeton N.J.: Princeton University Press 1974.

[24] Sun, T.C. On central and non-central limit theorems for non linear functions of a stationary Gaussian process. In: Dependence in Probability and Statistik.

[25] Surgailis, D. Convergence of sums of non-linear functions of moving averages to self-similar processes. (Russian). Dokl.Akad.Nauk SSSR 257 (1981), No 1, 51-54.

[26] Surgailis, D. Domains of attraction of self-similar multiple integrals. (Russian) Litovsk.Mat.Sb. 22 (1982), No 2, 195-201.

[27] Surgailis, D. On Poisson multiple stochastic integrals and associated equillibrium Markov processes. - In: Lecture Notes Contr.Inform.Sci., 49, Springer: Berlin 1983, pp. 233-248.

[28] Taqqu, M.S. Weak convergence to fractional Brownian motion and to Rosenblatt processes. Z.Wahrsch.verw.Gebiete 31 (1975), 287-302.

[29] Taqqu, M.S. Convergence of iterated processes of arbitrary Hermite rank. Z.Wahrsch.verw.Gebiete 50 (1979), 53-83.

[30] Yosida, K. Functional Analysis. Springer: Berlin-Göttingen 1965.

USING RENEWAL PROCESSES TO GENERATE LONG-RANGE DEPENDENCE AND HIGH VARIABILITY

by

Murad S. Taqqu*

and

Joshua B. Levy

1. Introduction

We explore here three types of convergence theorems involving the normalized partial sums of two random processes $W = W(t)$ and $V = V(t)$ indexed by the integers $t = \ldots, -1, 0, 1, \ldots$. $W(t)$ is a stationary renewal reward process with large inter-renewal intervals, while $V(t)$ is a non-stationary process that takes the value zero except at some rare instants t where it achieves extremely high values.

The two processes W and V, were originally introduced in B. Mandelbrot's 1969 treatise, "Long-run linearity, locally Gaussian processes, H-spectra, and infinite variances." The results which we present here capitalize on the initial groundwork established in that paper. We do not include proofs because these involve some delicate estimations. Some hints can be found in [3]. The proofs will appear elsewhere (but see also [2]). Using M independent and identically (i.i.d) copies W_m and V_m, $m = 1, \ldots, M$ of the processes W and V, we define

$$W^*(T,M) = \sum_{t=1}^{T} \sum_{m=1}^{M} W_m(t)$$

and

$$V^*(T,M) = \sum_{t=1}^{T} \sum_{m=1}^{M} V_m(t).$$

In the economic framework introduced by Mandelbrot, T denotes time, M is a model index, and the processes W^* and V^* are commodity prices. We study W^* and V^* in the cases where T is much larger than M ($T >> M$) and where M is much larger than T ($M >> T$). For instance, suppose that the inter-renewal times of the

* Research supported by the National Science Foundation grant ECS-80-15585.

underlying renewal processes have a stable (Pareto) distribution with parameter $1 < \alpha < 2$, and suppose that both T and M are large. Then $W^*(T,M)$ behaves like the Gaussian fractional Brownian motion when $T << M$ and like a Levy-stable process when $T >> M$. These two processes although both self-similar are very different. Fractional Brownian motion is a Gaussian process whose increments exhibit a long-run dependence [4], [5] whereas the Levy-stable process has infinite variance and independent increments [6].

2. The underlying sequences $\{U_k\}$ and $\{W_k\}$

The renewal reward process $W(t)$ will be introduced in the next section. Here, we define its inter-renewal times U_k and the corresponding rewards W_k. The assumption on the sequences $\{U_k\}$ and $\{W_k\}, k = \ldots, -1, 0, 1, \ldots$ are as follows:

(A1) $\{U_k\}$ is i.i.d., $P(U_0 = 1, 2, \ldots) = 1$,

(A2) $\{W_k\}$ is i.i.d., $EW_0 = 0, EW_0^2 < \infty$,

(A3) $\{U_k\}$ and $\{W_k\}$ are independent of each other.

U or U_0 will denote a generic U_k and W_0 will denote a generic W_k. To produce the desired convergence results, we stipulate one of the following distributional assumptions on U:

(A4) Either $EU^2 < \infty$ or $P(U \geq u) \sim u^{-\alpha}h(u), 1 < \alpha < 2$. (We write $U \epsilon \mathcal{D}(\alpha)$ in the second case.)

h belongs to the class $\mathcal{H}$ of slowly varying functions. A function $h \epsilon \mathcal{H}$ is slowly varying at ∞ if $\lim_{t \to \infty} \frac{h(xt)}{h(t)} = 1$ for all $x \epsilon [A, \infty)$. A constant function and $\log x$ are two examples of slowly varying functions.

If U satisfies the first part of (A4), then the Central Limit Theorem is in force: the distribution of $\frac{1}{\sqrt{n}} \sum_{k=1}^{n} (U_k - EU_k)$ converges to the normal distribution N with mean zero and variance $\text{Var } U_k = EU_k^2 - (EU_k)^2$. We say that U belongs to the domain of attraction of the normal law and we write $U \epsilon \mathcal{D}(N)$.

If U satisfies the second part of (A4), then U belongs to the domain of attraction of a stable distribution with index α: there exist centering constants $A_n(\alpha)$ and a slowly varying function $L(n)$ related to h (namely $L(n)^{-\alpha}h(n^{1/\alpha}L(n)x) \to 1$ as $n \to \infty$, $x > 0$) such that the distribution of the sequence $\frac{1}{n^{1/\alpha}L(n)}\sum_{k=1} U_k - A_n(\alpha)$ converges to the distribution of the stable random variable $\Lambda_\alpha \cdot \Lambda_\alpha$ can assume high values with high probability because the probability tails are of the form

$$P(\Lambda_\alpha \leq x) \sim c(-x)^{-\alpha} \text{ as } x \to -\infty;\ P(\Lambda_\alpha \geq x) \sim dx^{-\alpha} \text{ as } x \to \infty$$

where c and d are non-negative constants with $c + d > 0$. The stable random variable Λ_α has an infinite variance. In fact, for any $0 < \alpha < 2$, $E|\Lambda_\alpha|^\gamma = \infty$ but $E|\Lambda_\alpha|^\gamma < \infty$ for all $\gamma < \alpha$. For further details about stable distributions, see Chapter 2 of Ibragimov and Linnik [1], or the monograph of Mihnheer [6].

If

$$S_k = S_0 + \sum_{j=1}^{k} U_j \text{ for } k \geq 1; S_k = S_{-1} - \sum_{j=-k}^{-2} U_j \text{ for } k \leq -2,$$

then $\{S_k\}, k = \ldots -1, 0, 1, \ldots$ is a two-sided renewal sequence with S_k representing the time of the kth renewal. $S_0 \geq 0$ and $S_{-1} \leq -1$ must be chosen adequately to ensure that the renewal sequence is stationary. We take S_0 independent of the U_k's and distributed as

$$P(S_0 = u) = \mu^{-1}P(U \geq u + 1) \quad \text{for } u = 0, 1, \ldots$$

where $\mu = EU$ and we define S_{-1} by conditioning on S_0:

$$P(S_{-1} = -v|S_0 = u) = \frac{P(U = u + v)}{P(U \geq u + 1)}, u = 0, 1, 2, \ldots; v = 1, 2, \ldots$$

These definitions ensure that the renewal sequence $\{S_k\}$ is stationary. Note that $S_0 \geq 0$ and $S_0 - S_{-1} \geq 1$. The distribution of $S_k - S_{k-1}$ is the same as that of U except for $k = 0$.

3. The process $W(t)$ and $V(t)$

The two processes $W = W(t)$ and $V = V(t), t = \ldots, -1, 0, 1, \ldots$, are defined in terms of the sequences $\{U_k\}$ and $\{W_k\}$ as follows. $W(t)$ is the random variable W_k

when t lies within the kth renewal interval $(S_{k-1}, S_k]$. $V(t)$, however, is assigned the value zero except when a renewal occurs at t, in which case it assumes a value equal to the product U_kW_k if $S_k = t$. A sample path of $W = W(t), t = \ldots, -1, 0, 1, \ldots$, is thus a sequence of points located on segments parallel to the horizontal axis at a level W_k when $t \epsilon (S_{k-1}, S_k]$. Note that each segment ends at a renewal. On the other hand, a sample path of $V = V(t)$ is a sequence of points on the time axis except at renewal instants. As a consequence, when $k \geq 1$, the cumulative totals of W and V during $(S_{k-1}, S_k]$, $\sum_{t \epsilon (S_{k-1}, S_k]} W(t)$ and $\sum_{t \epsilon (S_{k-1}, S_k]} V(t)$, are both U_kW_k, although W achieves this by prescribing a constant amount W_k at each discrete t while V waits until the end of the interval to generate the entire total. Formally,

$$W(t) = \sum_{k=-\infty}^{\infty} W_k I(S_{k-1} < t \leq S_k)$$

$$V(t) = \sum_{k=-\infty}^{\infty} U_k W_k I(S_k = t)$$

for $t = \ldots, -1, 0, 1, \ldots$, with $I(A)$ denoting the indicator function of the set A. Figure 1 illustrates a sample path for both processes W and V.

If $t \geq 0$, then the above sums range over $0 \leq k \leq t$. For simplicity, we shall restrict our attention to this case.

Neither the W process nor the V process have independent marginals. However, W is stationary in that its finite-dimensional distributions are invariant with respect to time shifts. V is not stationary because the lack of independence between $\{U_kW_k\}$ and $\{S_k\}$ forces $V(t)$ and $V(t+h)$ to have different marginal distributions.

A more significant distinction between W and V arises under the second part of (A4) when $P(U \geq u) \sim u^{-\alpha}h(u), 1 < \alpha < 2$. For, although (A1) through (A3) imply

$$EW(t) = 0 = EV(t) \text{ and } \operatorname{Var} W(t) < \infty,$$

V has a finite variance only when $EU^2 < \infty$. When $V \epsilon \mathcal{D}(\alpha)$, V imitates the behavior of U in that $P(|V(t)| \geq x) \sim x^{-\alpha} h_t(x)$, for some $h_t \epsilon \mathcal{H}$.

To see this, recall that $V(t)$ is zero except at a countable number of renewal instants. High values of $|V|$ can then only occur at these renewal points. At renewals, $V(t)$ is distributed as $U_0 W_0$, and because $U \epsilon \mathcal{D}(\alpha)$, it is the random variable U, and not W_0 which controls the tail fluctuations of $U_0 W_0$. Indeed, if $U \epsilon \mathcal{D}(\alpha)$ and G denotes the distribution of $|W_0|$, then

$$
\begin{aligned}
P(|U_0\, W_0| \geq x) &= \int_0^\infty P(U \geq x w^{-1}) dG(w) \\
&\sim x^{-\alpha} h(x) \int_0^\infty w^\alpha dG(w) \qquad (3.1) \\
&= E|W_0|^\alpha x^{-\alpha} h(x) \quad \text{as } x \to \infty
\end{aligned}
$$

In addition, the constants of asymptoticity in the tail probabilities of $V(t)$ turn out to depend on the event $[S_0 = t]$ and not at all on the events $[S_k = t], 1 \leq k \leq t$. Indeed,

$$
P(|V(t)| \geq x) = P(|U_0\, W_0| \geq x, S_0 = t) + \sum_{k=1}^{t} P(|U_k\, W_k| \geq x, S_k = t) = (I) + (II)
$$

where, using (3.1),

$$
(I) = P(|U_0\, W_0| \geq x) P(S_0 = t) \sim E\, W_0|^\alpha P(S_0 = t) x^{-\alpha} h(x) \; .
$$

Since $\{S_k = t\}$ implies $\{U_k \leq t\}$ and since $P\{|W_0| \geq x\} = o(x^{-2})$ as $x \to \infty$, we conclude that the contribution of (II) is negligible:

$$
(II) \leq \sum_{k=1}^{t} P(|W_k| \geq x t^{-1}, S_k = t) = P(|W_0| \geq x t^{-1}) \sum_{k=1}^{t} P(S_k = t) = o(x^{-2} t^2).
$$

Therefore,

$$
P(|V(t)| \geq x) \sim E|W_0|^\alpha P(S_0 = t) x^{-\alpha} h(x) = x^{-\alpha} h_t(x). \qquad (3.2)
$$

Note that the tail behavior of $V(t)$ merely differs from the tail behavior of U by the multiplicative constant $E|W_0|^\alpha P(S_0 = t)$.

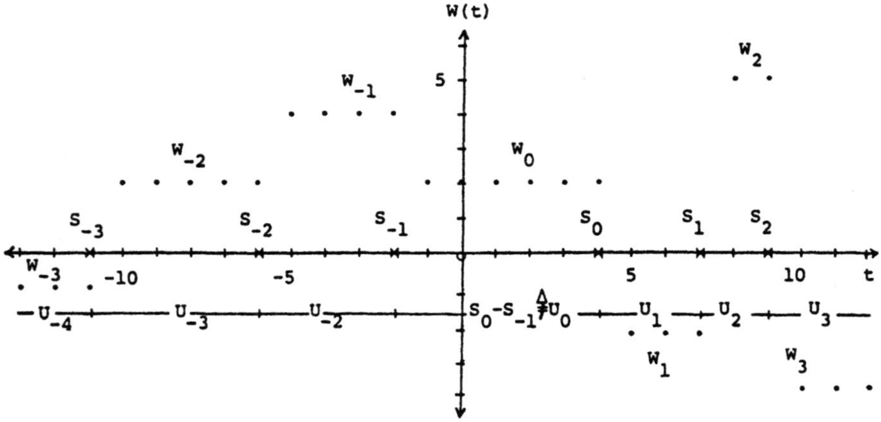
W(t)
5
W_2
W_-1
W_-2
W_0
S_-3
S_-2
S_-1
S_0
S_1
S_2
-10
-5
5
10
t
W_-3
U_-4
U_-3
U_-2
S_0-S_-1
Δ
U_0
U_1
U_2
U_3
W_1
W_3

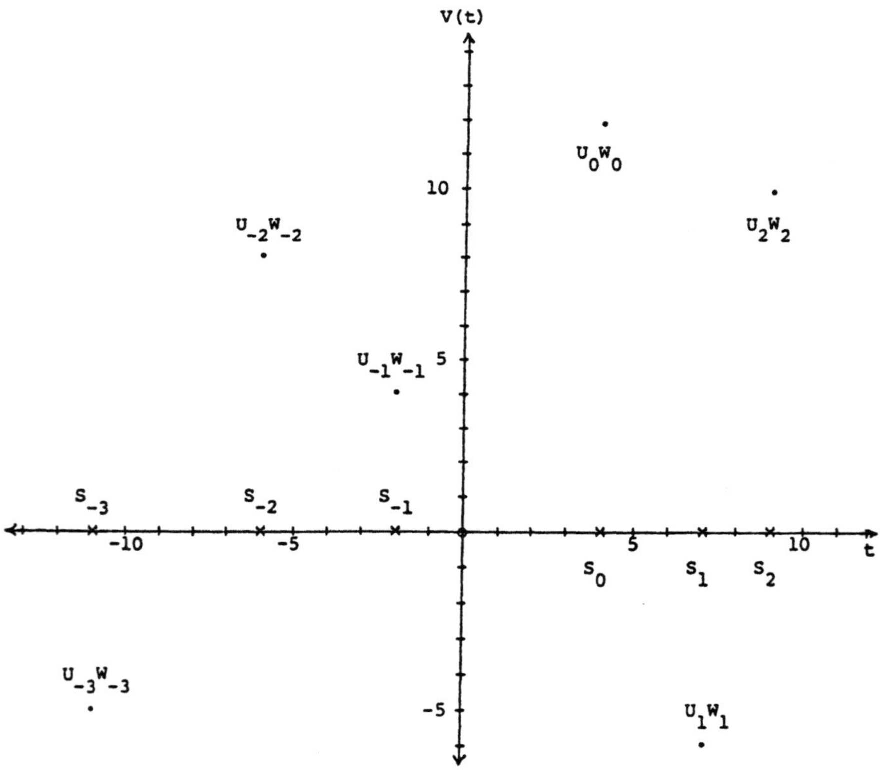
V(t)
10
U_0W_0
U_-2W_-2
U_2W_2
U_-1W_-1
5
S_-3
S_-2
S_-1
-10
-5
5
10
t
S_0
S_1
S_2
U_-3W_-3
-5
U_1W_1

Figure 1

It illustrates a sample path of the processes $W = W(t)$ and $V = V(t)$. There is an asymmetry between positive and negative times. When $W = W_{k'}$, the renewal interval has length U_k for $k \geq 1$ but has length U_{k-1} for $k \leq -1$. This has two related consequences: (a) If t is the time of the kth renewal, i.e. when $S_k = t$, then $V(t) = U_k W_k$. When $k \geq 1$, this means that $V(t)$ is equal to W_k times the length of the *preceding* inter-renewal time. When $k \leq -2, V(t)$ is equal to W_k times the length of the *succeeding* inter-renewal time. (b) The partial sums of $W(t)$ and $V(t)$ over the inter-renewal interval $(S_{k-1}, S_k]$ are identical to $U_k W_k$ when $k \geq 1$ but only equal in distribution when $k \leq -1$. When $k = 0$, however, these partial sums are not even equal in distribution because neither U_0 nor U_{-1} represent the length of an inter-renewal interval. In Figure 1, we have chosen arbitrarily U_0 to be 6 and U_{-1} to be 1.

4. The summed processes $W^*(M,T)$ and $V^*(M,T)$

For convenience, we let $X = X(t)$ denote either the $W = W(t)$ process or the $V = V(t)$ process and we set $X^*(T) = \sum_{t=1}^{T} X(t)$ with $X^*(0) = 0$. The processes W and V have relatively simple sample paths. To obtain a more random behavior, we introduce M i.i.d. copies of the underlying renewal reward process, hence M i.i.d. copies $X_m, m = 1, 2, \ldots M$ of X, and we define

$$X(t,M) = \sum_{m=1}^{M} X_m(t).$$

Accumulating the values of $X(t,M)$ over time, yields

$$X^*(T,M) = \sum_{t=1}^{T} X(t,M) = \sum_{t=1}^{T} \sum_{m=1}^{M} X_m(t) \quad \begin{matrix} T = 1,2,\ldots \\ M = 1,2,\ldots \end{matrix}$$

with $X^*(0,M) = 0$. (Note that $X^*(T,1) = X^*(T)$.)

The processes $X^*(T,M)$ (namely the processes $W^*(T,M)$ and $V^*(T,M)$) are the models of interest. To analyze their behavior for either large T or large M, we consider three types of asymptotic convergence.

With "L-lim" denoting convergence in the sense of the finite-dimensional distributions, we examine first

$$(i) \qquad L - \lim_{M\to\infty} \frac{1}{C(M)} (X^*(T_1,M), X^*(T_2,M), \ldots, X^*(T_J,M)).$$

Then,

$$(ii) \qquad L - \lim_{T\to\infty} \frac{1}{d(T)} X^*([Ty],M), \quad 0 \le y \le 1;$$

and, finally,

$$(iii) \qquad L - \lim_{M,T\to\infty} \frac{1}{a(T,M)} X^*([Ty],M), \quad 0 \le y \le 1.$$

The third case is a juxtaposition of the first two.

What happens when both the number of copies M and the time T grow simultaneously? We want to determine the combined effect of summing an increasing number of copies of X and accumulating through time the fluctuations of these copies.

5. The first type of convergence

We set $EW_0^2 \equiv \sigma^2$ and we let $C(\cdot)$ be the covariance function of W. Then

$$C(n) = E[W(t)W(t+n)] \text{ and } C(0) = \text{Var } W(t) = \sigma^2.$$

Define

$$C^{**}(T) = \sum_{t=1}^{T} \sum_{n=1}^{t} C(n) \quad T = 1, 2, \ldots$$

with the convention $C^{**}(0) = 0$. Note that $EX^*(T) = 0$ whether X is W or V.

We are interested in the convergence as $M \to \infty$ of the finite-dimensional distributions of the process $[C(M)]^{-1}X^*(T, M)$, that is of the vector

$$[C(M)^{-1}](X^*(T_1, M), \ldots, X^*(T_J, M))$$

for all finite $T_1, \ldots, T_J$.

Let the symbol $\triangleq$ denote equality of finite-dimensional distributions. The first theorem assumes that U has finite variance.

Theorem 1. *Suppose $EU^2 < \infty$. Then*

$$L - \lim_{M\to\infty} M^{-1/2} X^*(T, M) \triangleq Z_{X^*}(T)$$

where $Z_{X^*}(T), T = 0, 1, 2, \ldots$ is a mean-zero Gaussian process with $Z_{X^*}(0) = 0$ and covariances

$$R_{X^*}(S, T) = \text{Cov}[X^*(S), X^*(T)] \ .$$

Furthermore, when $X = W$, the increments of $Z_{W^*}(T)$ are stationary. The covariances of Z_{W^*} are

$$R_{W^*}(S, T) = \frac{1}{2}[\text{Var } W^*(s) + \text{Var } W^*(T) - \text{Var } W^*(T - S)] \tag{5.1}$$

where, for $T > 0$,

$$\text{Var } W^*(T) = \sigma^2 T + 2C^{**}(T-1).$$

When $X = V$, *the increments of* $Z_{V^*}(T)$ *are independent, non-stationary and the covariances of* Z_{V^*} *are*

$$R_{V^*}(S,T) = \text{Var } V^*(\min(S,T)) \tag{5.2}$$

where, for $T > 0$,

$$\text{Var } V^*(T) = \sigma^2\{EU^2 P(1 \le S_0 \le T) + \sum_{k=1}^{T} E(U_k^2 I(k \le S_k \le T))\}.$$

We remark that Z_{W^*} and Z_{V^*} are Gaussian with covariances identical respectively to those of W^* and V^*. Z_{W^*} has stationary increments, while the unusual nature of V^* induces $Z_{V^*}(S)$ and $Z_{V^*}(T)$ to have covariance $\text{Var}[V^*(\min(S,T))]$. Neither Z_{W^*} nor Z_{V^*} is discrete Brownian motion. However, in the case $EU^2 < \infty$, we shall use the fact that Var $V^*(T)$ is asymptotically proportional to T as $T \to \infty$, so that $T^{-1/2}Z_{V^*}([Ty])$, $y \epsilon [0,1]$, approaches Brownian motion. In fact, $T^{-1/2}V^*([Ty])$ also converges to Brownian motion as $T \to \infty$. Identical results will also hold for Z_{W^*} and for W^*.

The next theorem assumes that $U \epsilon \mathfrak{D}(\alpha)$. It also involves $C^{**}(T)$, although this function is different from the one in the previous theorem because the distribution of U is different.

Theorem 2. *Suppose* $P(U \ge u) \sim u^{-\alpha}h(u)$, $1 < \alpha < 2$. *Then*

$$L - \lim_{M\to\infty} M^{-1/2}W^*(T,M) \triangleq Z^{\alpha}_{W^*}(T) \tag{a}$$

where $Z^{\alpha}_{W^*}(T), T = 0,1,2,\ldots$ *is a mean-zero Gaussian process with* $Z^{\alpha}_{W^*}(0) = 0$ *and covariances given by the relation (5.1).*

$$L - \lim_{M\to\infty} M^{-1/\alpha}[L(M)]^{-1}V^*(T,M) \triangleq Z^{\alpha}_{V^*}(T) \tag{b}$$

where $Z^{\alpha}_{V^*}(T), T = 0,1,2,\ldots$ *is a mean-zero stable process with* $Z^{\alpha}_{V^*}(0) = 0$. *Its increments are independent, non-stationary, stable random variables satisfying (for* $T_1 \leq T_2$*)*

$$(5.3)\qquad \begin{aligned} P(Z^{\alpha}_{V^*}(T_2) - Z^{\alpha}_{V^*}(T_1) \leq -x) &\sim E(W_0^-)^{\alpha}\Big(\sum_{t=T_1+1}^{T_2} P(S_0 = t)\Big)(-x)^{-\alpha} \\ P(Z^{\alpha}_{V^*}(T_2) - Z^{\alpha}_{V^*}(T_1) \geq x) &\sim E(W_0^+)^{\alpha}\Big(\sum_{t=T_1+1}^{T_2} P(S_0 = t)\Big)x^{-\alpha} \end{aligned}$$

as $x \to \infty$.

We note that $Z^{\alpha}_{V^*}$ is not a Levy-stable process because its increments are not stationary. The delicate part in the proof of (b), is to show that the increments of $Z^{\alpha}_{V^*}$ are independent.

6. The second type of convergence

We focus here on the behavior of $X^*([Ty], M)$, $0 \leq y \leq 1$ as $T \to \infty$ and show that the results are identical for both $X = W$ and $X = V$. This is because for large T, the random variable $X^*(T,1) = X^*(T)$ is actually a sum of $X(t)$ over an exact number of renewal intervals plus some remainder terms which are asymptotically negligible for either W or V. When $X = V$, we note that $X^*(T)$ is approximately a random sum of i.i.d. random variables $U_k W_k$ which will have finite variance if $EU^2 < \infty$ and which will have tails similar to those of U if $U \epsilon \mathfrak{D}(\alpha)$.

Let $B(y)$ be Brownian motion, that is, a Gaussian process with stationary, independent increments satisfying $B(0) = 0$, $EB(y) = 0$ and $EB^2(y) = y$. Thus $B(ay) \triangleq a^{1/2}B(y)$. It appears as $T \to \infty$ when U has finite variance.

Theorem 3. *If* $EU^2 < \infty$, *then*

$$L - \lim_{T\to\infty} T^{-1/2}X^*([Ty], M) \triangleq (M\sigma^2\mu^{-1}EU^2)^{1/2}B(y).$$

Let $\Lambda_{\alpha}(y)$ be the Levy-stable process with parameter α, that is, a process with stationary independent increments, with $\Lambda_{\alpha}(y) \triangleq y^{1/\alpha}\Lambda_{\alpha}(1)$ and

$$P(\Lambda_{\alpha}(1) \leq -x) \sim E(W_0^-)^{\alpha}(-x)^{-\alpha},\ P(\Lambda_{\alpha}(1) \geq x) \sim E(W_0^+)^{\alpha}x^{-\alpha}$$

as $x \to \infty$. It appears as $T \to \infty$ when $U \epsilon \mathfrak{D}(\alpha)$.

Theorem 4. *If* $P(U \geq u) \sim u^{-\alpha} h(u), 1 < \alpha < 2$, *then*

$$L - \lim_{T\to\infty} T^{-1/\alpha} L(T)^{-1} X^*([Ty], M) \triangleq M^{1/\alpha} \mu^{-1/\alpha} \Lambda_\alpha(y), 0 \leq y \leq 1.$$

The slowly varying function L is related to h as in Section 2.

This theorem illustrates the fact that a finite variance stationary process $X = W$, satisfying Var $W^*(T) \to \infty$, is not necessarily attracted to a Gaussian law.

7. The third type of convergence

The third and final group of theorems concerns the case when both the number M of copies of X and the time T increase simultaneously to infinity. Specifically, we want to determine the renormalization factor $a(T, M)$ so that $L - \lim a(T, M)^{-1} X^*([Ty], M)$ converge to a non-trivial limit as both M and T tend to infinity. We consider three cases:

(i) T tends to infinity first, followed by M.

(ii) M tends to infinity first, followed by T.

(iii) M is a function $M(T)$ of T such that $M(T) \to \infty$ as $T \to \infty$.

In case (iii) we could interchange the roles of T and M so that $T = T(M) \to \infty$ as $M \to \infty$, and obtain identical results.

Theorem 5. *Suppose* $EU^2 < \infty$, $EW_0^2 \equiv \sigma^2 < \infty$, *and* $0 \leq y \leq 1$. *Then*

(i) and (ii) $$L - \lim_{M\to\infty} \lim_{T\to\infty} M^{-1/2} T^{-1/2} X^*([Ty], M) \triangleq (\sigma^2 \mu^{-1} EU^2)^{1/2} B(y)$$

(7.1)

$$\triangleq L - \lim_{T\to\infty} \lim_{M\to\infty} T^{-1/2} M^{-1/2} X^*([Ty], M)$$

(iii) If $M(T) \to \infty$ *as* $T \to \infty$, *then*

(7.2) $$L - \lim_{T\to\infty} [M(T)]^{-1/2} T^{-1/2} X^*([Ty], M(T)) \triangleq (\sigma^2 \mu^{-1} EU^2)^{1/2} B(y).$$

The first equality in (7.1) is an immediate consequence of Theorem 3. To understand the second equality in (7.1) note that $Z_{X^*}([Ty]) \triangleq L-\lim M^{-1/2}X^*([Ty],M)$ is Gaussian by Theorem 2, and that both $T^{-1/2}Z_{X^*}([Ty])$ and $T^{-1/2}X^*([Ty])$ converge to $(\sigma^2\mu^{-1}EU^2)^{1/2}B(y)$ because Var $Z_{X^*}(T)$ = Var $X^*(T) \sim (\sigma^2\mu^{-1}EU^2)T$ as $T \to \infty$. We also remark that the result (7.2) does not depend on how fast $M(T)$ diverges to infinity with T.

The next theorem includes important results. One of them involves the *fractional Brownian motion* process, which we now define. This process is denoted $B_H(y)$, $y \geq 0$, where H is a parameter restricted here to the interval $(\frac{1}{2}, 1)$. The process $B_H(y)$ satisfies $B_H(0) = 0$ and is Gaussian with mean zero and covariance

$$R(y,z) = \frac{1}{2}[y^{2H} + z^{2H} - |y-z|^{2H}].$$

As a consequence, the process $B_H(y)$ has stationary increments and satisfies $E[B_H(y)]^2 = y^{2H}$ and $B_H(ay) \triangleq a^H B_H(y)$ for all $a > 0$. The increments of the process $B_H(t)$ exhibit a long-range dependence. Refer to [4] and [5] for details.

Theorem 6. *Assume* $P(U \geq u) \sim u^{-\alpha}h(u), 1 < \alpha < 2$, *and let* $0 \leq y \leq 1$. *Let* $EM_0^2 \equiv \sigma^2 < \infty$.

(i) If T *tends to infinity first, followed by* M, *then*

$$L - \lim_{M\to\infty}\lim_{T\to\infty} M^{-1/\alpha}T^{-1/\alpha}[L(T)]^{-1}X^*([Ty],M) \triangleq \mu^{-1/\alpha}\Lambda_\alpha(y). \tag{7.3}$$

where $\Lambda_\alpha(y)$ *is the Levy-stable process appearing in Theorem 4.*

(ii) If M *tends to infinity first, followed by* T, *then*

$$L - \lim_{T\to\infty}\lim_{M\to\infty} T^{\frac{-(3-\alpha)}{2}}[h(T)]^{-1/2}M^{-1/2}W^*([Ty],M) \tag{7.4}$$

$$\triangleq (2\sigma^2)^{1/2}[\mu(\alpha-1)(2-\alpha)(3-\alpha)]^{-1/2}B_{\frac{3-\alpha}{2}}(y)$$

where $B_{\frac{3-\alpha}{2}}$ *is fractional Brownian motion with parameter* $H = \dfrac{3-\alpha}{2}$. *Moreover,*

$$L - \lim_{T\to\infty}\lim_{M\to\infty} M^{-1/\alpha}[L(M)]^{-1}V^*([Ty],M) \triangleq \Lambda_{\alpha,+}(y) \quad (y \neq 0)$$

where $\Lambda_{\alpha,+}(y)$ is a constant process equal (in distribution), for all y, to the stable random variable $\Lambda_{\alpha,+}$ satisfying

$$P(\Lambda_{\alpha,+} \le -x) \sim E(W_0^-)^\alpha (1-\mu^{-1})x^{-\alpha}$$

$$, P(\Lambda_{\alpha,+} \ge x) \sim E(W_0^+)^\alpha (1-\mu^{-1})x^{-\alpha}$$

as $x \to \infty$.

(iii) Let $M(T) \to \infty$ as $T \to \infty$ at a faster rate than $T^{\frac{4\alpha}{2-\alpha}+\gamma}$ for some $\gamma > 0$. Then for $y \neq 0$,

$$(7.5) \qquad L - \lim_{T\to\infty} \lim_{M(T)\to\infty} (M(T))^{-1/\alpha}(L(M(T)))^{-1}V^*([Ty], M) \triangleq \Lambda_{\alpha^+}(y)$$

where $\Lambda_{\alpha,+}(y)$ is the constant process defined above.

To understand the results of this last theorem, suppose first that $X = V$. We obtain the stable process $\mu^{-1/\alpha}\Lambda_\alpha(y)$ under $\lim_{M\to\infty}\lim_{T\to\infty}$ because of Theorem 4. That process appears as a limit when $T \to \infty$ and, since it is a stable process, its finite-dimensional distributions will not change when summing i.i.d. copies after renormalizing by $M^{1/\alpha}$. Under $\lim_{T\to\infty}\lim_{M\to\infty}$ we obtain the constant process $\Lambda_{\alpha,+}(y)$ when $y \neq 0$. (Because $V^*(0) = 0$, we get 0 when $y = 0$.) To see why, recall that, by Theorem 2, the process $Z_{V^*}^\alpha([Ty])$ appears as a limit when $M \to \infty$. That process has increments $Z_{V^*}^\alpha([Ty_j]) - Z_{V^*}^\alpha([Ty_{j-1}])$ that are independent, non-stationary, stable random variables with constants of asymptoticity involving the factor $t = \sum_{[Ty_{j-1}]+1}^{[Ty_j]} P(S_0 = t)$ (see (5.3)). That factor tends to 0 as $T \to \infty$ unless $y_{j-1} = 0$. Thus, the increment tends to 0 in probability, when $y_{j-1} \neq 0$. This suggests that $Z_{V^*}^\alpha([Ty])$ behaves as $Z_{V^*}(T)$ when $T \to \infty$, and so the limiting finite-dimensional distributions will not depend on y.

When $X = W$, the stable process $\mu^{-1/\alpha}\Lambda_\alpha(y)$ arises under $\lim_{M\to\infty}\lim_{T\to\infty}$ for reasons similar to the case $X = V$. However, under $\lim_{T\to\infty}\lim_{M\to\infty}$ the resulting limit is Gaussian. This is because the first limit $\lim_{M\to\infty}$ yields the *Gaussian* process $Z_{W^*}^\alpha(T)$ which has stationary increments and covariance (5.1) (Theorem 1). If adequately renormalized, the process remains Gaussian as $T \to \infty$, and will still possess stationary increments. To determine the normalization factor note that $\text{Var } Z^{\alpha}_{W^*}(T) = \text{Var } W^*(T) = \sigma^2 T + 2C^{**}(T-1)$ is asymptotically proportional to $T^{3-\alpha}h(t)$ because $P(U \geq u) \sim u^{-\alpha}h(u)$ as $u \to \infty$ and because

$$\begin{aligned} C(n) &= EW(t)\, W(t+n) \\ &= \sigma^2\mu^{-1}[1 - \sum_{u=1}^{n} P(U \geq u)] \\ &= \sigma^2\mu^{-1} \sum_{u=n+1}^{\infty} P(U \geq u) \\ &\sim \sigma^2\mu^{-1}(\alpha-1)^{-1}n^{-\alpha+1}h(n), \\ C^*(n) &\sim \sigma^2[\mu(\alpha-1)(2-\alpha)]^{-1}n^{2-\alpha}h(n), \\ C^{**}(n) &\sim \sigma^2[\mu(\alpha-1)(2-\alpha)(3-\alpha)]^{-1}n^{3-\alpha}h(n), \end{aligned}$$

as $n \to \infty$. The correct normalization for $Z^{\alpha}_{W^*}(Ty)$ is then $T^{\frac{3-\alpha}{2}}h(T)^{1/2}$, and the limit is fractional Brownian motion. The occurrence of fractional Brownian motion as the limit can be intuitively explained as follows. When $U \epsilon \mathcal{D}(\alpha)$, the process W can assume a constant value (namely W_k) over a period of time U_k which can be relatively long with high probability. Combinations of more and more copies of W^*, each of which is evaluated over longer periods of time, produces an overall sum which becomes more and more Gaussian and exhibits a high dependence, attributes shared by fractional Brownian motion.

Case (iii) of Theorem 6 is presently the most complex situation to analyze. In fact, we only obtain results for $X = V$ where for each T, $V^*(T)$ is attracted to the stable law. However, we require a particular rate of divergence of $M(T)$ to ensure a non-trivial limit. Consider for example the two following cases where that rate is not obtained: $M(T) \to M_0 < \infty$ and $M(T) = T \to \infty$. In either of these two cases, $M(T)^{-1/\alpha}L(M(T))^{-1}V^*(T,M)$ tends to infinity. Indeed, when $M(T) \to M_0 < \infty$,

$$V^*(T, M(T)) \sim T^{1/\alpha}L(T)\, M_0^{1/\alpha}\Lambda_\alpha,$$

and when $M(T) = T \to \infty$,

$$V^*(T, M(T)) \sim T^{2/\alpha}L(T)\Lambda_\alpha.$$

We are currently uncertain as to which distribution, Gaussian or non-normal stable, if any, qualifies as the appropriate limit law in the case $X = W$. The reason is that although $W^*(T, M)$ appropriately normalized tends to a normal as $M \to \infty$ with T fixed, it tends to a stable with $\alpha < 2$ if $T \to \infty$ with M fixed. However, at this stage it appears that either more delicate analysis is required or further conditions beyond Assumptions (A1) through (A4) are necessary.

References

[1] Ibragimov, I.A. and Linnik, Y.V. (1971). *Independent and Stationary Sequences of Random Variables*. Groningen: Wolters-Nordhoff.

[2] Levy, J.B. (1983). High variability and long-range dependence: a modeling approach based on renewal sequences. M.S. thesis: Cornell University.

[3] Mandelbrot, B.B. (1969). Long-run linearity, locally Gaussian processes, H-spectra, and infinite variances. *International Economic Review*, 10, 82-113.

[4] Mandelbrot, B.B. (1982). *The Fractal Geometry of Nature*. San Francisco: W.H. Freeman & Co.

[5] Mandelbrot, B.B. and Taqqu, M.S. (1979). Robust R/S analysis of long-run serial correlation. Proceedings of the 42nd Session of the International Statistical Institute, Manila. *Bulletin of the I.S.I.*, Vol. 48, Book 2, pp. 69-104.

[6] Mijnheer, J.L. (1975). *Sample Path Properties of Stable Processes*. Mathematical Centre Tracts 59. Amsterdam: Mathematisch Centrum.

Murad S. Taqqu
Department of Mathematics
Boston University
Boston, MA 02215

Joshua B. Levy
School of Business
State University of New York
Albany, NY 12222

SOJOURNS OF MULTIDIMENSIONAL GAUSSIAN PROCESSES

Makoto MAEJIMA

1. Introduction

We shall survey some recent results on sojourns of multidimensional stationary Gaussian processes with strongly dependent structures, given by Berman [2], Maejima [6],[7] and Taqqu [9].

Let $\{X(t) = (X_1(t), \cdots, X_p(t))', t \geqq 0\}$ be a measurable separable p-dimensional stationary Gaussian process, where x' denotes the transposed vector of x.

Assumptions and notation throughout the paper are the following. Suppose that

$$EX(0) = 0,$$

$$R(t) \equiv EX(0)X(t)' = (R_{ij}(t))_{1 \leqq i,j \leqq p},$$

$$R_{ij}(t) = \begin{cases} r(t), & \text{if } i = j, \\ \rho(t), & \text{if } i \neq j, \end{cases}$$

$r(t)$ and $\rho(t)$ are continuous,

$$r(0) = 1, \quad \rho(0) = \rho_0 \quad (0 \leqq \rho_0 < 1),$$

$$r(t) \sim t^{-\alpha}L(t) \quad \text{as} \quad t \to \infty, \tag{1.1}$$

$$\rho(t) \sim \rho_\infty t^{-\alpha}L(t) \quad \text{as} \quad t \to \infty \quad (0 \leqq \rho_\infty < 1), \tag{1.2}$$

where $0 < \alpha < 1/2$ and L is slowly varying at infinity.

Define

$$M(t) = \int_0^t I[X(s) \in D]\, ds, \quad t > 0,$$

where $I[\cdot]$ is the indicator function and $D \subset R^p$. Let

$$D(a_1, \cdots, a_p; u(t))$$

$$= \{(x_1, \cdots, x_p)' \mid \sum_{j=1}^{p} a_j^2 x_j^2 \leqq u(t)\}.$$

Our concern is asymptotic behaviours of $M(t)$ as $t \to \infty$ under suitable normalization. We survey the results in the following cases.

Case I (Taqqu [9]). $p = 2$, $D = D(a, b; 1)$, $0 < a \leqq b$.

Case II (Maejima [7]). $p \geqq 2$, $D = D(1, \cdots, 1; 1)$.

Case III (Berman [2]). $p \geqq 2$, $\rho(t) \equiv 0$,

(i) $D = D(1, \cdots, 1; u(t))^c$,
$u(t)$ is nondecreasing, $u(t) = o(\log t)$,

(ii) $D = D(1, \cdots, 1; v(t))$,
$v(t)$ is nonincreasing,
$v(t) \geqq Ct^{-\delta}$ with $0 < \delta < (2/p)\min(\alpha, 1-2\alpha)$.

Case IV (Maejima [6]). $p \geqq 2$, $\rho_0 = 0$,
$D = D(1/b, \cdots, 1/b; 1) - D(1/a, \cdots, 1/a; 1)$,
$0 < a < b$.

2. Preliminaries

In this section, we explain the basic idea to attack the problems, which is used in [6], [7] and [9]. Berman's approach [2] is different. However, if we assume (1.1) on r(t) from the first, we can discuss his case (Case III above) in the same framework. Our approach is based on the Hermite expansion for functionals of Gaussian processes, which has been successfully used in, for example, [1], [3], [4], [5] and [8].

Denote by F the class of real-valued functions of

p variables square-integrable with respect to $\Pi_{j=1}^{p}\phi(x_j)$, $\phi(x)$ being the standard normal density, and let $H_n(x)$, $n = 0, 1, 2, \cdots$ denote the n-th normalized Hermite polynomials defined by $H_n(x) = (-1)^n\phi(x)^{-1}(d^n/dx^n)\phi(x)$. Then $f_t \in F$, having the parameter $t > 0$, has an expansion

$$f_t(x_1,\cdots,x_p) = \sum_{n=0}^{\infty} \sum_{\Sigma_{j=1}^{p} n_j = n} c_t(n_1,\cdots,n_p) \prod_{j=1}^{p} H_{n_j}(x_j) \tag{2.1}$$

in the mean square sense, where

$$c_t(n_1,\cdots,n_p) = \{\prod_{j=1}^{p} n_j!\}^{-1} \int_{R^p} f_t(x_1,\cdots,x_p)$$

$$\{\prod_{j=1}^{p} H_{n_j}(x_j)\phi(x_j)dx_j\},$$

$$n_j \geqq 0, \quad \sum_{j=1}^{p} n_j = n.$$

Denote a matrix by

$$T = (t_{ij})_{1\leqq i,j\leqq p},$$

where

$$t_{1j} = \{p(1+(p-1)\rho_0)\}^{-1/2}, \quad 1 \leqq j \leqq p,$$

$$t_{ij} = \{i(i-1)(1-\rho_0)\}^{-1/2}, \quad 2 \leqq i \leqq p, \quad 1 \leqq j < i,$$

$$t_{ii} = -(i-1)^{1/2}\{i(1-\rho_0)\}^{-1/2}, \quad 2 \leqq i \leqq p,$$

$$t_{ij} = 0, \quad \text{otherwise},$$

and define $Y(t) = (Y_1(t), \cdots, Y_p(t))'$ by $Y(t) = TX(t)$.

T is simply one of the well-known diagonalizing matrices for the covariance matrix $R(0)$. Then

$$\tilde{R}(t) \equiv EY(0)Y(t)' = (\tilde{R}_{ij}(t))_{1 \leq i,j \leq p},$$

where

$$\tilde{R}_{ij}(t) = \begin{cases} \dfrac{r(t)+(p-1)\rho(t)}{1+(p-1)\rho_0}, & \text{if } i = j = 1, \\[2ex] \dfrac{r(t)-\rho(t)}{1-\rho_0}, & \text{if } i = j = 2, 3, \cdots, p, \\[2ex] 0, & \text{if } i \neq j. \end{cases}$$

Therefore, $\{Y_j(t)\}$, $1 \leq j \leq p$, are independent Gaussian processes.

We thus have

$$M(t) = \int_0^t I[Y(s) \in \tilde{D}]\, ds,$$

where

$$\tilde{D} = \{(y_1,\cdots,y_p)' \mid T^{-1}(y_1,\cdots,y_p)' \in D\},$$

and by (2.1)

$$M(t) = \sum_{n=0}^{\infty} \sum_{\Sigma_{j=1}^{p} n_j = n} c_t(n_1,\cdots,n_p) \int_0^t \prod_{j=1}^{p} H_{n_j}(Y_j(s))ds,$$

where

$$c_t(n_1,\cdots,n_p) = \{\prod_{j=1}^{p} n_j!\}^{-1} \int_{\tilde{D}} \{\prod_{j=1}^{p} H_{n_j}(x_j)\phi(x_j)dx_j\}.$$

Note that if

$$D = \{(x_1,\cdots,x_p)' \mid \sum_{j=1}^{p} x_j^2 \leqq u(t)\}$$

then

$$\tilde{D} = \{(y_1,\cdots,y_p)' \mid (1+(p-1)\rho_0)y_1^2+(1-\rho_0)\sum_{j=2}^{p} y_j^2 \leqq u(t)\}.$$

Therefore, if $\rho_0 = 0$, then $D = \tilde{D}$. (See [7] for the details.)

The following proposition will play an essential role in the proofs of the results.

Proposition 1. Let $f_t \in F$ and $m \geqq 1$. Suppose that

(a) $r(t) \sim t^{-\alpha}L(t)$, $\rho(t) \sim \rho_\alpha t^{-\alpha}L(t)$ for some α with

$0 < \alpha < 1/m$,

(b) $c_t(n_1,\cdots,n_p) = 0$ for $(n_1,\cdots,n_p)$ with

$\Sigma_{j=1}^{p} n_j = n \leqq m-1$

and

(c) $$\eta(t) \equiv \{\sum_{\Sigma_{j=1}^{p} m_j=m} c_t(m_1,\cdots,m_p)^2\}^{-1} \int_{R^p} f_t(x_1,\cdots,x_p)^2$$

$$\{\Pi_{j=1}^{p}\phi(x_j)dx_j\} = o(t^\gamma)$$

for some γ with $0 < \gamma < \min(\alpha, 1-m\alpha)$.

Write

$$K(t) = \int_0^t f_t(Y_1(s),\cdots,Y_p(s))\, ds$$

and

$$I(t) = \sum_{\Sigma_{j=1}^{p} m_j = m} c_t(m_1,\cdots,m_p) \int_0^t \{\prod_{j=1}^{p} H_{m_j}(Y_j(s))\}ds.$$

Then as $t \to \infty$

$$\{\mathrm{Var}K(t)\}^{-1/2}\{K(t\tau) - EK(t\tau)\}, \qquad \tau \in [0, \infty),$$

is asymptotically equal to

$$\{\mathrm{Var}I(t)\}^{-1/2} I(t\tau), \qquad \tau \in [0, \infty),$$

in the sense of all finite-dimensional distributions.

In case that $f_t \in F$ does not depend on t, we define m by

$$m = \min\{n \geq 1 \mid \text{there exists } c(n_1,\cdots,n_p) \neq 0 \text{ with } \Sigma_{j=1}^{p} n_j = n \text{ in the expansion (2.1)}\}.$$

We call this m the Hermite rank of $f \in F$. If $f \in F$ does not depend on t and has the Hermite rank m, then conditions (b) and (c) in Proposition 1 are automatically satisfied. (See Proposition 1 in [7].)

Proof of Proposition 1. The proof is the same as in that of Proposition 1 in [7]. As in [7], if we let

$$R(t) = \sum_{n=m+1}^{\infty} \sum_{\Sigma_{j=1}^{p} n_j = n} c_t(n_1,\cdots,n_p) \int_0^t \{\prod_{j=1}^{p} H_{n_j}(Y_j(s))\}ds,$$

then

$$\mathrm{Var}K(t\tau) = \mathrm{Var}I(t\tau) + \mathrm{Var}R(t\tau). \qquad (2.2)$$

Also,

$$\mathrm{VarI}(t\tau) \sim C(t)(t\tau)^{2-m\alpha}L(t\tau)^m,$$

where

$$C(t) = \frac{2}{(1-m\alpha)(2-m\alpha)} \sum_{\sum_{j=1}^{p} m_j = m} c_t(m_1,\cdots,m_p)^2$$

$$m_1!\left(\frac{1+(p-1)\rho_\infty}{1+(p-1)\rho_0}\right)^{m_1}\left\{\prod_{j=2}^{p} m_j!\left(\frac{1-\rho_\infty}{1-\rho_0}\right)^{m_j}\right\},$$

and for large t,

$$\mathrm{VarR}(t\tau) \leqq \mathrm{const.}\times \sum_{n=m+1}^{\infty} \sum_{\sum_{j=1}^{p} n_j = n} c_t(n_1,\cdots,n_p)^2 \prod_{j=1}^{p} n_j!$$

$$\int_0^{t\tau} (t\tau-s)s^{-(m+1)\alpha}L(s)^{m+1}ds$$

$$\leqq \mathrm{const.}\times\int_{R^p} f_t(x_1,\cdots,x_p)^2\left\{\prod_{j=1}^{p}\phi(x_j)dx_j\right\}$$

$$\{(t\tau)^{2-(m+1)\alpha}L(t\tau)^{m+1} + O(t\tau)\}.$$

Hence by condition (c),

$$\mathrm{VarR}(t\tau) = o(\mathrm{VarI}(t\tau))$$

and by (2.2)

$$\mathrm{VarK}(t\tau) \sim \mathrm{VarI}(t\tau).$$

Thus

$$\lim_{t\to\infty} E\left|\{\mathrm{VarK}(t)\}^{-1/2}\{K(t\tau) - EK(t\tau)\}\right.$$

$$\left. - \{\mathrm{VarI}(t)\}^{-1/2}I(t\tau)\right|^2$$

$$= \lim_{t\to\infty}\{\mathrm{Var} I(t)\}^{-1} E[R(t\tau)^2] = 0,$$

completing the proof.

3. Case I

The result in this section is given by Taqqu [9]. Let $p = 2$ and $D = D(a, b; 1)$, $0 < a \leq b$. Let $B_1(\cdot)$ and $B_2(\cdot)$ be two independent Gaussian random measures satisfying $EB_j(A_1)B_j(A_2) = |A_1 \cap A_2|$, $j = 1, 2$, for all Borel sets A_1 and A_2 in R, with $|\cdot|$ denoting Lebesgue measure.

Theorem 1. As $t \to \infty$,

$$Z_t(\tau) \equiv \{\mathrm{Var} M(t)\}^{-1/2}\{M(t\tau) - EM(t\tau)\} \overset{W}{\Rightarrow} \Delta_1(\tau)$$

in the sense of weak convergence in $C[0, \infty)$. The limiting process $\Delta_1(\tau)$ has the following representation:

$$\Delta_1(\tau) = \frac{K(\alpha)}{\sigma} \int_{R^2} g_\tau(y_1,y_2)\{K_1 \frac{1+\rho_\infty}{1+\rho_0} dB_1(y_1)dB_1(y_2)$$

$$+ K_2 \frac{1-\rho_\infty}{1-\rho_0} dB_2(y_1)dB_2(y_2)$$

$$+ K_3 \left(\frac{1-\rho_\infty^2}{1-\rho_0^2}\right)^{1/2} dB_1(y_1)dB_2(y_2)\},$$

where

$$g_\tau(y_1,y_2) = \int_0^\tau \prod_{i=1}^2 \{(s-y_i)^+\}^{-(\alpha+1)/2} ds,$$

$$K(\alpha) = \{(1-2\alpha)(1-\alpha)\}^{1/2}\{\int_0^\infty (x+x^2)^{-(\alpha+1)/2}\}^{-1},$$

$$\sigma^2 = 2K_1^2\left(\frac{1+\rho_\infty}{1+\rho_0}\right)^2 + 2K_2^2\left(\frac{1-\rho_\infty}{1-\rho_0}\right)^2 + K_3\,\frac{1-\rho_\infty^2}{1-\rho_0^2},$$

and K_1, K_2 *and* K_3 *are constants determined by* a, b *and* ρ_0, (*for their explicit forms, see* [6]).

Remark. In [6], the present author proved this theorem for the case $\tau = 1$, $\rho_\infty = 0$ and $a = b$, or for $\tau = 1$, $\rho(t) \equiv 0$ and $a < b$. Later Taqqu [9] proved the general case.

The proof of Theorem 1 is based on Proposition 1 and the following lemma.

Lemma 1 (Taqqu [9]). *Suppose that* (1.1) *and* (1.2) *are satisfied with* $0 < \alpha < 1/2$, *and let*

$$J(t) = \int_0^t \{K_1H_2(Y_1(s)) + K_2H_2(Y_2(s)) + K_3Y_1(s)Y_2(s)\}ds.$$

Then as $t \to \infty$

$$\{\mathrm{Var}J(t)\}^{-1/2}J(t\tau)$$

converges weakly in the space $C[0, \infty)$ *to* $\Delta_1(\tau)$.

4. *Case II*

In this section, we present a result in [7]. Let $p \geq 2$ and $D = D(1, \cdots, 1; 1)$. Let

$$A = \{(z_1,\cdots,z_{p-1})' \mid \sum_{j=1}^{p-1} z_j^2 \leq (1-\rho_0)^{-1}\},$$

$$B = \{(z_1,\cdots,z_{p-1})' \mid (1-\rho_0)^{-1}[1-(1+(p-1)\rho_0)z_1^2] \geq \sum_{j=2}^{p-1} z_j^2\},$$

$$u = u(z_1,\cdots,z_{p-1}) = \{\frac{1-(1-\rho_0)\Sigma_{j=1}^{p-1} z_j^{\,2}}{1+(p-1)\rho_0}\}^{1/2}$$

$$v = v(z_1,\cdots,z_{p-1})$$

$$= \{(1-\rho_0)^{-1}[1-(1+(p-1)\rho_0)z_1^{\,2}] - \sum_{j=2}^{p-1} z_j^{\,2}\}^{1/2},$$

$$c_1 = - \int\cdots\int_A u\phi(u)\phi(z_1)\cdots\phi(z_{p-1})dz_1\cdots dz_{p-1} < 0,$$

$$c_2 = \cdots = c_p = - \int\cdots\int_B v\phi(v)\phi(z_1)$$

$$\cdots\phi(z_{p-1})dz_1\cdots dz_{p-1} < 0.$$

Let $B_j(\cdot)$, $j = 1, \cdots, p$, be p independent Gaussian random measures satisfying $EB_j(A_1)B_j(A_2) = |A_1 \cap A_2|$.

__Theorem 2.__ __As__ $t \to \infty$

$$Z_t(\tau) \overset{W}{\Rightarrow} \Delta_2(\tau).$$

__The limiting process is__

$$\Delta_2(\tau) = \frac{K(\alpha)}{d} \int_{R^2} g_\tau(y_1,y_2)\{c_1(\frac{1+(p-1)\rho_\infty}{1+(p-1)\rho_0})dB_1(y_1)dB_1(y_2)$$

$$+ c_2(\frac{1-\rho_\infty}{1-\rho_0}) \sum_{j=2}^{p} dB_j(y_1)dB_j(y_2)\},$$

__where__

$$d^2 = 2\{c_1^{\,2}(\frac{1+(p-1)\rho_\infty}{1+(p-1)\rho_0})^2 + (p-1)c_2^{\,2}(\frac{1-\rho_\infty}{1-\rho_0})^2\}.$$

The proof is based on Proposition 1 and the following

lemma.

Lemma 2 (Maejima [7]). *Suppose that* (1.1) *and* (1.2) *are satisfied with* $0 < \alpha < 1/2$ *and let*

$$J(t) = \int_0^t \sum_{j=1}^{p} c_j H_2(Y_j(s))\, ds.$$

Then as $t \to \infty$

$$\{\mathrm{Var} J(t)\}^{-1/2} J(t\tau)$$

converges weakly in the space $C[0, \infty)$ *to* $\Delta_2(\tau)$.

5. *Case III*

In this section, we explain the main results in Berman [2]. Our approach is different from his original one and based on Proposition 1.

5.1. Let $p \geq 2$ and $D = D(1, \cdots, 1; u(t))^c$.

Theorem 3. *Suppose that* $\rho(t) \equiv 0$, $u(t)$ *is nondecreasing and* $u(t) = o(\log t)$. *Then as* $t \to \infty$

$$Z_t(\tau) \overset{d}{\Rightarrow} \Delta_3(\tau),$$

where

$$\Delta_3(\tau) = K(\alpha)(2p)^{-1/2} \sum_{j=1}^{p} \int_{R^2} g_\tau(y_1, y_2)\, dB_j(y_1)\, dB_j(y_2),$$

and $\overset{d}{\Rightarrow}$ *denotes the convergence of all finite-dimensional distributions.*

Remark. The process

$$\int_{R^2} g_\tau(y_1, y_2)\, dB_1(y_1)\, dB_2(y_2)$$

is called the Rosenblatt process. So, $\Delta_2(\tau)$ is the weighted sum of independent Rosenblatt processes and $\Delta_3(\tau)$ is the sum of p independent Rosenblatt processes.

<u>Proof of Theorem 3</u>, (based on Proposition 1). Let

$$f_t(x_1,\cdots,x_p) = I[\sum_{j=1}^{p} x_j^2 \geqq u(t)]. \tag{5.1}$$

Since $\rho(t) \equiv 0$, we do not have to consider $\{Y(t)\}$. We apply Proposition 1 with $m = 2$ to f_t in (5.1). Since the region is symmetric with respect to each axis,

$$c_t(1,0,\cdots,0) = c_t(0,1,0,\cdots,0)$$

$$= \cdots = c_t(0,\cdots,0,1) = 0.$$

We next check condition (c). We have

$$c_t(2,0,\cdots,0) = \frac{1}{2}\int\cdots\int_{\Sigma_{j=1}^{p}x_j^2 \geqq u(t)} H_2(x_1)\phi(x_1)\phi(x_2)$$

$$\cdots\phi(x_p)\,dx_1\cdots dx_p$$

$$= -\int\cdots\int_{\Sigma_{j=2}^{p}x_j^2<u(t)} \phi(x_2)\cdots\phi(x_p)\,dx_2\cdots dx_p$$

$$\int_0^{\{u(t)-\Sigma_{j=2}^{p}x_j^2\}^{1/2}} H_2(x_1)\phi(x_1)\,dx_1$$

$$= \int\cdots\int_{\Sigma_{j=2}^{p}x_j^2<u(t)} \phi(x_2)\cdots\phi(x_p)\{u(t) - \sum_{j=2}^{p} x_j^2\}^{1/2}$$

$$\phi(\{u(t)-\sum_{j=2}^{p} x_j^2\}^{1/2})\,dx_2\cdots dx_p$$

$$= (2\pi)^{-p/2} e^{-u(t)/2} \int\cdots\int_{\Sigma_{j=2}^{p} x_j^2 < u(t)} \{u(t) - \sum_{j=2}^{p} x_j^2\}^{1/2}$$

$$dx_2 \cdots dx_p$$

$$= \frac{1}{2} (2\pi)^{-p/2} e^{-u(t)/2} C_1(p) u(t)^{p/2},$$

where

$$C_1(p) = \pi^{p/2}/\Gamma((p/2)+1),$$

and

$$c_t(2,0,\cdots,0) = c_t(0,2,0,\cdots,0) = \cdots = c_t(0,\cdots,0,2).$$

We also see that $c_t(m_1,\cdots,m_p) = 0$ if $m_i = m_j = 1$ for some $1 \leqq i < j \leqq p$ and $m_k = 0$ for all $k \neq i, j$.

On the other hand,

$$\int_{R^p} f_t(x_1,\cdots,x_p)^2 \{\prod_{j=1}^{p} \phi(x_j) dx_j\}$$

$$= \int\cdots\int_{\Sigma_{j=1}^{p} x_j^2 \geqq u(t)} \{\prod_{j=1}^{p} \phi(x_j) dx_j\}$$

$$\sim C_2(p) u(t)^{(p-1)/2} e^{-u(t)/2}.$$

Hence in (c)

$$\eta(t) \sim C \frac{u(t)^{p-1} e^{-u(t)/2}}{u(t)^{p} e^{-u(t)}} = C \frac{e^{u(t)/2}}{u(t)} \leqq Ct^{\varepsilon}$$

for any $\varepsilon > 0$, by the assumption that $u(t) = o(\log t)$. Hence (c) is fulfilled.

Consequently by Proposition 1, $Z_t(\tau)$ has the same limiting distribution as

$$\tilde{I}_t(\tau) \equiv \{\mathrm{Var}I(t)\}^{-1/2} I(t\tau),$$

where

$$I(t) = \tfrac{1}{2}(2\pi)^{-p/2} e^{-u(t)/2} C_1(p) u(t)^p \sum_{j=1}^{p} \int_0^t H_2(X_j(s))ds$$

and

$$\mathrm{Var}I(t) \sim \frac{p}{2(1-2\alpha)(1-\alpha)} (2\pi)^{-p} e^{-u(t)} C_1(p)^2 u(t)^{2p} t^{2-2\alpha} L(t)^2.$$

Hence

$$\tilde{I}_t(\tau) \sim \frac{\{(1-2\alpha)(1-\alpha)\}^{1/2}}{(2p)^{1/2} t^{1-\alpha} L(t)} \sum_{j=1}^{p} \int_0^{t\tau} H_2(X_j(s))ds,$$

and thus the non-central limit theorem ([3], [8]) concludes the theorem.

5.2. Let $p \geqq 2$ and $D = D(1, \cdots, 1; v(t))$.

<u>Theorem 4</u>. <u>Suppose that</u> $\rho(t) \equiv 0$, $v(t)$ <u>is nonincreasing and</u> $v(t) \geqq Ct^{-\delta}$ <u>with</u> $0 < \delta < (2/p)\min(\alpha, 1-2\alpha)$. <u>Then as</u> $t \to \infty$

$$Z_t(\tau) \overset{d}{\Rightarrow} \Delta_3(\tau).$$

<u>Remark</u>. Berman [2] proved Theorem 4 under weaker condition that $0 < \delta < (2/p)(1-\alpha)$ by a different approach.

<u>Proof</u>. Let

$$f_t(x_1, \cdots, x_p) = I[\sum_{j=1}^{p} x_j^2 \leqq v(t)].$$

It is enough to check (c) in Proposition 1. We have

$$c_t(2,0,\cdots,0) = -\frac{1}{2}(2\pi)^{-p/2}e^{-v(t)/2}C_1(p)v(t)^{p/2}$$

and

$$\int\cdots\int_{\Sigma_{j=1}^{p} x_j^2 \leqq v(t)} \{\prod_{j=1}^{p}\phi(x_j)dx_j\} \sim C_1(p)v(t)^{p/2}.$$

Hence in (c)

$$\eta(t) \sim \frac{Cv(t)^{p/2}}{v(t)^p e^{-v(t)}} \leqq Cv(t)^{-p/2} = o(t^{p\delta/2}),$$

completing the proof.

6. Case IV

Let $D = D(1/b, \cdots, 1/b; 1) - D(1/a, \cdots, 1/a; 1)$.
6.1. A result in [6] is extended to the case of general dimension as follows.

Theorem 5. Suppose that $\rho(t) \equiv C$, $a < p^{1/2} < b$ *and* $a^p\phi(a) \neq b^p\phi(b)$. *Then as* $t \to \infty$

$$Z_t(\tau) \overset{W}{\Rightarrow} \Delta_3(\tau).$$

Proof. Let

$$f(x_1,\cdots,x_p) = I[a^2 \leqq \sum_{j=1}^{p} x_j^2 \leqq b^2].$$

Note that $c(m_1,\cdots,m_p) = 0$ if $m_i = 1$ for some $1 \leqq i \leqq p$ and $m_j = 0$ for all $j \neq i$. Moreover

$c(2,0,\cdots,0)$

$$= \int \cdots \int_{a^2 \leq \Sigma_{j=1}^{p} x_j^2 \leq b^2} H_2(x_1)\phi(x_1)\phi(x_2)\cdots\phi(x_p)\,dx_1\cdots dx_p$$

$$= \frac{1}{2} C_1(p)(2\pi)^{-p/2}\{a^p e^{-a^2/2} - b^p e^{-b^2/2}\} \neq 0$$

by $a^p\phi(a) \neq b^p\phi(b)$, and

$$c(2,0,\cdots,0) = c(0,2,0,\cdots,0) = \cdots = c(0,\cdots,0,2).$$

This completes the proof of the finite dimensional convergence by Proposition 1. The tightness of $\{Z_t(\tau)\}$ can be proved in the same manner as in [7].

6.2. In [6], the present author considered the case $p = 2$ satisfying $\rho_0 = 0$ and $a^2\phi(a) = b^2\phi(b)$. Then we saw that $Z_t(\tau)$ has the same limiting distribution as $\mathfrak{J}_t(\tau) = \{\mathrm{Var} J(t)\}^{-1/2} J(t\tau)$, where

$$J(t) = C\int_0^t \{H_4(Y_1(s)) + H_4(Y_2(s))$$

$$+ 2H_2(Y_1(s))H_2(Y_2(s))\}ds$$

if it exists. Here $C = (1/64)(a^2-b^2)a^2\exp(-a^2/2)$. Taqqu's argument [9] suggests us that the limiting distribution of $\mathfrak{J}_t(\tau)$ exists if $r(t) \sim t^{-\alpha}L(t)$, $0 < \alpha < 1/4$ and it seems to be

$$\Delta_4(\tau) = \mathrm{const.}\times\int_{R^4} g_\tau(y_1,\cdots,y_4)$$

$$\{dB_1(y_1)dB_1(y_2)dB_1(y_3)dB_1(y_4)$$

$$+ dB_2(y_1)dB_2(y_2)dB_2(y_3)dB_2(y_4)$$

$$+ 2dB_1(y_1)dB_1(y_2)dB_2(y_3)dB_2(y_4)\}$$

with integrand

$$g_\tau(y_1,\cdots,y_4) = \int_0^\tau \prod_{i=1}^{4} \{(s-y_i)^+\}^{-(\alpha+1)/2} ds.$$

The problem can also be treated in the general dimension.

7. Additional comment

As we have seen in [6], in the situation of section 6, the assupmtion $\rho_0 \neq 0$ might give us the different type of limiting distributions, even if all other conditions are the same. Furthermore, the constant K_3 in Theorem 1 in section 3 is zero if $a = b$ or $\rho(t) \equiv 0$ and nonzero otherwise, and the processes $\Delta_1(\tau)$ with zero K_3 and with nonzero K_3 are of different type. From this point of view, the author believes it worthwile to study how the limiting distributions depend on the dependence of the components of vector-valued Gaussian processes, in the sojourn time problems we have discussed in this paper.

Reference

[1] Berman, S.M. (1979) : High level sojourns for strongly dependent Gaussian processes. Z. Wahrsch. 50, 223-239.

[2] Berman, S.M. (1984) : Sojourns of vector Gaussian processes inside and outside spheres. Z. Wahrsch. 66, 529-542.

[3] Dobrushin, R.L. and Major, P. (1979) : Non-central limit theorems for non-linear functionals of Gaussian fields. Z. Wahrsch. 50, 27-52.

[4] Maejima, M. (1981) : Some sojourn time problems for strongly dependent Gaussian processes. Z. Wahrsch. 57, 1-14.

[5] Maejima, M. (1982) : Some limit theorems for sojourn times of strongly dependent Gaussian processes. Z. Wahrsch. 60, 359-380.

[6] Maejima, M. (1985) : Some sojourn time problems for 2-dimensional Gaussian processes. To appear in J. Multivar. Anal.

[7] Maejima, M. (1985) : Sojourns of multidimensional Gaussian processes with dependent components. To appear in Yokohama Math. J.

[8] Taqqu, M.S. (1979) : Convergence of integrated processes of arbitrary Hermite rank. Z. Wahrsch. 50, 53-83.

[9] Taqqu, M.S. (1984) : Sojourn in an elliptical domain. Technical Report No.630, School of Operations Research and Industrial Engineering, Cornell University.

Department of Mathematics
Keio University
3-14-1, Hiyoshi
Kohoku-ku, Yokohama 223
Japan

Hausdorff dimension of sample paths for self-similar processes

by

Norio Kôno

1. Introduction. First we consider a real continuous function f defined on $I = [0,1]$ which satisfies a uniform Hölder condition:

$$(1) \qquad |f(t) - f(s)| \leq K|t-s|^{\alpha} \quad t,s \in I \quad (0 < \alpha \leq 1).$$

Then, we have the upper bound of the Hausdorff dimension of the graph $G = \{(t,f(t));\ t \in I\}$ i.e. $d_H(G) \leq 2 - \alpha$, where $d_H(\cdot)$ means the Hausdorff dimension.

Now we consider a one dimensional Brownian motion path $B(t, \omega)$. We know that for almost all ω, $B(t, \omega)$ satisfies a uniform Hölder condition when $\alpha < 1/2$, which means that the Hausdorff dimension of the graph $G_\omega = \{(t,B(t,\omega);t \in I\}$ is bounded above by 3/2 a.s. But recall that to obtain the uniform Hölder condition for the Brownian motion paths usually we need the exponential moment condition

$$(2) \qquad E[\exp\{\tfrac{1}{4}(B(t) - B(s))^2/|t-s|\}] \leq M < +\infty.$$

Surprisingly enough, to obtain the Hausdorff dimension of the graph of self-simialr stationary increment processes we do not require an exponential moment condition like (2). Strictly speaking, we do not assume even "self-similarity". We only require "incremental stationarity" with some additional conditions. We also discuss the existence of local times under the similar conditions.

2. Results. Let $\{X(t, \omega);\ 0 \leq t \leq 1\}$ be a R^d-valued separable and measurable stochastic process, where R^d is the d-dimensional Euclidean space with the usual norm $\|\ \|$. Set $G_\omega = \{(t,X(t, \omega));\ 0 \leq t \leq 1\}$ and $R_\omega =$

$\{X(t, \omega);\ 0 \leq t \leq 1\}$.

Theorem 1. We assume that there exist three positive constants α, β and M with $\beta > \max(d, 1/\alpha)$ such that

$$(3) \qquad E[||X(t) - X(s)||^{\beta}] \leq M\,|t-s|^{\alpha\beta}$$

holds for all $0 \leq t, s \leq 1$. Then, we have

(i) $\quad d_H(G_\omega) \leq 1 + d(1-\alpha)$ a.s. and

(ii) $\quad d_H(R_\omega) \leq \min(d, 1/\alpha)$ a.s.

Remark 1. The condition (3) implies necessarily $\alpha < 1$ except for the degenerate case, and by the Kolmogorov's theorem we can choose $\{X(t)\}$ as path continuous version, so do we.

Theorem 2. We assume that there exist three positive constants a, α, d' such that

$$(4) \qquad P(||X(t) - X(s)|| \leq |t-s|^{\alpha}x) \leq ax^{d'}$$

holds for all $x \geq 0$. Then, we have

(i) $\quad d_H(G_\omega) \geq \max[1, \min\{1+d'(1-\alpha), 1/\alpha\}]$ a.s. and

(ii) $\quad d_H(R_\omega) \geq \min(d', 1/\alpha)$ a.s.

Corollary. If $\{X(t)\}$ satisfies the conditions (3) and (4) with $d = d'$, then we have $d_H(G_\omega) = 1+d(1-\alpha)$ a.s. if $\alpha d < 1$ and $d_H(R_\omega) = \min(d, 1/\alpha)$ a.s.

Remark 2. If a process $\{X(t)\}$ satisfies both the conditions (3) and (4) of Theorem 1 and 2, and if $\alpha d < 1$, then d' must be less than or equal to d.

As for the existence of local times, applying the

results by D. Geman [2] or S. Berman [1], we have the following theorem.

Theorem 3. Let $\{X(t); 0 \le t \le 1\}$ be a real valued separable measurable stochastic process. Assume that $\{X(t)\}$ satisfies one of the following two conditions:

(i) There exist three positive constants a, d, α with $0 < \alpha < 1$ and $d \ge 1$ such that

$$(5) \qquad P(|X(t) - X(s)| \le |t-s|^{\alpha} x) \le ax^{d}$$

holds for all $x \ge 0$. Or,

(ii) there exist two positive constances α and M with $0 < \alpha < 1$ such that

$$(6) \qquad \int_{-\infty}^{\infty} E[e^{iu(X(t)-X(s))|t-s|^{-\alpha}}]du \le M$$

holds for all $0 \le t, s \le 1$. Then, there exists a square integrable local time with respect to the Lebesgue measure a.s.

Remark 3. Taking account of Berman's remark [1, p.280], if $\{X(t)\}$ has a square integrable local time, the sample paths cannot satisfy a local Hölder condition of order greater than 2 at any point a.s.

3. Proofs of the Theorems. First we derive a lemma which may be of independent interest.

Lemma 1. If there exist three positive constants α, β and M with $\alpha\beta > 1$ and $\beta \ge 1$ such that

$$(7) \qquad E[||X(t) - X(s)||^{\beta}] \le M\, |t-s|^{\alpha\beta}$$

holds for all $0 \le t, s \le 1$, then for any t we have

$$(E[\sup_{0\le h\le T} ||X(t+h) - X(t)||^{\beta}])^{1/\beta} \le K\, T^{\alpha},$$

where

$$K = 2^{-1/\beta} M^{1/\beta} \Sigma_{n=1}^{\infty}\, 2^{-(\alpha-1/\beta)n}.$$

Proof. Recall that by (7) we can choose $\{X(t)\}$ having continuous sample paths. Set

$$h = T\, \Sigma_{k=1}^{\infty} \varepsilon_k 2^{-k}, \quad h_n = \Sigma_{k=1}^{n}\, \varepsilon_k 2^{-k} T, \quad h_0 = 0,$$

where $\varepsilon_k = 0$ or 1. Then, we have

$$X(t+h,\omega) - X(t,\omega) = \Sigma_{n=1}^{\infty} (X(t+h_n,\omega) - X(t+h_{n-1},\omega))$$

$$= T^{\alpha} \Sigma_{h_n \ne h_{n-1}} 2^{-n\alpha} (X(t+h_n,\omega) - X(t+h_{n-1},\omega))(h_n - h_{n-1})^{-\alpha}$$

and

$$\sup_{0\le h\le T} ||X(t+h,\, \omega) - X(t,\, \omega)||$$

$$\le T^{\alpha} \Sigma_{n=1}^{\infty} 2^{-n\alpha} \max_{0\le k\le 2^{n-1}-1} ||X(t+2^{-n}T+k2^{-n+1}T,\omega)$$

$$- X(t+k2^{-n+1}T,\, \omega)||\, 2^{n\alpha} T^{-\alpha}.$$

It follows by the Minkovski inequality that

$$(E[\sup_{0\le h\le T} ||X(t+h) - X(t)||^{\beta}])^{1/\beta}$$

$$\le T^{\alpha} \Sigma_{n=1}^{\infty} 2^{-n\alpha} (E[(\max_{0\le k\le 2^{n-1}-1} ||X(t+2^{-n}T+k2^{-n+1}T)$$

$$- X(t+k2^{-n+1}T)||\, 2^{n\alpha} T^{-\alpha})^{\beta}])^{1/\beta}$$

$$\le T^{\alpha} \Sigma_{n=1}^{\infty} 2^{-n\alpha} (\Sigma_{k=0}^{2^{n-1}-1} E[||X(t+2^{-n}T+k2^{-n+1}T)$$

$$- X(t+k2^{-n+1}T)||^{\beta} / (2^{-n\alpha\beta} T^{\alpha\beta})])^{1/\beta}$$

$$\le 2^{-1/\beta} M^{1/\beta} T^{\alpha} \Sigma_{n=1}^{\infty} 2^{-(\alpha-1/\beta)n}.$$

Using this lemma we start proving Theorem 1.

Proof of Theorem 1, (i). Set $N_\varepsilon = [1/\varepsilon]$ $(0 < \varepsilon < 1)$,

$M_i(\varepsilon) = \sup_{i\varepsilon \le t \le (i+1)\varepsilon} ||X(t, \omega) - X(i\varepsilon, \omega)||$ $\quad 0 \le i \le N_\varepsilon$,

$B_i = \{x \in R^d; ||x - X(i\varepsilon, \omega)|| \le M_i(\varepsilon)\}$,

S_i = a minimal ε-net of B_i, and

$B_i(y) = \{(t, x) \in [0, 1] \times R^d ; i\varepsilon \le t \le (i+1)\varepsilon, ||x - y|| \le \varepsilon\}$ for $y \in S_i$. Since we have the diameter of $B_i(y) \le \sqrt{5}\,\varepsilon$, the cardinal number of $S_i \le c_1(M_i(\varepsilon))^d$ and

$$G_\omega \subset \bigcup_{0\le i\le N_\varepsilon,\ y\in S_i} B_i(y),$$

it follows for γ-dimensional Hausdorff outer measure Λ^γ that

$$\begin{aligned} E[\Lambda^\gamma(G_\omega)] &\le c_2 E[\Sigma_{i=0}^{N_\varepsilon} \Sigma_{y\in S_i} \varepsilon^\gamma] \\ &\le c_2\varepsilon^{\gamma-d} E[\Sigma_{i=0}^{N_\varepsilon} M_i^d(\varepsilon)] \\ &\le c_2\varepsilon^{\gamma-d} \Sigma_{i=0}^{N_\varepsilon} (E[M_i^\beta(\varepsilon)])^{d/\beta} \\ &\le c_3\varepsilon^{\gamma-d-1+\alpha d} \quad \text{(by Lemma 1).} \end{aligned}$$

Therefore, if $\gamma > 1+d(1-\alpha)$ we have $E[\Lambda^\gamma(G_\omega)] = 0$. This completes the proof of Theorem 1, (i).

The proof for the second part of Theorem 1 is almost the same way as the first one. Since we have $R_\omega \subset \bigcup_{0\le i\le N_\varepsilon} B_i$, it follows for $\beta > \gamma > 1/\alpha$ that

$$\begin{aligned} E[\Lambda^\gamma(R_\omega)] &\le \Sigma_{i=0}^{N_\varepsilon} E[M_i^\gamma(\varepsilon)] \\ &\le \Sigma_{i=0}^{N_\varepsilon} E[M_i^\beta(\varepsilon)]^{\gamma/\beta} \\ &\le c_4\varepsilon^{-1+\alpha\gamma} \to 0 \quad \text{as} \quad \varepsilon \downarrow 0. \end{aligned}$$

By the trivial inequality $d_H(R_\omega) \le d$, we have $d_H(R_\omega) \le \min$

$(d, 1/\alpha)$ a.s.

Proof of Theorem 2,(i). To obtain the lower bound of the Hausdorff dimension, it is sufficient to show that there exists a probability measure μ on G_ω such that for $0 < \gamma < d_H$

$$\int_{G_\omega \times G_\omega} r_{pq}^{-\gamma} d\mu(p)d\mu(q) < +\infty$$

holds, where r_{pq} is the distance in R^{d+1} between two points p and q in G_ω. Now we choose the μ as the image measure of the Lebesgue measure on $[0, 1]$ by the map $t \in [0, 1] \to (t, X(t, \omega)) \in G_\omega$, i.e. we have

$$\int_{G_\omega \times G_\omega} r_{pq}^{-\gamma} d\mu(p)d\mu(q)$$
$$= \int_0^1 \int_0^1 (|t-s|^2 + ||X(t, \omega) - X(s, \omega)||^2)^{-\gamma/2} \, dtds.$$

Setting $F_{t,s}(x) = P(||X(t) - X(s)|| \leq |t-s|^\alpha x)$, we have

$$E[\int_0^1\int_0^1(|t-s|^2 + ||X(t) - X(s)||^2)^{-\gamma/2}dtds]$$
$$= \int_0^1\int_0^1 (\int_0^\infty(|t-s|^2 + |t-s|^{2\alpha}x^2)^{-\gamma/2}dF_{t,s}(x)) \, dtds$$
$$= \int_0^1\int_0^1\gamma(|t-s|^{2\alpha}\int_0^\infty x(|t-s|^2 + |t-s|^{2\alpha} x^2)^{-\gamma/2-1}$$
$$\times F_{t,s}(x)dx) \, dtds.$$

Now let us estimate the inner term:

$$I = |t-s|^{2\alpha}\int_0^\infty x(|t-s|^2 + |t-s|^{2\alpha}x^2)^{-\gamma/2-1}F_{t,s}(x)dx.$$

We devide into the following three cases.

(i) The case $\alpha \geq 1$.

$$I \leq |t-s|^{2\alpha}\int_0^\infty x(|t-s|^2 + |t-s|^{2\alpha}x^2)^{-\gamma/2-1}dx$$
$$= |t-s|^{-\gamma}\int_0^\infty y(1+y^2)^{-\gamma/2-1}dy.$$

(ii) The case $0 < \alpha < 1$ and $\alpha d' < 1$.
In this case, take γ such as $1+d'(1-\alpha) > \gamma > d'$ and devide I into two parts as follows:

$$I = |t-s|^{2\alpha}(\int_0^{|t-s|^\delta} + \int_{|t-s|^\delta}^\infty) = I_1 + I_2,$$

where $\delta = (1-\alpha)(\gamma-d')/\gamma$. Then we have

$$I_1 \le a|t-s|^{2\alpha}\int_0^{|t-s|^\delta}(|t-s|^2 + |t-s|^{2\alpha}x^2)^{-\gamma/2-1}x^{1+d'}\,dx$$

$$\le a|t-s|^{-\gamma+d'(1-\alpha)}\int_0^\infty(1+y^2)^{-\gamma/2-1}y^{1+d'}dy \quad \text{and}$$

$$I_2 \le |t-s|^{2\alpha}\int_{|t-s|^\delta}^\infty x(|t-s|^2 + |t-s|^{2\alpha}x^2)^{-\gamma/2-1}dx$$

$$\le |t-s|^{-\gamma}\int_{|t-s|^{\delta+\alpha-1}}^\infty y(1+y^2)^{-\gamma/2-1}dy$$

$$\le c_5|t-s|^{-\gamma+d'(1-\alpha)} \quad (\text{recall } \delta < 1-\alpha).$$

(iii) The case $0 < \alpha < 1$ and $\alpha d' \ge 1$.
In this case, take γ such as $\gamma < d'$ and devide I into the following two parts:

$$I = |t-s|^{2\alpha}(\int_0^1 + \int_1^\infty) = I_1 + I_2.$$ Then we have

$$I_1 \le a|t-s|^{2\alpha}\int_0^1(|t-s|^2 + |t-s|^{2\alpha}x^2)^{-\gamma/2-1}x^{1+d'}dx$$

$$= a|t-s|^{-\gamma+d'(1-\alpha)}\int_0^{|t-s|^{\alpha-1}}(1+y^2)^{-\gamma/2-1}y^{1+d'}dy$$

$$\le c_6|t-s|^{-\alpha\gamma} \quad \text{and}$$

$$I_2 \le |t-s|^{2\alpha}\int_1^\infty x(|t-s|^2 + |t-s|^{2\alpha}x^2)^{-\gamma/2-1}dx$$

$$= |t-s|^{-\gamma}\int_{|t-s|^{\alpha-1}}^\infty y(1+y^2)^{-\gamma/2-1}dy$$

$$\le c_7|t-s|^{-\alpha\gamma}.$$

In the above all cases, choose γ such as $0 < \gamma < \max[1, \min\{1+d'(1-\alpha), 1/\alpha\}]$ then we get

$$E[\int_0^1\int_0^1(|t-s|^2 + ||X(t)-X(s)||^2)^{-\gamma/2}dtds] < +\infty.$$

The standard argument yields the proof of Theorem 2, (i).

Proof of Theorem 2,(ii). By the similar way as before, we have for $\gamma < \min(d', 1/\alpha)$

$$E[\int_0^1\int_0^1||X(t)-X(s)||^{-\gamma}dtds]$$

$$= \int_0^1\int_0^1|t-s|^{-\alpha\gamma}(\int_0^\infty x^{-\gamma}dF_{t,s}(x))dtds$$

$$= \gamma\int_0^1\int_0^1|t-s|^{-\alpha\gamma}(\int_0^\infty x^{-\gamma-1}F_{t,s}(x)dx)dtds$$

$$\leq c_8 \int_0^1\int_0^1|t-s|^{-\alpha\gamma}dtds < +\infty.$$

Proof of Theorem 3,(i). By Theorem A of [2], it is enough to check that for any $\phi \in K$

$$\int_0^1\int_0^1 \phi(X(t, \omega) - X(s, \omega))dtds < +\infty \quad \text{a.s.},$$

where $K = \{\phi;\ \phi(x) = \int_{|x|}^\infty y^{-1}|f(y)|dy,\ f$ belongs to a countable dense subset of $L^1(0, \infty)\}$.

To do so we calculate the followings.

$$I = E[\int_0^1\int_0^1 \phi(X(t, \omega) - X(s, \omega))dtds]$$

$$= E[\int_0^1\int_0^1(\int_{|X(t,\omega)-X(s,\omega)|}^\infty y^{-1}|f(y)|dy)dtds]$$

$$= \int_0^1\int_0^1 E[\int_{|X(t)-X(s)||t-s|^{-\alpha}}^\infty u^{-1}|f(|t-s|^\alpha u)|du]dtds,$$

but

$$E[\int_{|X(t)-X(s)||t-s|^{-\alpha}}^\infty u^{-1}|f(|t-s|^\alpha u)|du]$$

$$= \int_0^\infty(\int_x^\infty u^{-1}|f(|t-s|^\alpha u)|du)dF_{t,s}(x)$$

$$\leq a|t-s|^{-\alpha}\int_0^\infty|f(y)|dy + \int_0^\infty|f(y)|dy$$

$\leq c_9|t-s|^{-\alpha}.$

Therefore $I < +\infty$ for $0 < \alpha < 1$.

The second part of Theorem 3 follows immediately by Lemma 3.1 of [1].

Institute of Mathematics
Yoshida College
Kyoto University
Kyoto, 606 Japan

References

[1] S.M. Berman; Local times and sample funciton properties of stationary Gaussian processes. Trans. Amer. Math. Soc. 137(1969), 277-299.

[2] D. Geman and J. Horowitz; Local times for real and random functions. Duke Math. J. 43(1976),809-828.

[3] S. Orey; Gaussian sample functions and the Hausdorff dimension of level crossings. Z. Wahr. Geb. 15 (1970), 249-256.

[4] S. J. Taylor; The Hausdorrf α-dimensional measure of Brownian paths in n-space. Camb. Phil. Soc. 49 (1953), 31-39.

[5] __________; The α-dimensional measure of the graph and set of zeros of a Brownian path. Ibid. 51 (1955), 265-274.

[6] __________; On the connexion between Hausdorff measures and generalized capacity. Ibid. 57(1961), 524-531.

[7] I. A. Ibragimov; On smoothenss conditions for trajectories on random functions. Theory of Prob. App. 28(1984), 240-262.

Almost sure approximations for U-statistics

Herold Dehling, Universität Göttingen

1. Introduction:

U-statistics have been introduced by Hoeffding in 1948 [13] as generalizations of the sample mean. Let $X_1, X_2 \ldots$ be a sequence of i.i.d real-valued random variables.

Definition: For a symmetric function $h: \mathbf{R}^m \to \mathbf{R}$, $m \geq 1$, we define the U-statistic $U_n(h)$ by

(1.1) $$U_n(h) = \binom{n}{m}^{-1} \sum_{1 \leq i_1 < \ldots < i_m \leq n} h(X_{i_1}, \ldots, X_{i_m})$$

h is called the kernel of the U-statistic.

From a statistical viewpoint, $U_n(h)$ is an unbiased estimator of the parameter $\theta = Eh(X_1, \ldots, X_m)$. If the family of distributions under consideration is rich enough, $U_n(h)$ is the minimum variance unbiased estimator of θ.

The kernel $h: \mathbf{R}^m \to \mathbf{R}$ is called degenerate if

(1.2) $$\int h(x_1, \ldots, x_m) dF(x_i) = 0$$

where F is the c.d.f. of X_1.

We define the following norm for kernels:

(1.3) $$\|h\|_p = (E|h(X_1, \ldots, X_m)|^p)^{1/p}$$

For brevity we shall denote by $L_p(m, dF)$ the space of all kernels $h: \mathbf{R}^m \to \mathbf{R}$ with $\|h\|_p < \infty$. If dF is Lebesgue-measure on $[0, 1]$, we write instead of $L_p(m, dF)$ for short $L_p(m)$.

Hoeffding (1948) proved the following decomposition theorem:

Theorem 1.1: *(i) There exist degenerate kernels $h_c: \mathbf{R}^c \to \mathbf{R}$, $1 \leq c \leq m$ such that for every $n \geq m$:*

(1.4) $$U_n(h) - \theta = \sum_{c=1}^{m} \binom{m}{c} U_n(h_c)$$

(ii) If $h_d \not\equiv 0$ and $h_c \equiv 0$ for $1 \leq c < d$ for some $d < m$, then $n^{d/2}\binom{m}{d}U_n(h_d)$ and $n^{d/2}(U_n(h) - \theta)$ are stochastically equivalent,i.e. their difference converges to 0 in probability.

Hence the asymptotic distribution of $U_n(h) - \theta$ is determined by the first non-vanishing term in the expansion (1.4) and it suffices to study the asymptotics of $U_n(h)$ for degenerate h. Especially if $0 < \varsigma_1 = \text{Var } h_1 < \infty$, $\sqrt{n}(U_n(h) - \theta)$ is asymptotically normal with mean zero and variance $m^2\varsigma_1$. This was observed by Hoeffding. Roughly speaking, one can say that all limit results (asymptotic normality, law of the iterated logarithm, invariance principles,...) for U-statistics with $\varsigma_1 > 0$ follow from the corresponding results for sequences of i.i.d. random variables and (usually fairly crude) estimates of the size of $U_n(h_c)$, $c \geq 2$. This applies e.g. to the almost sure invariance principles given by Sen [26].

We shall from now on restrict our attention to degenerate kernels having finite 2nd moments. Moreover, by considering the kernel
$h^*(u_1, \ldots, u_m) := h(F^{-1}u_1, \ldots, F^{-1}u_m)$, we may from now on assume that the X_i's are uniformly distributed on $[0,1]$ and that $h : [0,1]^m \to \mathbf{R}$. Weak convergence in this case has been studied by von Mises [20], Filippova [12], a distribution invariance principle has been obtained by Denker, Grillenberger and Keller [7] and independently by Mandelbaum and Taqqu [19]. In the more general setup of symmetric statistics weak convergence has been studied by Dynkin and Mandelbaum [11].

Except for the case $m = 2$ the limit laws of $U_n(h)$ cannot be explicitly given in the sense that one could write down their distribution function or characteristic function. Instead, the limit laws are given as multiple stochastic integrals

$$\int_0^1 \cdots \int_0^1 h(x_1, \ldots, x_m) 1_{\{x_i \neq x_j \text{ for } i \neq j\}} K(dx_1, t) \ldots K(dx_m, t) \tag{1.5}$$

with respect to the Kiefer process. Here by Kiefer-process we denote a mean-zero two-parameter Gaussian process $K(s,t)$, $0 \leq x \leq 1$, $t \geq 0$ with the following covariance structure:

$$E\ K(s,t)\ K(s',t') = s(1-s') \cdot (t \wedge t') \text{ if } s \leq s' \tag{1.6}$$

In the case $m = 2$ the distribution of the stochastic double integral (for fixed t) can be explicitly computed. Note that by homogeneity it suffices to consider the stochastic integral with respect to $K(x,1) = B(x)$, where B is the Brownian bridge process on $[0,1]$.

The kernel h induces a bounded linear operator $T_h : L_2(dx) \to L_2(dx)$ by

$$T_h f(x) = \int_0^1 h(x,y) f(y) dy$$

T_h is a Hilbert-Schmidt operator and has eigenvalues $\lambda_i, i = 1, 2 \ldots$ satisfying $\sum_{i=1}^{\infty} \lambda_i^2 < \infty$. If $W_1, W_2 \ldots$ are independent standard normal random variables then

$\sum_{i=1}^{\infty} \lambda_i(W_i^2 - 1)$ has the same distribution as $\int_0^1 \int_0^1 h(x,y)\ 1_{\{x \neq y\}} B(dx)B(dy)$ (Neuhaus [23], for a short proof see Denker [6]).

Alternatively, one can compute the characteristic function:

$$E(\exp \text{it} \int_0^1 \int_0^1 h(x,y)B(dx)B(dy)) = \frac{1}{\sqrt{D(2it,h)}}$$

where $D(z,h)$ is the Fredholm determinant of the kernel $h(x,y)$, i.e.:

$$D(z,h) = 1 + \sum_{k=1}^{\infty} \frac{(-z)^k}{k!} \int_0^1 \dots \int_0^1 \begin{vmatrix} h(t_1,t_1) & \dots & h(t_1,t_k) \\ h(t_k,t_1) & \dots & h(t_k,t_k) \end{vmatrix} dt_1 \dots dt_k$$

(von Mises [20] and Kac-Siegert [14])

The distribution invariance principle in the formulation of Denker, Grillenberger and Keller is the following.

Theorem 1.2.: *Let $h \in L_2(m)$ be a degenerate kernel. Then the $D[0,1]$-valued random element $\{n^{m/2}U_{[nt]}(h) : 0 \le t \le 1\}$ converges in distribution to*

$$\left\{\int_0^1 \dots \int_0^1 h(x_1,\dots,x_m)\ 1_{\Delta^c} K(dx_1,t)\dots K(dx_m,t),\ 0 \le t \le 1\right\}$$

where Δ is the set

$$\Delta = \{(x_1,\dots,x_m)\colon \exists i \neq j \text{ with } x_i = x_j\} \tag{1.7}$$

The work ([3],[4],[5]) surveyed in this article was motivated by two classical refinements of the Central Limit Theorem for i.i.d. random variables, namely the Law of the Iterated Logarithm and Strassen's Almost Sure Invariance Principle:

Let $X_1, X_2, \dots$ be an i.i.d. sequence of random variables with $EX_1 = 0$ and $EX_1^2 = 1$. Then:

(i) (Law of the Iterated Logarithm)

$$\lim_{n\to\infty} \begin{matrix}\sup\\ \inf\end{matrix} \frac{X_1 + \dots + X_n}{(2n \log\log n)^{1/2}} = \pm 1 \quad \text{a.s.} \tag{1.8}$$

(ii) (Almost Sure Invariance Principle) If the probability space is rich enough, there exists Brownian Motion $\{W_t, t \ge 0\}$ such that

$$X_1 + \dots + X_n - W_n = o((n \log\log n)^{1/2}) \quad \text{a.s.} \tag{1.9}$$

For sequences having higher moments, (1.9) was strengthened by Komlos, Major, Tusnady [15] and Major [17]. Generalizations to dependent random variables have been obtained by several authors (see Philipp and Stout [25]). Moreover, random variables with values in Banach spaces have been considered (Kuelbs and Philipp [16], Philipp [24], Morrow and Philipp [22], Dehling [2]).

We shall prove results in the spirit of (1.8) and (1.9) for U-statistics. Without being explicitly stated, the ideas developed in the above-mentioned papers enter into these proofs.

In the next section we define multiple stochastic integrals with respect to Kiefer processes. Section 3 deals with U-statistics of kernels having some smoothness properties and gives the method of proof for almost sure invariance principles in the very simplest case. Section 4 contains the Law of the Iterated Logarithm for U-statistics and in section 5 this result is used to derive the Almost Sure Invariance Principle for general kernels. The last section gives applications, e.g. to the LIL for multiple Wiener integrals.

2. Multiple stochastic integrals with respect to Kiefer-processes.

In this section we shall, following the approach of Filippova [12], give an outline of the definition of the multiple stochastic integral

$$\int_0^1 \cdots \int_0^1 h(x_1, \ldots, x_m)\ K(dx_1, t) \ldots K(dx_m, t) \tag{2.1}$$

for kernels h satisfying $|||h|||_2 < \infty$, where $|||\quad|||_p$ is defined by:

$$|||h|||_p = \sum_{\substack{Q \text{partition} \\ \text{of} \{1,\ldots,m\}}} \|h_Q\|_p \tag{2.2}$$

Here for any partition $Q = \{Q_1, \ldots, Q_q\}$ of $\{1, \ldots, m\}$ the kernel $h_Q : [0,1]^q \to R$ is given by

$$h_Q(y_1, \ldots, y_q) = h(x_1, \ldots, x_m), \text{ where } x_i = y_j \text{ iff } i \in Q_j$$

As far as the distribution of (2.1) for a fixed t is concerned, it suffices because of the homogeneity of the Kiefer-process to consider the integral for $t = 1$, which is equivalent to integration with respect to the Brownian Bridge B_t on $[0, 1]$. That one can even choose a version of (2.1) that is continuous in t has been proved by Denker, Grillenberger and Keller [7].

First we shall define the stochastic integral for step-functions given by:

$$h(x_1, \ldots, x_m) = \sum_{1 \le i_1, \ldots, i_m \le s} h_{i_1 \ldots i_m} \prod_{j=1}^{m} 1(t_{i_j}, t_{i_j+1}](x_j) \tag{2.3}$$

where $h_{i_1 \ldots i_m} \in R$ and $0 = t_0 < t_1 < \ldots < t_s < t_{s+1} = 1$

For such an h we put:

(2.4)
$$\int_0^1 \cdots \int_0^1 h(x_1,\ldots,x_m)\; B(dx_1)\ldots B(dx_m) =$$
$$= \sum_{1\le i_1,\ldots,i_m\le s} h_{i_1\ldots i_m} \prod_{j=1}^{m} \left(B(t_{i_j+1}) - B(t_{i_j})\right)$$

This definition can in the usual way be extended to all kernels h with $|||h|||_2 < \infty$ using the following two lemmas:

Lemma 2.1: *The set of simple kernels of the form (2.3) is dense in* $(L_2(m), |||\cdot|||_2)$.

Lemma 2.2: *For h as in (2.3) we have:*

(2.5)
$$E\left|\int_0^1 \cdots \int_0^1 h(x_1,\ldots,x_m)\; B(dx_1)\ldots B(dx_m)\right|^2 \le C|||h|||_2$$

for some absolute constant C.

The proofs of these lemmas are given e.g. in Denker [6], pp. 69-73.

3. "Smooth" kernels; method of proof.

Note that Strassens almost sure invariance principle does not imply the Central Limit Theorem nor a distribution-type invariance principle. For this the remainder term in (1.9) would have to be $o(n^{1/2})$. But, as was shown by Major [18] this can only be obtained under sharper moment conditions; e.g. $E|X_1|^{2+\delta} < \infty$ for some $\delta > 0$. Similarly, in almost sure approximations of U-statistics, results that imply also the distribution-type invariance principle can only be obtained under additional requirements on the kernel function h.

First we need more than 2nd moments:

(3.1)
$$||h||_{2+\delta} < \infty \quad \text{for some } \delta > 0$$

Moreover, we have to impose a "smoothness" restriction on the kernel, which in its most general form may be stated as follows:

(3.2) There is a refining sequence of partitions $(\alpha(r), r \ge 1)$ of $[0,1]$ with $\alpha(r) = \{A_i(r) : 1 \le i \le 2^r\}$ and $h_{i_1\ldots i_m} \in R$, $1 \le i_1,\ldots,i_m \le 2^r$ and constants $C, \gamma > 0$ such that:

$$||h - \sum_{1\le i_1,\ldots,i_m\le 2^r} h_{i_1\ldots i_m} 1_{A_{i_1}(r)\times\ldots\times A_{i_m}(r)}||_2 \le C2^{-r\gamma}$$

This is satisfied, if the kernel h has some Hölder-continuity or bounded variation property (see [3]). Especially, all classical kernels satisfy (3.2).

Theorem 3.1: *Let $X_1, X_2, \ldots$ be a sequence of independent random variables, uniformly distributed on $[0,1]$ and let $h : [0,1]^m \to R$ be a kernel satisfying (3.1) and (3.2). Then, after a possible enlargening of the probability space, there exists a standard Kiefer-process $\{K(s,t),\ 0 \le s \le 1, t \ge 0\}$ such that:*

(3.3)

$$\frac{n!}{(n-m)!}U_n(h) - \int_0^1 \ldots \int_0^1 h1_{\Delta^c}(x_1,\ldots,x_m)\ K(dx_1,n)\ldots K(dx_m,n) << n^{\frac{m}{2}-\lambda}$$

for some $\lambda > 0$.

Note that, since the random variables X_i have continuous distribution, we may replace $U_n(h)$ by $U_n(h1_{\Delta^c})$. Moreover, we find $U_n(h1_{\Delta^c}) = \frac{(n-m)!}{n!}V_n(h1_{\Delta^c})$ where V_n is the von-Mises-statistic, which for a degenerate kernel g is defined by:

(3.4)
$$V_n(g) = \sum_{1 \le i_1,\ldots,i_m \le n} g(X_{i_1}\ldots,X_{i_m})$$

In addition, if (3.2) holds, we can in a similar way approximate $h1_{\Delta^c}$ in the $|||\cdot|||_2$-norm (defined in (2.2)) by simple, degenerate kernels.

The basic idea of the proof of Theorem 3.1 is probably best understood from the proof for degenerate kernels with finitely many values. In addition, we restrict ourself to the case $m = 2$, so that:

(3.5)
$$h(x,y) = \sum_{1 \le \mu,\nu \le d} h_{\mu\nu} 1_{A_\mu \times A_\nu}(x,y)$$

where $\{A_\mu,\ 1 \le \mu \le d\}$ is a partition of $[0,1]$.

To avoid confusion, we make the following remark: For kernels as in (3.5) one could give a proof of Theorem 3.1 relying on the multidimensional analogue of Strassens invariance principle for i.i.d. random variables. We shall not follow this, since this approach is of no use for non-simple kernels.

The main ingredients of our proof are the following results, which for reference we state here:

(a) (Yurinskii [30]) Let $\xi_1, \xi_2, \ldots, \xi_n$ be independent R^d-valued random variables with mean 0 and $E\|\xi_i\|^3 < \infty$. If μ_n is the law of $n^{-1/2}\sum_{i \le n}\xi_i$ and ν_n the Gaussian law on R^d with mean 0 and the same covariance matrix as μ_n, then the following estimate for the Prohorov-distance d_p of μ_n and ν_n holds:

$$d_p(\mu_n,\nu_n) \le cd^{1/3}n^{-1/9}(n^{-1}\sum_{i \le n} E\|\xi_i\|^3)^{1/3}$$

(b) (Strassen [29]) Let μ and ν be two probability measures on the separable metric space (S,d) with $d_p(\mu,\nu) \le \epsilon$. Then there exists a law Q on $S \times S$ with marginals μ and ν satisfying

$$Q(\{(x,y) : d(x,y) > \epsilon\}) \le \epsilon$$

(c) (Skorohod [27], Dudley, Philipp [10]) Let S and T be Polish spaces and Q a law on $S \times T$ with marginal μ on S. Let $(\Omega, \mathfrak{A}, P)$ be a probability space and $X : \Omega \to S$ a random variable with distribution μ. Assume that there exists a random variable $U : \Omega \to [0,1]$, uniformly distributed on $[0,1]$ and independent of X. Then there exists a random variable $Y : \Omega \to T$ such that (X, Y) has distribution Q.

Proof of Theorem 3.1: Choose $\lambda > 0$, $\alpha > 10$ with $\lambda < \min(\frac{1}{\alpha+1}(\frac{1}{2} - \frac{1}{2+\delta}), \frac{1}{\alpha}(\frac{\alpha}{9} - \frac{10}{9}))$ and put $t_k = [k^{\alpha+1}]$, $H_k = [t_k, t_{k+1}) \cap \mathbf{Z}, n_k = \text{card } H_k \approx (\alpha+1)k^\alpha$
Consider the d-dimensional random vector

(3.6) $$Y_k = (n_k^{-1/2} \sum_{i \in H_k} (1_{A_\mu}(X_i) - |A_\mu|);\ 1 \leq \mu \leq d)$$

where $|\ \ |$ denotes the Lebesgue-measure. This random vector has mean 0 and covariance matrix $C = (c_{\mu\nu})$ where

$$c_{\mu\nu} = \begin{cases} |A_\mu|(1 - |A_\mu|) & \text{if } \mu = \nu \\ -|A_\mu||A_\nu| & \text{if } \mu \neq \nu \end{cases}$$

Then by Yurinskii's theorem we find:

$$d_p(L(Y_k); N(0,C)) << d^{1/3} n_k^{-1/9}$$

Applying Theorems (b) and (c) together, we can find an $N(0,C)$-distributed random variable Z_k, depending only on the σ-field generated by Y_k and U_k such that

(3.7) $$P(\{|Y_k - Z_k| \geq cd^{1/3} n_k^{-1/9}\}) \leq cd^{1/3} n_k^{-1/9}$$

Here, $(U_k, k \geq 1)$ is an i.i.d. sequence of random variables, uniformly distributed on $[0,1]$ and independent of $(X_k, k \geq 1)$. That such a sequence has been constructed is meant by the phrase "after a possible enlargening of the probability space."

Altogether we find an i.i.d. sequence $\{Z_k\}$ with $N(0,C)$-distribution, satisfying (3.7). Now let $K'(s,t)$ be any standard Kiefer process and define

$$Z_k'(\mu) = n_k^{-\frac{1}{2}} \int 1_{A_\mu}(x)(K'(dx, t_{k+1}) - K'(dx, t_k))\ ,\ 1 \leq \mu \leq d$$

It is easily seen that (Z_k) and (Z_k') have the same law. Now we can apply the Skorohod-Dudley-Philipp theorem to the joint law Q of K' and Z_k' and find a Kiefer process $K(s,t)$, such that

(3.8) $$Z_k = n_k^{\frac{1}{2}} (\int 1_{A_\mu}(x)(K(dx, t_{k+1}) - K(dx, t_k)).\ 1 \leq \mu \leq d)$$

An application of the Borel-Cantelli-Lemma together with (3.7) gives:

(3.9) $$|Y_k - Z_k| << n_k^{-\frac{1}{9}}$$

Hence we get for the kernel h defined in (3.5):

$$
\begin{aligned}
V_{t_\ell}(h) - & \int\int h(x,y)\; K(dx,t_\ell)\; K(dy,t_\ell) \\
&= \sum_{1\le\mu,\nu\le d} h_{\mu\nu} \sum_{j,k\le \ell} (n_j n_k)^{\frac{1}{2}} (Y_j(\mu)Y_k(\nu) - Z_j(\mu)Z_k(\nu)) \\
&\le \Big(\sum_{1\le\mu,\nu\le d} h_{\mu\nu}^2\Big)^{1/2} \sum_{j,k\le\ell} \{ n_j^{\frac12} |Y_j - Z_j| \cdot n_k^{\frac12}|Z_k| + n_k^{\frac12}|Y_k - Z_k| \cdot n_j^{\frac12}|Y_j - Z_j| \\
&\quad + n_k^{1/2}|Y_k - Z_k| \cdot n_j^{\frac12}|Z_j|\}
\end{aligned}
$$

Since $\{Z_j, j\ge 1\}$ is an i.i.d. Gaussian sequence, we have $|Z_j| << \log j$ and hence we get using (3.9) and the definition of t_k

$$
\begin{aligned}
V_{t_\ell}(h) - & \int_0^1\int_0^1 h(x,y) K(dx,t_\ell) K(dy,t_\ell) \\
&\le \Big(\sum_{1\le\mu,\nu\le d} h_{\mu\nu}^2\Big)^{1/2} \sum_{j,k} \Big(n_j^{\frac12} n_j^{-\frac19} n_k^{\frac12} n_k^{-\frac19} + 2 n_j^{\frac12} n_j^{-\frac19} n_k^{\frac12} \log k \Big) \\
&\le \Big(\sum_{1\le\mu,\nu\le d} h_{\mu\nu}^2\Big)^{1/2} \cdot 3 \Big(\sum_{j\le\ell} n_j^{\frac12-\frac19}\Big)\Big(\sum_{k\le\ell} n_k^{\frac12}\log k\Big) \\
&<< \Big(\sum_{1\le\mu,\nu\le d} h_{\mu\nu}^2\Big)^{1/2} \cdot t_\ell^{\frac89+\frac{10}{9\alpha}}
\end{aligned}
$$

At this point we have proved the almost sure invariance principle along the subsequence $t_\ell, \ell = 1,2,\ldots$. Hence what remains to be done is to "break into the blocks" via a maximal inequality. To make use of the martingale structure of U-statistics we write:

$$
\begin{aligned}
V_n(h) &= 2 \sum_{1\le i<j\le n} h(X_i,X_j) + \sum_{i=1}^n h(X_i,X_i) \\
&= 2U_n'(h) + \sum_{i=1}^n h(X_i,X_i),
\end{aligned}
$$

where we put $U_n'(h) = \binom{n}{m} U_n(h)$.

We need the following moment estimates which can be proved easily along the lines of Doob [9], pp. 226-227.

$$
(3.10) \qquad E|\sum_{\substack{n_1<i\le n_1+l_1\\ n_2<j\le n_2+l_2}} \sum_{i<j} h(X_i,X_j)|^{2+\delta} \le C(l_1 l_2)^{1+\delta/2} \int_0^1\int_0^1 |h(x,y)|^{2+\delta} dx dy
$$

$$
(3.11) \qquad E|\sum_{i=1}^n (h(X_i,X_i) - \int_0^1 h(x,x)dx)|^{2+\delta} \le n^{1+\frac{\delta}{2}} \int_0^1 |h(x,x)|^{2+\delta} dx
$$

Then we get using the martingale maximal inequality:

$$P\{\max_{t_k<n\le t_{k+1}} |V_n - V_{t_k}| > t_k^{1-\lambda}\}$$
$$\le P\{\max_{t_k<n\le t_{k+1}} |U_n'(h) - U_{t_k}'(h)| > \frac{1}{4}t_k^{1-\lambda}\}$$
$$+ P\{\max_{t_k<n\le t_{k+1}} |\sum_{j=t_k+1}^{n} (h(X_j,X_j) - \int_0^1 h(x,x)dx)| > \frac{1}{2}t_k^{1-\lambda} - n_k \cdot |\int_0^1 h(x,x)dx|\}$$
$$<< t_k^{-(1-\lambda)(2+\delta)} \cdot (n_k t_k)^{1+\frac{\delta}{2}} + t_k^{-(1-\lambda)(2+\delta)} \cdot n_k^{1+\frac{\delta}{2}}$$
$$<< k^{-1-\epsilon}$$

Now we can apply the Borel-Cantelli lemma to obtain (3.3)

4. The functional LIL for U-statistics

In this section we shall first prove a crude upper bound in the LIL for U-statistics. The basic fact used here is that for degenerate h the U-statistic $\{U_n'(h); n \ge 1\}$ is a martingale. Using a martingale LIL (Theorem 4.2) one can relate the a.s. behavior of $U_n'(h)$ to that of the sum of conditional variances

$$V_n = \sum_{j=1}^{n} E[(U_j'(h) - U_{j-1}'(h))^2 | L_{j-1}],$$

when L_j is the σ-field generated by $X_1, \dots, X_j$. To get a bound on the growth of this is the main problem in the proof. We can write the conditional variances in the following way:

$$\begin{aligned} &E[(U_n'(h) - U_{n-1}'(h))^2 | L_{n-1}] \\ &= E[\{\sum_{1\le i_1<\dots<i_{m-1}<n} h(X_{i_1}, \dots, X_{i_{m-1}}, X_n)\}^2 | L_{n-1}] \\ &= \int_0^1 \{\sum_{1\le i_1<\dots<i_{m-1}<n} h(X_{i_1}, \dots, X_{i_{m-1}}, t)\}^2 dt \\ &= \|U_{n-1}'(\tilde{h})\|_{L_2(dx)}^2 \end{aligned} \tag{4.1}$$

Here $\tilde{h}: [0,1]^{m-1} \to L_2(dx)$ is defined by

$$\tilde{h}(x_1, \dots, x_{m-1})(\cdot) = h(x_1, \dots, x_{m-1}, \cdot) \tag{4.2}$$

Hence we have related the almost sure behavior of $U_n'(h)$ to the almost sure behavior of $U_{n-1}'(\tilde{h})$, where $\tilde{h}$ is a Hilbert-space valued kernel of degree $m-1$. If we set up the whole theory for H-valued kernels from the beginning, we can proceed by induction on m.

Theorem 4.1:[4] *Let $h\colon [0,1]^m \to H$ be a measurable function with values in the separable Hilbert space $(H, |\cdot|)$. Let $\{X_j, j \geq 1\}$ be a sequence of independent random variables uniformly distributed on $[0,1]$. Suppose h is degenerate and*

$$E|h(X_1, \ldots, X_m)|^{2+\delta} < \infty$$

Then

$$\limsup_{n\to\infty} (\log\log n)^{-\frac{m}{2}} U_n'(h) \leq \frac{8^{\frac{m}{2}}}{(m!)^{1/2}} \cdot (E|h(X_{1,\ldots,}X_m)|^2)^{1/2} \text{ a.s.} \tag{4.3}$$

The main ingredient in the proof of Theorem 4.1, sketched above and in detail given below, is the following bounded LIL for Hilbert-space valued martingales, due to Dehling, Denker and Philipp [4].

Theorem 4.2: *Let $\{X_n, L_n, n \geq 1\}$ be a square-integrable martingale difference sequence with values in a separable Hilbert space $(H, |\cdot|)$. We denote the conditional expectation operator $E_n(\cdot) = E_n(\cdot|L_{n-1})$. Suppose that there is an adapted sequence $(M_n, L_{n-1}, n \geq 1)$ with $M_n \to \infty$ a.s. and*

$$\limsup_{n\to\infty} M_n^{-1} \sum_{j\leq n} E_j |X_j|^2 \leq 1 \tag{4.4}$$

Let f be a non-decreasing function such that $f(x)(\log\log x)^\alpha / x$ is non-increasing for some $\alpha > 17$ and that

$$\sum_{n\geq 1} (f(M_n))^{-1} E_n\{|X_n|^2 1\{|X_n|^2 > f(M_n)\}\} < \infty \text{ a.s.} \tag{4.5}$$

Then we have with probability 1:

$$\limsup_{n\to\infty} (2M_n \log\log M_n)^{-1/2} |\sum_{j\leq n} X_j| \leq 2 \tag{4.6}$$

Taking Theorem 4.2 for granted, it is fairly easy to prove Theorem 4.1. First we need the following moment estimates for $U_n'(h)$:

Lemma 4.3: *Let $U_n'(h)$ be a U-statistic with H-valued kernel h. Then*

$$E|U_n'(h)|^2 = \binom{n}{m} E|h(X_1, \ldots, X_m)|^2 \tag{4.7}$$

$$E|U_n'(h)|^{2+\delta} \leq A(m,\delta) n^{m(1+\delta/2)} E|h(X_1, \ldots, X_m)|^{2+\delta} \tag{4.8}$$

where the constant $A(m,\delta)$ depends only on m and δ.

The proof of (4.7) is just a straightforward computation, (4.8) is proved by a modification of the standard proof in the case $m = 1$, given e.g. in Doob [pp. 226-227].

Proof of Theorem 4.1: Let $u_n = U_n'(h) - U_{n-1}'(h)$ $= \sum_{1\le i_1<\ldots<i_{m-1}<n} h(X_{i_1},\ldots,X_{i_{m-1}},X_n)$ denote the martingale difference sequence of $U_n'(h)$. Using (4.1) and the induction hypothesis we obtain:

$$\limsup_{n\to\infty} (n\log\log n)^{-m+1} E_n u_n^2 \le \frac{8^{m-1}}{(m-1)!}\|h\|_2^2$$

If we put in Theorem 4.2 $M_n = \frac{8^{m-1}}{m!} n^m (\log\log n)^{m-1}\|h\|_2^2$ and $f(x) = \frac{x}{\log x}$ then we get from (4.6):

$$\limsup_{n\to\infty} (n\log\log n)^{-\frac{m}{2}} |U_n'(h)| \le \left(\frac{8^{\frac{m}{2}}}{(m!)^{1/2}}\right)\cdot \|h\|_2 \text{ a.s.}$$

once we have proved that (4.5) holds. But this follows immediately from the Beppo-Levi-theorem and the following estimation:

$$\begin{aligned} &f(M_n)^{-1}E\{|u_n|^2 1\{|u_n|^2 > f(M_n)\}\} \\ &\qquad\le f(M_n)^{-1-\delta/2}E|u_n|^{2+\delta} \\ &\qquad<< M_n^{-(1+\delta/2)}E(E(|u_n|^{2+\delta}|X_n))\cdot(\log M_n)^{1+\frac{\delta}{2}} \\ &\qquad<< n^{-m(1+\delta/2)} n^{(m-1)(1+\delta/2)}(\log n)^{1+\delta/2} \\ &\qquad<< n^{1-\delta/4} \end{aligned}$$

In the remaining part of this paper we shall refine Theorem 4.1 to identify the lim sup in (4.3) exactly. We will even prove a functional version of the LIL, i.e. we consider the almost sure behavior of the following $D[0,1]$-valued random elements:

$$Z_n^h(t) := (2n\log\log n)^{-\frac{m}{2}}\frac{n!}{(n-m)!}U_{[nt]}(h) \tag{4.9}$$

Theorem 4.4: (Dehling, [5]) *Let $h\in L_{2+\delta}(m)$ for some $\delta>0$. Then with probability 1 $\{Z_n^h, n\ge 1\}$ is relatively compact and has the following set of limit points:*

$$\begin{aligned} K_h = \{&x\in C[0,1]: \\ &x(t) = \int_0^1\ldots\int_0^1 h(x_1,\ldots,x_m)\int_0^t f(x_1,s)ds\ldots\int_0^t f(x_m,s)ds dx_1\ldots dx_m, \\ &\text{where } \int_0^1\int_0^1 f^2(x,s)dxds\le 1\} \end{aligned} \tag{4.10}$$

To prove this result, we need the following approximation lemma for degenerate kernels, proved in [5].

Lemma 4.5: *Let $h \in L_2(m)$ be a symmetric degenerate kernel. Then for any $\epsilon > 0$ there exist $N \geq 1$, $\lambda_1, \ldots, \lambda_N \in \mathbf{R}$ and $g_1, \ldots, g_N \in L_2^0[0,1] = \{g \in L_2[0,1] : \int_0^1 g dx = 0\}$ such that $g(x_1, \ldots, x_m) = \sum_{i=1}^{N} \lambda_i g_i(x_1) \cdot \ldots \cdot g_i(x_m)$ satisfies $\|h - g\|_2 \leq \epsilon$*

From this and (4.3) it follows via a standard argument that it is sufficient to prove Theorem 4.4 for simple kernels

$$g(x_1, \ldots, x_m) = \sum_{i=1}^{N} \lambda_i g_i(x_1) \cdot \ldots \cdot g_i(x_m) \tag{4.11}$$

Let us recall first the functional LIL for $\mathbf{R}^d$-valued random variables:

If $Y_1, Y_2, \ldots$ is an i.i.d. sequence of $\mathbf{R}^d$-valued mean-zero random variables having finite 2nd moments and with partial sums $S_n = \sum_{i=1}^n Y_i$, then

$$Z_n(t) := (2n \log\log n)^{-\frac{1}{2}} S_{[nt]}, \ 0 \leq t \leq 1$$

is almost surely relatively compact in $D_d[0,1]$, the space of all functions $f : [0,1] \to \mathbf{R}^d$ that are right-continuous and have left limits. If C is the covariance matrix of Y_1 and A some $d \times d$ matrix satisfying $A^T A = C$, then the set of limit points of Z_n is given by:

$$K_C = \{x \in D_d[0,1] : \ x(t) = \int_0^t y(s) \cdot A ds, \ \int_0^1 \|y\|^2 ds \leq 1\}$$

If we define $\tilde{Z}_n^h$ as in (4.9) only with $\frac{n!}{(n-m)!} U_n$ replaced by V_n, i.e. including diagonals, then it is easily seen that for g as in (4.11) $\tilde{Z}_n^g - Z_n^g \longrightarrow 0$ as $n \to \infty$ a.s. and that $\tilde{Z}_n^g = \sum_{i=1}^N \lambda_i (Z_n^{g_i})^m$. Since the map $\Phi : D_N[0,1] \to D[0,1]$, defined by $\Phi(x_1, \ldots, x_N) = \sum_{i=1}^N \lambda_i x_i^m$ is continuous, we can apply the N-dimensional FLIL and get:

With probability 1, $\tilde{Z}_n^g$ is relatively compact and has the following set of limit points:

$$\{\sum_{i=1}^{N} \lambda_i x_i^m : x(t) = \int_0^t y(s) \cdot A \ ds, \ y : [0,1] \to \mathbf{R}^N \text{ and} \int_0^1 \|y(s)\|^2 ds \leq 1\}$$

where A is any $N \times N$-matrix satisfying $A^T A = (E g_i(X_1) g_j(X_1))_{1 \leq i,j \leq N}$. Hence it remains to show that K_g coincides with the above set. From (4.10) we get:

$$K_g = \{\sum_{i=1}^{N} \lambda_i x_i^m : x_i(t) = \int_0^t \int_0^1 g_i(x) f(x,s) dx ds, \ \int_0^1 \int_0^1 f^2(x,s) dx ds \leq 1\}$$

Let $e_1, \ldots, e_N$ be a complete ONS for the subspace of L_2 generated by $g_1, \ldots, g_N$. Then we may assume without loss of generality that $f(\cdot, s) \in \text{span}\,(e_1, \ldots, e_N)$ for all $s \in [0,1]$. Denoting the inner product in L_2 by $\langle \cdot, \cdot \rangle$, we obtain

$$K_g = \{\sum_{i=1}^{N} \lambda_i x_i^m : x_i(t) = \sum_{j=1}^{N} \langle g_i, e_j \rangle \int_0^t \langle f, e_j \rangle ds, \sum_{j=1}^{N} \int_0^1 \langle f, e_j \rangle^2 ds \leq 1\}$$

$$= \{\sum_{i=1}^{N} \lambda_i x_i^m : x(t) = (\int_0^t \langle f, e_j\rangle ds, j = 1, \dots N) A_g; \int_0^1 \sum_{j=1}^{N} \langle f, e_j\rangle^2 ds \le 1\}$$

where A_g is defined by $(A_g)_{j,k} = \langle g_k, e_j\rangle$ and hence

$$(A_g^T A_g)_{i,k} = \sum_j \langle g_i, e_j\rangle\langle g_k, e_j\rangle = \langle g_i, g_k\rangle = E(g_i(X_1) g_k(X_1)).$$

5. The almost sure invariance principle for general (non-smooth) kernels

In Section 3 we proved the almost sure invariance principle for kernels having only finitely many values. Now it is easy to see that each kernel $h \in L_2(m)$ can be approximated in L_2-norm by these simple kernels, i.e. for every $\epsilon > 0$ there exists $h_\epsilon \in L_2(m)$ having only finitely many values such that

$$\|h - h_\epsilon\|_2 \le \epsilon \tag{5.1}$$

We can approximate $U_n(h_\epsilon)$ by a stochastic integral with respect to the Kiefer-process and we have a bound for $U_n(h - h_\epsilon)$ by Theorem 4.1. What is needed in addition is a similar bound for the stochastic integral of $h - h_\epsilon$ with respect to the Kiefer process.

Theorem 5.1:[4] *Let $h \in L_{2+\delta}(m)$ for some $\delta > 0$ be a degenerate kernel vanishing on all diagonals. Then*

$$\begin{aligned} &\limsup_{n\to\infty} (n \log\log n)^{-\frac{m}{2}} \int_0^1 \dots \int_0^1 h(x_1, \dots, x_m) K(dx_1, n) \dots K(dx_m, n) \\ &\le (8m!)^{1/2} \|h\|_2 \end{aligned} \tag{5.2}$$

Putting these results together, we immediately obtain the following:

Proposition 5.2: *Let $h \in L_{2+\delta}(m)$ for some $\delta > 0$ be a degenerate kernel vanishing on all diagonals. Then for for every $\epsilon > 0$ there exists a standard Kiefer process $K^{(\epsilon)}(s, t)$ such that:*

$$\begin{aligned} &\limsup_{n\to\infty} (n \log\log n)^{-\frac{m}{2}} \\ &\cdot \left|\frac{n!}{(n-m)!} U_n(h) - \int_0^1 \cdots \int_0^1 h(x_1, \dots, x_m) K^{(\epsilon)}(dx_1, n) \dots K^{(\epsilon)}(dx_m, n)\right| \\ &\le \epsilon \end{aligned} \tag{5.3}$$

Hence it remains to find a "limiting" Kiefer-process $K(s, t)$ satisfying (5.3) with $\epsilon = 0$. This requires some tricky ideas, originally due to P. Major [18] and for the present case carried out in [3] and [4]. We finally get the following result:

Theorem 5.3: *Under the conditions of Proposition 5.2 there exists a Kiefer-process $K(s,t)$ such that*

(5.4)

$$\frac{n!}{(n-m)!}U_n(h) - \int_0^1 \dots \int_0^1 h(x_1,\dots,x_m)K(dx_1,n)\dots K(dx_m,n) = o((n\log\log n)^{\frac{m}{2}})$$

6. Applications, Examples:

In general, almost sure invariance principles like Strassen's classical theorem can be used to carry over limit results that can be fairly easily proved for the limiting Gaussian process to the original partial sum process. In our case we use the reverse implication to get a functional LIL for multiple stochastic integrals with respect to the Kiefer process, where a direct proof seems to be considerably more difficult. We get as a simple Corollary of Theorems 4.4 and 5.3 the following result:

Theorem 6.1: *Let $h \in L_{2+\delta}(m)$ be a degenerate kernel vanishing on Δ. Then*

$$(2n\log\log n)^{-\frac{m}{2}} \int_0^1 \dots \int_0^1 h(x_1,\dots,x_m)K(dx_1,nt)\dots K(dx_m,nt), \ 0 \le t \le 1$$

is with probability 1 relatively compact in $D[0,1]$ and has K_h as its set of limit points.

A similar theorem has been proved by Mori and Oodaira [21]. Theorems 4.4 and 6.1, of course, also imply the ordinary LIL.

Corollary 6.2: *Let $h \in L_{2+\delta}(m)$ be a degenerate kernel. Then*

$$\begin{aligned}\limsup_{n\to\infty}\ & (2n\log\log n)^{-\frac{m}{2}} \frac{n!}{(n-m)!}U_n(h) \\ &= \limsup_{n\to\infty}\ (2n\log\log n)^{-\frac{m}{2}} \int_0^1 \dots \int_0^1 h(x_1,\dots,x_m)1_{\Delta^c}K(dx_1,n)\dots K(dx_m,n) \\ &= C(h)\ , \quad \text{where}\end{aligned}$$

$$C(h) = \sup\{\int_0^1 \dots \int_0^1 f(x_1)\dots f(x_m)h(x_1,\dots,x_m)dx_1\dots dx_m : \int_0^1 f^2 dx \le 1\}$$

In special cases $C(h)$ can be explicitly computed.

Example 6.3: If $(I_\ell, \ell \le L)$ is a partition of $[0,1]$ and $p_\ell = |I_\ell|$,we define

$$h(x,y) = \sum_{\ell \le L} p_\ell^{-1}(1_{I_\ell}(x) - p_\ell)(1_{I_\ell}(y) - p_\ell)$$

h is a degenerate kernel of dimension 2 and its associated von-Mises-statistic is just the statistic for the χ^2-test of goodness of fit. We compute:

$$\int\int f(x)h(x,y)f(y)dxdy = \sum_{\ell\le L} p_\ell^{-1}(\int_0^1 f(x)(1_{I_\ell}(x) - p_\ell)dx)^2$$

We may without loss of generality assume that $\int_0^1 f(x)dx = 0$ and that $f \equiv f_k \equiv$ const. on I_k and get hence

$$= \sum_{\ell \leq L} p_\ell^{-1}(f_\ell p_\ell)^2 = \sum_{\ell \leq L} p_\ell f_\ell^2 = \int_0^1 f^2(x)dx$$

Therefore we have $C(h) = 1$ in this case.

If $m = 2$, the constant $C(h)$ can be computed from the eigenvalues of the operator T_h. It is easily seen that $C(h) = \max\{\lambda_k : \lambda_k \text{ eigenvalue of } T_h\}$

Example 6.4: Consider the Cramer-von-Mises-Smirnoff statistic

$$G_n = \int_0^1 n \cdot w(x)(F_n(x) - x)^2 dx$$

where F_n is the e.d.f. of a uniformly $[0,1]$-distributed sample. G_n is a von-Mises-statistic with kernel:

$$h(x,y) = \int_0^1 w(u)(1_{\{x \leq u\}} - u)(1_{\{y \leq u\}} - u)du$$

One can show that $h(x,y)$ is the Green function of the Sturm-Liouville-type differential equation

$$\frac{d}{dx}(P(x) \cdot Z') + \lambda Z = 0, \quad P(x) = \frac{1}{w(x)} \tag{6.1}$$

under the boundary condition $P(x) \cdot Z' \to 0$ as $x \to 0$ or $x \to 1$ (de Wet and Venter [8]).

Since the eigenvalues of the Sturm-Liouville equation have been extensively studied in the literature, one can in this way compute $C(h)$ for a large class of kernels h.

As an example, consider $w(x) = \dfrac{1}{x(1-x)}$. The solutions of (6.1) with this weight-function are the Jacobi-Polynomials. Their eigenvalues are

$$\lambda_k = \frac{1}{k(k+1)}, \quad k \geq 1$$

and hence $C(h) = 1/2$, a result obtained earlier by Csaki [1].

References

[1] Csáki, E.: On the standardized empirical distribution function. *Coll. Mathem. Soc. János Bolyai*, **32**. Non-parametric Statistical Inference, pp. 123-138. Budapest (1980).

[2] Dehling, H.: Limit theorems for sequences of weakly dependent Banach space valued random variables. *Z. Wahrscheinlichkeitstheorie verw. Gebiete* **63**, 393-432 (1983).

[3] Dehling, H.; Denker, M.; Philipp, W.: Invariance principles for von Mises and U-statistics. *Z. Wahrscheinlichkeitstheorie verw. Gebiete* **67**, 139-167 (1984).

[4] Dehling, H.: Denker, M.; Philipp, W.: A bounded law of the iterated logarithm for Hilbert space valued martingales and its application to U-statistics. *Z. Wahrscheinlichkeitstheorie verw. Gebiete* (to appear).

[5] Dehling, H.: The functional law of the iterated logarithm for von Mises functionals, preprint (1984).

[6] Denker, M.: Asymptotic Theory in Nonparametric Statistics. *Braunschweig-Wiesbaden: Vieweg Verlag* (1985).

[7] Denker, M.; Grillenberger, Chr.; Keller, G.: A note on invariance principles for von Mises statistics. *Metrika* (to appear).

[8] de Wet, T.; Venter, J.H.: Asymptotic distributions for quadratic forms with applications to tests of fit. *Ann. Statistics* **1**, 380-387 (1973).

[9] Doob, J.L.: *Stochastic processes.* New York: Wiley (1953).

[10] Dudley, R.M.; Philipp, W.: Invariance principles for sums of Banach space valued random elements and empirical processes. *Z. Wahrscheinlichkeitstheorie verw. Gebiete* **62**, 509-552.

[11] Dynkin, E.B.; Mandelbaum, A.: Symmetric statistics, Poisson point processes and multiple Wiener integrals. *Ann. Statistics* **11**, 739-745 (1983).

[12] Filippova, A.A.: Mises' theorem on the asymptotic behavior of functionals of empirical distribution functions and its statistical applications. *Theory Probability Appl.* **7**, 24-57 (1961).

[13] Hoeffding, W.: A class of statistics with asymptotically normal distribution. *Ann. Math. Statistics* **19**, 293-325 (1948).

[14] Kac, M.; Siegert, A.J.F.: An explicit representation of a stationary Gaussian process. *Ann. Math. Statistics* **18**, 438-442 (1947).

[15] Komlos, J.; Major, P.; Tusnady, G.: An approximation of partial sums of independent RV's and the sample d.f. I and II. *Z. Wahrscheinlichkeitstheorie verw. Gebiete* **32**, 111-131 (1975) and **34**, 33-58 (1976).

[16] Kuelbs, J.; Philipp, W.: Almost sure invariance principles for partial sums of mixing B-valued random variables. *Ann. Probability* **8**, 1003-1036 (1980).

[17] Major, P.: The approximation of partial sums of independent RV's. *Z. Wahrscheinlichkeitstheorie verw. Gebiete* **35**, 213-220 (1976).

[18] Major,P.: Approximation of partial sums of i.i.d. r.v.'s when the summands have only two moments. *Z. Wahrscheinlichkeitstheorie verw. Gebiete* **35**, 221-229 (1976).

[19] Mandelbaum, A.; Taqqu, M.S.: Invariance principles for symmetric statistics. *Ann. Statistics* **12**, 483-496 (1984).

[20] von Mises, R.: On the asymptotic distribution of differentiable statistical functions. *Ann. Math. Statistics* **18**, 309-348 (1947).

[21] Mori, T.; Oodaira, H.: The law of the iterated logarithm for self-similar processes represented by multiple Wiener integrals. preprint (1985).

[22] Morrow,G.J.; Philipp, W.: An almost sure invariance principle for Hilbert space valued martingales. *TAMS* **273**, 231-251 (1982).

[23] Neuhaus, G.: Functional limit theorems for U-statistics in the degenerate case. *J. Multivariate Analysis* **7**, 424-439 (1977).

[24] Philipp, W.: Almost sure invariance principles for sums of B-valued random variables. Probability in Banach spaces II, *Lecture Notes in Math.* **709**, 171-193, Berlin-Heidelberg-New York: Springer Verlag (1979).

[25] Philipp, W.; Stout, W.F.: Almost sure invariance principles for partial sums of weakly dependent random vectors. *Memoirs of the AMS* **161**, (1975).

[26] Sen, P.K.: Almost sure behavior of U-statistics and von Mises differentiable statistical functions. *Ann. Statistics* **2**, 387-396 (1974).

[27] Skorohod, A.V.: On a representation of random variables. *Theory Probability Appl.* **21**, 628-632 (1976).

[28] Strassen, V.: An almost sure invariance principle for the law of the iterated logarithm. *Z. Wahrscheinlichkeitstheorie verw. Gebiete* **3**, 211-226 (1964).

[29] Strassen, V.: The existence of probability measures with given marginals. *Ann. Math. Statistics* **36**, 423-439 (1967).

[30] Yurinskii, V.V.: On the error of the Gaussian approximation for convolutions. *Theory Probability Appl.* **22**, 236-247 (1977).

Institut für Mathematische Stochastik, Universität Göttingen, Lotzestraße 13
D-3400 Göttingen,Fed. Rep. of Germany

A BIBLIOGRAPHICAL GUIDE TO SELF-SIMILAR PROCESSES AND LONG-RANGE DEPENDENCE

by

Murad S. Taqqu*

Self-similar processes are of interest in probability, statistics, physics, turbulence and in modelling random phenomena with long-range dependence. Since the applications are diverse, references are scattered in the literature. The purpose of this bibliographical guide is to *bring together* many of the important references to the subject. Relevant references to some related topics are also included. Although this is definitely not a comprehensive bibliography, we hope that it can be a useful tool for researchers in this fast moving area.

Self-similarity is a scaling property of the finite-dimensional distributions. A stochastic process $X(t)$ is self-similar with parameter H if $X(at)$ and $a^H X(t)$

* Research supported by the National Science Foundation grant ECS-84-08524.

have the same finite-dimensional distributions for all $a > 0$. Brownian motion for example is self-similar with $H = 1/2$. There are other self-similar Gaussian processes and also non-Gaussian processes, some with finite variance and some with infinite variance.

For an *introduction* to self-similar processes see [241], [239], [218], [185], [186], [250]. Integer-valued self-similar processes are discussed in [249], extremal self-similar processes in [251], operator self-similar processes in [115] and Markovian ones in [119]. Many of the seminal ideas about self-similarity are due to Benoit Mandelbrot. See for example the paper [160] which develops the properties of fractional Brownian motion, a family of Gaussian self-similar processes. Refer also to the book [158] and to its extensive bibliography.

Important papers of *historical interest* are [104], where fractional Brownian motion was first introduced; that process is also studied in [88], [255] and [160]. In [198], Rosenblatt discusses an example of a sequence with long-range dependence that does not converge to a Gaussian limit.

As noted in [118], self-similar processes arise in (weak) *limit theorems*. Many of the results involve also non-linear filtering, Hermite expansions and polynomial chaos. See Sun and Ho [218], Taqqu [239] and Major [140] for a survey. The literature about limit theorems is rather extensive. See the papers of Taqqu [235], [237], [238], Fox and Taqqu [52], [53], [55], Dobrushin and Major [40], [41], [42], Major [140]-[143], [15], Rosenblatt [198]-[202], Surgailis [219]-[224], Giraitis [59], [61], Ho and Sun [80]-[82]. See also [86], [4], [5], [16].

These results should be contrasted with what happens when there is a short-range dependence (see [92], [191], [14], [14a], [216], [217], [167], [59b], [61], [61b]). For *limits of stable random variables with long-range dependence*, see [4a], [246], [102], [120], [128], [2b], [2c], [133]-[135], [139], [101].

Scaling limits for *generalized processes* are discussed in [38], [39], [41], [49], [50].

The paper [243] is a recent survey on the *law of the iterated logarithm* for self-similar processes and it contains many references ([22], [51], [106]-[109], [116], [117], [181], [189], [236]). For further extensions see [182] and [242]. Notes on the *Hausdorff dimension* of paths can be found in [110], [98a].

Levy's Brownian motion (see [123a], [24a], [171a]) is Brownian motion with many parameters. It is an *isotropic self-similar random field*. For its extension to fractional Brownian motion with many parameters, see [1], Chap. 8 and [63]. See also [158], p. 356.

The *sojourn time* of multidimensional self-similar Gaussian processes in various regions of space is discussed in [138]. See also [8], [9], [131], [132], [136], [137], [240].

Self-similar processes and processes that are attracted to them are characterized by scaling exponents. There are few rigorous results about *statistical estimation* techniques. See [161] for a slightly dated survey and [30]. See also [6], [10], [29], [50a], [54], [58], [66], [67], [68], [73], [98], [112], [113], [154], the papers of Mandelbrot and Wallis [162]-[165], [172], [177]-[179], [190], [192],

[234].

Applications of self-similar processes and long-range dependence include modelling in *economics* [146a], [146b], [147], [150], [152], [69], [70], [67], [37], [245], *turbulence* [105], [153a], [155], [56], [77], *weather* [197], [129], [130], [130a], *communications* [149].

There have been many applications to *geophysics* and *hydrology*, particularly related to the so-called "Hurst phenomenon." See [121] for a review and [161] for a discussion. Hurst's original work in [89]-[91], was given a new interpretation by Mandelbrot [148] which suggested the use of self-similar processes and led Mandelbrot to coin the term "self-similar." See also [27], [65], [58], [123], the papers of Mandelbrot [151]-[153], [156], [157], [162]-[165], [78], [79], [172], [173], [174], [179], [187], [215]. [248].

Many of the preceding papers deal also with *simulation* ([151], [27], [172]), *long-range dependence*, and various variations of processes attracted to a self-similar process, for example *fractional ARMA*. For fractional ARMA see also [67], [68], [83]-[85], [58], [124], [125], [54].

Self-similar processes are of interest in *physics*, particularly in connection with the so-called "renormalization group theory" and "critical phenomena" at phase transitions. The connection with limit theorems is discussed in [97], [23], [184], the ϵ-expansion in [212], [11a], fractals in [184]. Some recent papers are collected in [57], [211a], [193a]. Related models are discussed in [87], [180], [211] [228]-[230], [159], [159a]. Dyson's hierarchical model is considered in [46],

[12], [28]. Self-affine fractals are discussed in [159b], [159c].

An interesting problem involving *collision of particles* was first treated in [76], [214], [62] and solved in [44]. See also [231], [232], [144], [103], [43].

The notion of *infrared and ultraviolet catastrophes* first suggested by Mandelbrot is discussed in [239] and [49].

The literature on $1/f$ *noises* is very extensive. See for example [47] and [254]. For the role of $1/f$ in the variations of quarz clocks, see [192]. An interesting non-random dynamical model for $1/f$ noise is discussed in [166] and an approach based on finitely additive probabilities is suggested in [72]. The use of generalized functions for 1/f noise is considered in [239].

Stable processes are self-similar and many other self-similar processes can be subordinated to stable ones, for example fractional stable motion (see [246]). Because stable processes with index $\alpha < 2$ have infinite variance, one obtains the so-called "Noah effect" discussed in [162]. A useful paper about the domain of attraction of multidimensional distributions is in [207]. There has been recently many interesting developments related to stable processes that are of particular relevance to self-similar processes. [252] contains a good review. See also [176], [71], [122], [17]-[20], [21a], [21b], [74], [75], [126]. Resnick [196] uses point processes representations to establish limit theorems. See also [31]. For moving averages in stable variables see [4a] and [2b], [2c], [25], [26], [31], [32], [134], [135]. See also [127]. For stable generalized self-similar processes, see [1].

Many self-similar processes are represented by *multiple Wiener-Itô integrals*. For L^2 integration, there is the original approach of Wiener [253], modified by Itô [194], [95] and the approach of Neveu [183], summarized in the appendix of [146]. Engel [48] considers the role of the diagonal when the underlying processes are infinitely divisible. Perez-Abreu [193] offers a general perspective, and also includes a treatment of multiple Wiener-Itô integrals with respect to generalized processes. L^2 multiple Wiener-Itô integrals are also defined in [140], [99] and treated in [121c], [194], [195], [13], [244]. Various applications can be found in [208], [188], [33], [93], [167], [168], [59b], [24], [225], [226]. Hida [77a], [77b] develops a "causal calculus." An interesting discussion on the problems related to Wiener's polynomial chaos can be found in McKean [171b].

Non-L^2 multiple integrals are much harder to define and much effort is now being directed towards that goal. See [209] for a compendium of recent papers. See also [222], [227], [233], [18], [21], [169], [170], [111], [203]-[206], [114].

Multiple integrals appear also in the limits of *U-statistics*. See [45], [146], [60], [34], [35]. Surveys are found in [36], [37a]. For U statistics in stable random variables, see [145], [3].

The following papers discuss useful techniques: [7], [11a], [100], [175], [210], [213], [247], [256].

References

[1] Adler, R.J. (1981). *The Geometry of Random Fields*. Chichester: Wiley.

[2a] Astrauskas, A. (1982). Stable self-similar fields. *Lithuanian Mat. Journal* **22**(3), 215-221.

[2b] Astraukas, A. (1983a). Limit theorems for sums of linearly generated random variables. *Lithuanian Mat. Journal* 23(2), 127-134.

[2c] Astraukas, A. (1983b). Limit theorems for quadratic forms of linear processes. *Lithuanian Mat. Journal* 23(4), 355-361.

[3] Avram, F. and Taqqu, M.S. (1984). Symmetric polynomials of random variables attracted to an infinitely divisible law. School of Operations Research and Industrial Engineering Technical Report No. 622, Cornell University. To appear in *Z. Wahrscheinlichkeitstheorie verw. Geb.*.

[4] Avram, F. and Taqqu, M.S. (1984). Generalized powers of strongly dependent random variables. School of Operations Research and Industrial Engineering Technical Report, No. 643, Cornell University.

[4a] Avram, F. and Taqqu, M.S. (1985). Weak convergence of moving averages with infinite variance. Appears in this volume.

[5] Bentkus, V. and Surgailis, D. (1981). On some classes of self-similar random fields. *Lithuanian Mat. Journal* 21(2), 53-66.

[6] Beran. J. and Künsch, H. (1985). Location estimators for processes with long-range dependence. Preprint.

[7] Berman, S.M. (1978). A new kind of tightness of probability measures and its applications to integral functionals of stochastic processes. *Z. Wahrscheinlichkeitstheorie verw. Geb.* 42, 155-162.

[8] Berman, S.M. (1979). High level sojourns for strongly dependent Gaussian processes. *Z. Wahrscheinlichkeitstheorie verw. Geb.* 50, 223-236.

[9] Berman, S.M. (1984). Sojourns of vector Gaussian processes inside and outside spheres. *Z. Wahrscheinlichkeitstheorie verw. Geb.* 66, 529-542.

[10] Bhattacharya, R.N., Gupta, V.K. and Waymire, E. (1982). The Hurst effect under trends. Preprint. To appear in the *Journal of Applied Probability*.

[11a] Bleher, P.M. (1980). ϵ-expansion for automodel random fields. In: *Multicomponents Random Systems*, pp. 59-117. Advances in Prob. and Related Topics 6, New York: Dekker.

[11b] Bleher, P.M. (1981). Inversion of Toeplitz matrices. *Trans. Moscow Math. Soc.*, Issue 2, 201-229.

[12] Bleher, P.M. and Sinai, Ja.G. (1973). Investigation of the critical point in models of the type of Dyson's Hierarchical Model. *Commun. Math. Phys.* 33, 23-42.

[13] Borell, C. (1984). On polynomial chaos and integrability. *Prob. and Math. Stat.* 3, 191-203.

[14] Bradley, R.C. (1985). Basic properties of strong mixing conditions. Appears in this volume.

[14a] Bradley, R.C., Bryc, W., and Janson, S. (1985). Remarks on the foundations of measures of dependence. Tech. Report 105, Center for Stoch. Processes, Univ. of North Carolina.

[15] Breuer, P. and Major, P. (1983). Central limit theorems for non-linear functionals of Gaussian fields. *Journal of Multivariate Analysis* 13, 425-441.

[16] Burton, R.M. and Waymire, E. (1984). Scaling limits for point random fields. *Journal of Multivariate Analysis*, 15, 237-251.

[17] Cambanis, S. (1982). Complex symmetric stable variables and processes. Center for Stoch. Processes, University of North Carolina, Tech. Report 10.

[18] Cambanis, S. and Miller, G. (1980). Some path properties of p-th order and symmetric stable processes. *Ann. Prob.* 6, 1148-1156.

[19] Cambanis, S. and Miller, G. (1981). Linear problems in pth order and stable processes. *SIAM J. Appl. Math.* 41, 43-69.

[20] Cambanis, S. and Soltani, A.R. (1984). Prediction of stable processes: Spectral and moving average representation. *Z. Wahrscheinlichkeitstheorie verw. Geb.* 66, 593-612.

[21] Cambanis, S., Rosinski, J. and Woyczynski, W.A. (1985). Convergence of quadratic forms in p-stable random variables and θ_p-radonifying operators. *Ann. Prob.* **13**, 885-897.

[21a] Cambanis, S., Hardin, C.D. and Weron, A. (1985). Innovations and Wold decompositions of stable sequences. Tech. Report 106, Center for Stoch. Processes, Univ. of North Carolina.

[21b] Cambanis, S., and Miamee, J.C. (1985). On prediction of harmonizable stable processes. Tech. Report 110, Center for Stoch. Processes, Univ. of North Carolina.

[22] Carmona, R. and Kôno, N. (1976). Convergence en loi et lois du logarithme itéré pour les vecteurs gaussiens. *Z. Wahrscheinlichkeitstheorie und Verw. Geb.* 36, 241-267.

[23] Cassandro, M. and Jona-Lasinio, G. (1978). Critical point behavior and probability theory. *Adv. in Physics* 27, 913-941.

[24] Chambers, D. and Slud, E. (1985). Central limit theorems for nonlinear functionals of stationary Gaussian processes. Preprint.

[24a] Chentsov, N.N. (1957). Lévy's Brownian motion for several parameters and generalized white noise. *Th. Prob. Appl.* **2**, 265-266.

[25] Cline, D.B.H. (1985). Convolution tails, product tails and on domains of attraction. Preprint.

[26] Cline, D.B.H. and Brockwell, P.J. (1985). Linear prediction of ARMA processes with infinite variance. *Stoch. Proc. Appl.* 19, 281-296.

[27] Chi, M., Neal, E., and Young, G.K. (1973). Practical application of fractional Brownian motion and noise to synthetic hydrology. *Water Resour. Res.* 9, 1523-1533.

[28] Collet, P. and Eckmann, J.P. (1978). A renormalization group analysis of the hierarchical model in statistical mechanics. *Lecture Notes in Physics* 74. New York: Springer Verlag.

[29] Cox, D.R. (1981). Statistical analysis of time series: some recent developments. *Scand. J. Statist.* 8, 93-115.

[30] Cox, D.R. (1984). Long-range dependence: a review. *Statistics: An Appraisal:* H.A. David and H.T. David, eds., pp. 55-74. Iowa State Univ. Press.

[31] Davis, R. and Resnick, S. (1985). Limit theory for moving averages with regularly varying tail probabilities. *Ann. Prob.* 13, 179-195.

[32] Davis, R. and Resnick, S. (1985). Limit theory for the sample correlation function of moving averages. Appears in this volume.

[33] Dawson, D.A. and Salehi, H. (1980). Spatially homogeneous random evolutions. *J. Multivariate Anal.* 10, 141-180.

[34] Dehling, H., Denker, M. and Philipp, W. (1984). Invariance principles for von Mises and U-statistics. *Z. Wahrscheinlichkeitstheorie verw. Geb.* 67, 139-167.

[35] Dehling, H. (1985). The functional law of the iterated logarithm for von-Mises functionals and multiple Wiener integrals. Preprint.

[36] Dehling, H. (1985). Almost sure approximations for U-statistics. Appears in this volume.

[37] Dempster, A.P. (1979). A modelling/Bayesian approach to seasonal economic time series analysis. Abstract.

[37a] Denker, M. (1985). *Asymptotic distribution theory in nonparametric statistics*. Braunschweig/Wiesbaden: Fiedr. Vieweg & Sohn.

[38] Dobrushin, R.L. (1979). Gaussian and their subordinated self similar random generalized fields. *Ann. Probability* 7, 1-28.

[39] Dobrushin, R.L. (1980). Automodel generalized random fields and their renormalization group. In *Multicomponent Random Systems*, p. 153-198. Advances in Prob. and Related Topics 6, New York: Dekker.

[40] Dobrushin, R.L. and Major, P. (1979). Limit behavior of some self-similar fields. Preprint.

[41] Dobrushin, R.L. and Major, P. (1979). Non-central limit theorems for non-linear functionals of Gaussian fields. *Z. Wahrscheinlichkeitstheorie verw. Geb.* 50, 27-52.

[42] Dobrushin, R.L. and Takahashi, Y. (1979). Gaussian random fields with scale limits. Preprint.

[43] Dürr, D., Goldstein, S. and Lebowitz, J.L. (1983). Stochastic processes originating in deterministic microscopic dynamics. *J. Stat. Physics* 30, 519-526.

[44] Dürr, D., Goldstein, S. and Lebowitz, J.L. (1985). Asymptotics of particle trajectories in infinite one-dimensional systems with collisions. Preprint.

[45] Dynkin, E.B. and Mandelbaum, A. (1983). Symmetric statistics, Poisson point processes and multiple Wiener integrals. *Ann. Stat.* 11, 739-745.

[46] Dyson, F.J. (1971). An Ising ferromagnet with discontinuous long-range order. *Commun. Math. Physics* 21, 269-283.

[47] Eberhard, J.W. and Horn, P.M. (1978). Excess 1/f noise in metals. *Phys. Rev. B* 18, 6681-6693.

[48] Engel, D.D. (1982). The multiple stochastic integral. *Memoirs Am. Math. Soc.* 38, #265.

[49] Enss, V. (1978). Renormalization group limits for Wick polynomials of Gaussian processes. *Reports on Math. Phys.* 13, 87-99.

[50] Enss, V. (1980). Scaling limits of generalized random processes. In *Quantum Fields-Algebras, Processes*, L. Streit ed., Proceedings of the Symposium "Bielefeld Encounters in Physics and Mathematics II, 1978, pp. 107-124. Wien: Springer-Verlag.

[50a] Fortus, M.I. (1983). Canonical correlations and canonical functionals for a stationary random process with a power spectral density. *Soviet Math. Doklady* **28**(1), 280-283.

[51] Fox, R. (1981). Upper functional laws of the iterated logarithm for non-Gaussian self-similar processes. Technical Report No. 509, School of Operations Research & Industrial Engineering, Cornell University.

[52] Fox, R. and Taqqu, M.S. (1983). Central limit theorems for quadratic forms in random variables having long-range dependence. School of Operations Research and Industrial Engineering Technical Report No. 590, Cornell University.

[53] Fox, R. and Taqqu, M.S. (1984a). Multiple stochastic integrals with dependent integrators. School of Operations Research and Industrial Engineering Technical Report No. 599, Cornell University. To appear in the *Journal of Multivariate Analysis*.

[54] Fox, R. and Taqqu, M.S. (1984b). Large-sample properties of parameter estimates for strongly dependent stationary Gaussian time series. (Old title: Maximum likelihood type estimator for the self-similarity parameter in Gaussian sequences.) School of Operations Research and Industrial Engineering Technical Report No. 592, Cornell University. To appear in *Annals of Statistics*.

[55] Fox, R. and Taqqu, M.S. (1985). Non-central limit theorems for quadratic forms in random variables having long-range dependence. *Annals of Probability* 13, 428-446.

[56] Frisch, U., Sulem, P.L. and Nelkin, M. (1978). A simple dynamical model of intermittent fully developed turbulence. *J. Fluid Mech.* 87, 719-736.

[57] Frölich, J. (1983). Editor. *Scaling and Self-similarity in Physics: Renormalization in Statistical Mechanics and Dynamics*. Progress in Physics, Vol. 7. Boston: Birkhaüser.

[58] Geweke, J. and Porter-Hudak, S. (1984). The estimation and application of long memory time series models. *J. Time Series Anal.* 4, 221-238.

[59] Giraitis, L. (1983). On the convergence of some non-linear transformation of Gaussian sequences to self-similar processes. *Lithuanian Mat. Journal* 23(1), 31-39.

[59b] Giraitis, L. (1984). Asymptotic distribution of spectral estimates of Ito-Wiener integrals. *Lithuanian Mat. Journal* **24**(3), 98-104.

[60] Giraitis, L. and Surgailis, D. (1984). A limit theorem for triangular array of symmetric statistics. Preprint.

[61] Giraitis, L. and Surgailis, D. (1985). Central limit theorem and other limit theorems for functionals of Gaussian processes. Preprint.

[61b] Giraitis, L., and Surgailis, D. (1985). Multivariate Appell polynomials and the central limit theorem. Appears in this volume.

[62] Gisselquist, R. (1973). A continuum of collision process limit theorems. *Ann. of Prob.*1,231-239.

[63] Goldman, A. (1984a). Estimations analytiques concernant the movement brownian fractionnaire à plusieurs parametres. *C.R. Acad. Sci. Paris*, 298,
Serie I, No.5, 91-93.

[64] Goldman, A. (1984b). Decomposition de McKean, prediction et exacte mesure de Hausdorff des trajectoires du mouvement brownien à plusieurs parametres. Preprint. Submitted *Annals Prob.*

[65] Gomide, F.L.S. (1978). Markovian inputs and the Hurst phenomenon. *J. Hydrology* 37, 23-45.

[66] Graf, H.P. (1983). Long-range correlations and estimation of the self-similarity parameter. Thesis, ETH, Zürich.

[67] Granger, C.W.J. (1980). Long memory relationships and aggregation of dynamic models. *J. Econometrics* 14, 227-238.

[68] Granger, C.W.J. and Joyeux, R. (1980). An introduction to long-memory time series and fractional differencing. *J. Time Series Anal.* 1, 15-30.

[69] Greene, M.T. and Fielitz, B.D. (1977). Long-term dependence in common stock returns. *J. Financial Economics* 4, 339-349.

[70] Greene, M.T. and Fielitz, B.D. (1979). The effect of long-term dependence on risk return models of common stocks. *Operations Research* 27, 944-951.

[71] Greenwood, P.E. (1969). The variation of a stable path is stable. *Z. Wahrscheinlichkeitstheorie verw. Geb.* 14, 140-148.

[72] Grize, Y.L. (1984). Towards a stationary continuous lower probability based model for flicker noise. Thesis, Cornell University.

[73] Hampel, F.R. (1978). Modern trends in the theory of robustness. *Math. Operationsforsch. Statist. Scr. Statist.* 9, 425-442.

[74] Hardin Jr., C.D. (1981). Isometries on subspaces of L^p. *Indiana Univ. Math. J.* 30, 449-465.

[75] Hardin Jr., C.D. (1982). On the spectral representation of symmetric stable processes. *J. Mult. Anal.* 12, 385-401.

[76] Harris, T.E. (1965). Diffusion with "collisions" between particles. *J. Appl. Probability* 2, 323-338.

[77] Helland, K.N. and Van Atta, C.W. (1978). The "Hurst phenomenon" in grid turbulence. *J. Fluid Mech.* 85, part 3, 573-589.

[77a] Hida, T. (1970). *Stationary Stochastic Processes*. Princeton: Princeton Univ. Press.

[77b] Hida, T. (1980). *Brownian Motion*. New York: Springer Verlag.

[78] Hipel, K.W. and McLeod, A.I. (1978). Box-Jenkins modelling in the geophysical sciences. Preprint. Dept. of Systems Design, University of Waterloo, Ontario. To appear in *Future Trends in Geomathematics*.

[79] Hipel, K.W. and McLeod, A.I. (1979). Perspectives in stochastic hydrology. *Time Series Proceedings, International Meeting, Nottingham University*, pp. 73-102, Amsterdam: North-Holland.

[80] Ho, H.-C. (1984). Limit theorems for non-linear functions of a stationary Gaussian process. Thesis, Wayne State University.

[81] Ho, H.C. and Sun, T.C. (1984a). A central limit theorem for non-instantaneous filters of a stationary Gaussian process. Preprint.

[82] Ho, H.C. and Sun, T.C. (1984b). Limit theorems for non-linear functionals of a stationary Gaussian process. Preprint.

[83] Hoskings, J.R.M. (1981). Fractional differencing. *Biometrika*, 68, 1, 165-176.

[84] Hoskings, J.R.M. (1982a). Some models in persistence in time series. In: *Time Series Analysis: Theory and Practice* 1, 641-653. Amsterdam: North-Holland.

[85] Hoskings, J.R.M. (1982b). Modelling persistence in hydrological time series using fractional differencing. Preprint.

[86] Hsiao, C.T. (1981). Central limit theorem for stationary and dependent random vectors. Preprint.

[87] Hughes, B.D., Shlesinger, M.F. and Montroll, E.W. (1981). Random walks with self-similar clusters. *Proc. Nat. Acad. Sci. USA* 78, 3287-3291.

[88] Hunt, G.A. (1951). Random Fourier transforms. *Trans. Amer. Math. Soc.* 71, 38-69.

[89] Hurst, H.E. (1951). Long-term storage capacity of reservoirs. *Trans. Amer. Soc. Civil Engineers* 116, 770-808.

[90] Hurst, H.E. (1955). Methods of using long-term storage in reservoirs. *Proceedings of the Institution of Civil Engineers*, Part I, 519-577.

[91] Hurst, H.E., Black, R.P. and Simaika, Y.M. (1965). *Long-term Storage: An Experimental Study*. London: Constable.

[92] Ibragimov, I.A. and Linnik, Yu.V. (1971). *Independent and stationary sequences of random variables*. Groningen: Walters-Noordhoff.

[93] Isobe, E. and Sato, S. (1983). Wiener-Hermite expansion of a process generated by an Itô stochastic differential equation. *J. Appl. Prob.* 20, 754-765.

[94] Itô, K. (1951). Multiple Wiener Integral. *J. Math. Soc. Japan* 3, 157-169.

[95] Itô, K. (1956). Spectral type of the shift transformation of differential processes with stationary increments. *Trans. Amer. Math. Soc.* 81, 106-134.

[96] Johnston, S. (1981). Fractalfest: some prickly problems in plasma physics. Preprint.

[97] Jona-Lasinio, G. (1977). Probabilistic approach to critical phenomena. In: *New Developments in Quantum Field Theory and Statistical Mechanics, Cargese 1976* (ed: M. Levy and P. Mitter), pp. 419-446. New York: Plenum.

[98] Jonas, A.B. (1983). Persistent memory random processes. Thesis, Harvard University.

[98a] Kahane, J.P. (1985). *Some random series of functions* Cambridge: Cambridge Univ. Press.

[99] Kallianpur, G. (1980). *Stochastic filtering theory*. New York: Springer Verlag.

[100] Karlin, S. and Rinott, Y. (1982). Applications of ANOVA type decompositions for comparisons of conditional variance statistics including Jackknife estimates. *Annals of Stat.* 10, 485-501.

[101] Kasahara, Y. and Maejima, M. (1985). Functional limit theorems for weighted sums of i.i.d. random variables. Preprint.

[102] Kesten, H. and Spitzer, F. (1979). A limit theorem related to a new class of self-similar processes. *Z. Wahrscheinlichkeitstheorie und Verw. Geb.* 50, 5-25.

[103] Kipnis, C., Lebowitz, J.L., Presutti, E. and Spohn, H. (1983). Self-diffusion for particles with stochastic collisions in one dimension. *J. Stat. Physics* 30, 107-121.

[104] Kolmogorov, A.N. (1940). Wienersche Spiralen und einige andere interessante Kurven in Hilbertschen Raum. *C.R. (Doklady) Acad. Sci. USRR (N.S.)* 26, 115-118.

[105] Kolmogorov, A.N. (1941). Local structure of turbulence in an incompressible liquid for very large Reynolds numbers. *C.R. (Doklady) Acad. Sci.*

URSS (N.S.) 30, 299-303. Reprinted in Friedlander and Topper 1961, 151-155.

[106] Kôno, N. (1983). Classical limit theorems for dependent random variables having moment conditions. *Fourth USSR-Japan Symp. on Prob. and Math. Stat.* at Tbilisi. Lecture Notes in Math. 1021. New York: Springer-Verlag.

[107] Kôno, N. (1983). Approximate upper functions of Gaussian processes. *J. Math. Kyoto Univ.* 23, 195-209.

[108] Kôno, N. (1983). Iterated log type strong limit theorems for self-similar processes. *Proc. Japan Acad.*, Ser.A, 59, 85-87.

[109] Kôno, N. (1983). Propriétés quasi-partout de fonctions aléatoires gaussiennes. Preprint. IRMA, Math., Univ. Louis Pasteur, True R. Descartes, 67084 Strasbourg, France.

[110] Kôno, N. (1985). Hausdorff dimension of sample paths for self-similar processes. Appears in this volume.

[111] Krakowiak, W. and Szulga, J. (1984). Random multilinear forms. Preprint.

[112] Künsch, H. (1981). Thermodynamics and statistical analysis of Gaussian random fields. *Z. Wahrscheinlichkeitstheorie verw. Geb.* 58, 407-421.

[113] Künsch, H. (1985). Discrimination between small trends and long-range dependence. Preprint.

[114] Kwapien, S. and Woyczynski, W.A. (1985). Decoupling of martingale transforms and stochastic integrals for processes with independent increments.
Preprint.

[115] Laha, R.G. and Rohatgi, V.K. (1982). Operator self-similar stochastic processes in R_d. *Stoch. Proc. Appl.*, 12, 73-84.

[116] Lai, T.L. and Stout, W. (1978). The law of the iterated logarithm and upper-lower class tests for partial sums of stationary Gaussian sequences. *Ann. Prob.* 6, 731-750.

[117] Lai, T.L. and Stout, W. (1980). Limit theorems for sums of dependent random variables. *Z. Wahrscheinlichkeitstheorie verw. Geb.* 51, 1-14.

[118] Lamperti, J.W. (1962). Semi-stable stochastic processes. *Trans. Amer. Math. Soc.* 104, 62-78.

[119] Lamperti, J.W. (1972). Semi-stable Markov processes I. *Z. Wahrscheinlichkeitstheorie verw. Geb.* 22, 205-225.

[120] Lang, R. and Xank, N.X. (1983). Strongly correlated random fields as observed by a random walker. *Z. Warhscheinlichkeitstheorie verw. Geb.* 64, 327-340.

[121] Lawrence, A.J. and Kottegoda, N.T. (1977). Stochastic modelling of river-flow time series. *J. Royal Statis. Soc. A.* 140, Part 1, 1-47.

[122] Lepage, R., Woodroofe, M. and Zinn, J. (1981). Convergence of a stable distribution via order statistics. *Ann. Prob.* 9, 624-632.

[123] Lettenmaier, D.P. and Burgess, S.J. (1977). Operational assessment of hydro-logical models of long-term persistence. *Water Resour. Res.* **13**, 113-124.

[123a] Lévy, P. (1948). *Processus Stochastiques et Mouvement Brownien.* Paris: Gauthier-Villars.

[124] Li, W.K. (1981). Topics in time series modelling. Thesis, University of Western Ontario.

[125] Li, W.K. and McLeod, A.I. (1983). Fractional time series modelling. Preprint, University of Western Ontario.

[126] Linde, W. (1983). *Infinitely Divisible and Stable Measures on Banach Spaces.* Teubner-Texte zur Mathematik. Band 58, Leipzig.

[127] Logan, B.F., Mallows, C.L., Rice, S.O. and Shepp, L.A. (1973). Limit distributions of self-normalized sums. *Ann. Probability* 1, 788-809.

[128] Lou, J.-H. (1985). Some properties of a special class of self-similar processes. *Z. Wahrscheinlichkeitstheorie und verw. Geb.* 68, 493-502.

[129] Lovejoy, S. (1982). Area-perimeter relation for rain and cloud areas. *Science* 216, 185-187.

[130] Lovejoy, S. and Schertzer, D. (1982). Buoyancy, shear, scaling and fractals. Preprint.

[130a] Lovejoy, S., Mandelbrot, B.B. (1985). Fractal properties of rain, and a fractal model. *Tellus*, **37A**, 209-232.

[131] Maejima, M. (1981). Some sojourn time problems for strongly dependent Gaussian processes. *Z. Wahrscheinlichkeitstheorie verw. Geb.* 57, 1-14.

[132] Maejima, M. (1982). Some limit theorems for sojourn times of strongly dependent Gaussian processes. *Z. Wahrscheinlichkeitstheorie verw. Geb.* 60, 359-380.

[133] Maejima, M. (1982). On the weak convergence to a self-similar process without finite variance. Preprint.

[134] Maejima, M. (1983). On a class of self-similar processes. *Z. Wahrscheinlichkeitstheorie verw. Geb.* 62, 235-245.

[135] Maejima, M. (1983). A self-similar process with nowhere bounded sample paths. *Z. Wahrscheinlichkeitstheorie verw. Geb.* 65, 115-119.

[136] Maejima, M. (1985). Some sojourn time problems for two-dimensional Gaussian processes. *J. Multivar. Anal.* (To appear.)

[137] Maejima, M. (1985). Sojourns of multidimensional Gaussian processes with dependent components. Preprint.

[138] Maejima, M. (1985). Sojourn of multidimensional Gaussian processes. Appears in this volume.

[139] Maejima, M. and Rachev, S.T. (1984). An ideal metric and the rate of convergence to a self-similar process. Preprint.

[140] Major, P. (1981a). Multiple Wiener-Itô Integrals. *Springer Lecture Notes in Mathematics* 849. New York: Springer-Verlag.

[141] Major, P. (1981b). Limit theorems for non-linear functionals of Gaussian sequences. *Z. Wahrscheinlichkeitstheorie verw. Geb.* 57, 129-158.

[142] Major, P. (1982a). On renormalizing Gaussian fields. *Z. Wahrscheinlichkeitstheorie verw. Geb.* 59, 515-533.

[143] Major, P. (1982b). On a class of self-similar fields. Preprint.

[144] Major, P. and Szasz D. (1980). On the effect of collisions on the motion of an atom in R^1. *Ann. Probability* 8, 1068-1078.

[145] Malevich, T.L. and Abdalimov, B. (1977). Stable limit distributions for U-statistics. *Th. of Prob. and Applic.* 22, 370-377.

[146] Mandelbaum, A. and Taqqu, M.S. (1983). Invariance principle for symmetric statistics. *The Annals of Statistics* 12, 483-496.

[146a] Mandelbrot, B.B. (1960). The Pareto-Lévy law and the distribution of income. *International Economic Review* 1, 79-106.

[146b] Mandelbrot, B.B. (1961). Stable Paretian random functions and the multiplicative variation of income. *Econometria* 29, 517-543.

[147] Mandelbrot, B.B. (1963). The variation of certain speculative prices. *Journal of Business* 36, 394-419. Reprinted in P.H. Cootner, ed., *The Random Character of Stock Market Prices* (Cambridge, Mass: MIT Press, 1964), 297-337.

[148] Mandelbrot, B.B. (1965). Une classe de processus stochastiques homothétiques à soi; application à la loi climatologique de H.E. Hurst. *Comptes Rendus* 240, 3274-3277.

[149] Mandelbrot, B.B. (1965). Self-similar error clusters in communications systems and the concept of conditional systems and the concept of conditional stationarity. *I.E.E.E. Trans. on Communications Technology*, COM-13, 71-90.

[150] Mandelbrot, B.B. (1969). Long-run linearity, locally Gaussian processes, H-spectra and infinite variances. *International Econom. Rev.* 10, 82-113.

[151] Mandelbrot, B.B. (1971). A fast fractional Gaussian noise generator. *Water Resour. Res.* 7, 543-553.

[152] Mandelbrot, B.B. (1972a). Statistical methodology for non-periodic cycles: from the covariance to R/S analysis. *Ann. Econom. Soc. Measurement* 1, 259-290.

[153] Mandelbrot, B.B. (1972b). Broken line process derived as an approximation to fractional noise. *Water Resour. Res.* 8, 1354-1356.

[153a] Mandelbrot, B.B. (1974). Intermittent turbulence in self similar cascades; divergence of high moments and dimension of the carrier. *J. Fluid Mechanics* 62, 331-358.

[154] Mandelbrot, B.B. (1975). Limit theorems on the self-normalized range for weakly and strongly dependent processes. *Z. Wahrscheinlichkeitstheorie verw. Geb.* 31, 271-285.

[155] Mandelbrot, B.B. (1976). Intermittent turbulence and fractal dimension: kurtosis and the spectral exponent $5/3+B$. In —it Turbulence and Navier Stokes Equations, R. Teman, ed., Springer Lecture Notes in Math., Vol. 565, 121-145.

[156] Mandelbrot, B.B. (1979a). On scaling fractal facets of the Hurst effect, and a paper by Gomide. Preprint.

[157] Mandelbrot, B.B. (1979b). Stationarity, ergodic, settled and vagrant time series and the Hurst effect. Preprint.

[158] Mandelbrot, B.B. (1982). *The Fractal Geometry of Nature*. San Francisco: W.H. Freeman.

[159] Mandelbrot, B.B. (1984). Fractals in physics: squig clusters, diffusions, fractal measures, and the unicity of fractal dimensionality. *J. of Stat. Phys.* 34, 895-930.

[159a] Mandelbrot, B.B. (1984). Squig sheets and some other squig fractal constructions. *J. Statistical Physics*, 36, 519-545.

[159b] Mandelbrot, B.B. (1985). Self-affine fractals and fractal dimension. To appear in *Scripta Physica*.

[159c] Mandelbrot, B.B. (1985). Self-affine fractal sets, Parts I,II,III. To appear in *Fractals in Physics* (Trieste 1985) Ed. by Pietronero, L. and Tosatti, E., North Holland Publishing Company.

[160] Mandelbrot, B.B. and Van Ness, J.W. (1968). Fractional Brownian motions, fractional noises and applications. *SIAM Review* 10, 422-437.

[161] Mandelbrot, B.B. and Taqqu, M.S. (1979). Robust R/S analysis of long-run serial correlation. Proceedings of the 42nd Session of the International Statistical Institute, Manila. *Bulletin of the I.S.I.*, Vol. 48, Book 2, pp. 69-104.

[162] Mandelbrot, B.B. and Wallis, J.R. (1968). Noah, Joseph and operational hydrology. *Water Resour. Res.* 4, 909-918.

[163] Mandelbrot, B.B. and Wallis, J.R. (1969a). Computer experiments with fractional Gaussian noises. Parts 1,2,3. *Water Resour. Res.* 5, 228-267.

[164] Mandelbrot, B.B. and Wallis, J.R. (1969b). Some long-run properties of geophysical records. *Water Resour. Res.* 5, 321-340.

[165] Mandelbrot, B.B. and Wallis, J.R. (1969c). Robustness of the rescaled range R/S in the measurement of noncyclic long-run statistical dependence. *Water Resour. Res.* 5, 967-988.

[166] Manneville, P. (1980). Intermittency, self-similarity and 1/f spectrum in dissipative dynamical systems. *Le Journal de Physique* 41, 1235-1242.

[167] Maruyama, G. (1976). Nonlinear functionals of Gaussian stationary processes and their applications. Proc. 3rd Japan-USSR Symposium on Probability Th. (Tashkent 1975). Springer *Lecture Notes in Math.*, Vol. 550, 375-378.

[168] Maruyama, G. (1982). Applications of the multiplication of the Itô-Wiener expansions to limit theorems. *Proc. Japan Acad.* 58A, 388-390.

[169] McConnell, T. and Taqqu, M.S. (1984). Double integration with respect to stable processes. School of Operations Research and Industrial Engineering Technical Report No. 599, Cornell University.

[170] McConnell, T. and Taqqu, M.S. (1984). Decoupling inequalities for multilinear forms in symmetric random variables. To appear in *Annals of Probability*.

[171a] McKean, H.P. (1963). Brownian motion with a several dimensional time. *Th. Prob. Appl.* 8, 357-378.

[171b] McKean, H.P. (1973). Wiener's theory of non-linear noise. In *Stochastic Differential Equations, SIAM-AMS Proc.* 6, 191-209.

[171c] McKean, H.P. (1973). Geometry of differential space. *Ann. Prob.* **1**, 197-206.

[172] McLeod, A.I. and Hipel, K.W. (1978). Preservation of the rescaled adjusted range. Parts 1,2,3. *Water Resour. Res.* 14, 491-518.

[173] McLeod, A.I. and Hipel, K.W. (1983). Perspectives in long-memory modelling. Preprint.

[174] Mejia, J.M., Rodriguez-Iturbe, I. and Dawdy, D.R. (1972). Streamflow simulation 2. The broken line process as a potential model for hydrological simulation. *Water Resour. Res.* 8, 931-941.

[175] Meyer, P.A. (1966), ed. Séminaire de probabilités X, Université de Strasbourg. *Lecture notes in Math.* 511. Heidelberg: Springer-Verlag.

[176] Mijnheer, J.L. (1975). *Sample Path Properties of Stable Processes*. Mathematical Centre Tracts 59. Amsterdam: Mathematisch Centrum.

[177] Mohr, D. (1981). Modelling data as a fractional Gaussian noise. Ph.D. dissertation, Princeton University.

[178] Mohr, D. (1981a). Estimating the parameters of a fractional Gaussian noise. Preprint.

[179] Mohr, D. (1981b). A spectral estimate of Hurst's parameter. Preprint.

[180] Montroll, E.W. and Shlesinger, M.F. (1982). On 1/f noise and other distributions with long tails. *Proc. Nat. Acad. Sci. USA* 79, 3380-3383.

[181] Mori, T. and Oodaira, H. (1984). The law of the iterated logarithm for self-similar processes represented by multiple Wiener integrals. Preprint.

[182] Mori, T. and Oodaira, H. (1985). The functional iterated logarithm law for stochastic processes represented by multiple Wiener integrals. Preprint.

[183] Neveu, J. (1968). *Processus Aléatoires Gaussiens*. Les Presses de l'Université de Montréal, Montréal.

[184] Newman, C.M. (1981). Self-similar random fields in mathematical physics. In: *Measure Theory and Applications*, Goldin, G. and Wheeler, R. eds.

[185] O'Brien, G.L. and Vervaat, W. (1983). Marginal distributions of self-similar processes with stationary increments. *Z. Wahrscheinlichkeitstheorie verw. Geb.* 64, 129-138.

[186] O'Brien, G.L. and Vervaat, W. (1985). Self-similar processes with stationary increments generated by point processes. *Ann. Prob.* 13, 28-52.

[187] O'Connell, P.E. (1977). ARIMA models in synthetic hydrology. In *Mathematical Models for Surface Water Hydrology*, T.A. Ciriani, V. Maione, J.R. Wallis, ed., pp. 51-68, New York: J. Wiley & Sons.

[188] Ocone, D. (1983). Multiple integral expansions for nonlinear filtering. *Stochastics* 10, 1-30.

[189] Oodaira, H. (1973). The law of the iterated logarithm for Gaussian processes. *Ann. Prob.* 1, 954-967.

[190] Parzen, E. (1983). Time series, model identification, memory and quantiles. Preprint.

[191] Peligrad, M. (1985). Recent advances in the central limit theorem and its weak invariance principle for mixing sequences of random variables (a survey). Appears in this volume.

[192] Percival, D.B. (1983). The statistics of long memory processes. Thesis, University of Washington.

[193] Perez-Abreu, V. (1985). Product stochastic measures, multiple stochastic integrals and their extensions to nuclear space valued processes. Tech. Report 107, Center for Stoch. Processes, Univ. of North Carolina, Chapel Hill.

[193a] Pietrorino, L. and Tosatti, E. (1986) *Fractals in Physics (Trieste 1985)*. To be published by North Holland.

[194] Plikusas, A. (1980). Estimation of cumulants and large deviations for certain nonlinear transformations of a stationary Gaussian process. *Lithuanian Mat. Journal* 20(2), 150-156.

[195] Plikusas, A. (1981). Some properties of the multiple Itô integral. *Lithuanian Mat. Journal* 21(2), 184-191.

[196] Resnick, S.I. (1983). Point processes, regular variation and weak convergence. Preprint: To appear in the *Journal of Applied Probability*.

[197] Richardson, L.F. (1922). *Weather Prediction by Numerical Process*. (Reprinted in 1965.) London: Cambridge Univ. Press.

[198] Rosenblatt, M. (1961). Independence and dependence. *Proc. 4th Berkeley Sympos. Math. Statist. Probab.*, 411-443. Berkeley: Univ. Calif. Press.

[199] Rosenblatt, M. (1976). Fractional integrals of stationary processes and the central limit theorem. *J. Appl. Probability* 13, 723-732.

[200] Rosenblatt, M. (1979). Some limit theorems for partial sums of quadratic forms in stationary Gaussian variables. *Z. Wahrscheinlichkeitstheorie verw. Geb.* 49, 125-132.

[201] Rosenblatt, M. (1981). Limit theorems for Fourier transforms of functionals of Gaussian sequences. *Z. Wahrscheinlichkeitstheorie verw. Geb.* 55, 123-132.

[202] Rosenblatt, M. (1983). Stochastic processes with short-range and long-range dependence. Preprint.

[203] Rosinski, J. (1985). On stochastic integral representation of stable processes with sample paths in Banach spaces. Preprint.

[204] Rosinski, J. and Woyczynski, W.A. (1984). Moment inequalities for real and vector p-stable stochastic processes. Preprint.

[205] Rosinski, J. and Woyczynski, W.A. (1984). Products of random measures, multilinear random forms and multiple stochastic integrals. *Proc. Measure Theory Conf. Oberwolfach 1983*. Springer's Lecture Notes in Math.

[206] Rosinski, J. and Woyczynski, W.A. (1984). On Itô stochastic integration with respect to the p-stable motion: inner clock, integrability of sample paths, double and multiple integrals. Preprint.

[207] Rvaceva, E.L. (1962). On domains of attraction of multi-dimensional distributions. *Selected Translations in Math. Stat. and Prob.* IMS-AMS, 2, 183-205.

[208] Segall, A. and Kailath, T. (1976). Orthogonal functionals of independent-increment processes. *IEEE Trans. Information Theory*, Vol. IT-22, 287-298.

[209] Seminar notes on multiple stochastic integration, polynomials chaos and their applications (1985). Preprint No. 85-34. Dept. of Math. and Stat. Case Western Reserve Univ., Cleveland, OH 44106.

[210] Shiryayev, A.N. (1981). Martingales: recent developments, results and applications. *International Statistical Review* 49, 199-233.

[211] Shlesinger, M.F. and Hughes, B.D. (1981). Analogs of renormalization group transformations in random processes. *Physica* 190A, 597-608.

[211a] Shlesinger, M.F., Mandelbrot, B.B., and Rubin, R.J. (1984). *Fractals in Physics (Gaithersburg 1983)*. Appears in *J. Statistical Physics*, Vol. 34.

[212] Sinai, Ya. G. (1976). Self-similar probability distributions. *Theory Probability Appl.* 21, 64-80.

[213] Slepian, D. (1972). On the symmetrized Kronecker power of a matrix and extensions of Mehler's formula for Hermite polynomials. *SIAM J. Math. Anal.* 3, 606-616.

[214] Spitzer, F. (1969). Uniform motion with elastic collision of an infinite particle systems. *J. Math. Mech.* 18, 973-990.

[215] Srikanthan, R. and McMahon, T.A. (1979). Generation of skewed annual flows using fast fractional Gaussian noise generator. Preprint.

[216] Sun, T.C. (1963). A central limit theorem for non-linear functions of a normal stationary process. *J. Math. Mech.* 12, 945-978.

[217] Sun, T.C. (1965). Some further results on central limit theorems for non-linear functions of a normal stationary process. *J. Math. Mech.* 14, 71-85.

[218] Sun, T.C. and Ho, H.C. (1985). On central and non-central limit theorems for non-linear functions of a stationary Gaussian process. Appears in this volume.

[219] Surgailis, D. (1979). On self-similar stable random fields; Abstract listed in the Colloquium on "Random Fields: Rigorous Results in Statistical Mechanics and Quantum Field Theory," Janos Bolyai Math. Society, Esztergom, Hungary.

[220] Surgailis, D. (1981a). Convergence of sums of nonlinear functions of moving averages to self-similar processes. *Soviet Mathematics* 23, 247-250.

[221] Surgailis, D. (1981b). On infinitely divisible self-similar random fields. *Z. Wahrscheinlichkeitstheorie verw Geb.* 58, 453-477.

[222] Surgailis, D. (1981c). On L^2 and non-L^2 multiple stochastic integration. *Stochastic Differential Systems*. Lecture Notes in Control and Information Sciences **36** New York: Springer-Verlag.

[223] Surgailis, D. (1981d). Convergence of sums of nonlinear functions of moving averages to self-similar processes. *Soviet Math. Doklady* 23, 247-250.

[224] Surgailis, D. (1983). Zones of attraction of self-similar multiple integrals. *Lithuanian Mat. Journal* 22, 327-340.

[225] Surgailis, D. (1983). On Poisson multiple stochastic integrals and associated equilibrium Markov processes. Theory and Applications of Random Fields. Lecture Notes in Control and Information Sciences **49**, New York: Springer-Verlag.

[226] Surgailis, D. (1984). On multiple Poisson stochastic integrals and associated
Markov semigroups. *Probability and Math. Stat.* 3, 217-239.

[227] Surgailis, D. (1985). On the multiple stable integral. Preprint.

[228] Suzuki, M. (1983a). Phase transition and fractals. *Progress of Theoretical Physics* 69, 65-76.

[229] Suzuki, M. (1983b). Long-time tail relaxation noise. In: *Noise in Physical Systems and 1/f Noise*. (Savelli, M., Leroy, G., Nougier, J.P., eds.), 73-79. Elsevier.

[230] Suzuki, M. (1984). Finite-size scaling for transient similarity and fractals. *Prog. Theor. Phys.* 71, 1397-1400.

[231] Szasz, D. (1981). Joint diffusion on the line. In *Random fields: rigorous results in statistical mechanics and quantum field theory*. Colloquia Mathematica Societatis Janos Bolyai, Vol. 27. Amsterdam: North-Holland.

[232] Szatzschneider, W. (1975). A more deterministic version of Harris-Spitzer's "Random constant velocity" model for infinite systems of particles. Probability Winter School, Karpacz, Poland 1975. *Springer Lecture Notes in Math.* 472, 157-167.

[233] Szulga, J. and Woyczynski, W.A. (1983). Existence of a double random integral with respect to stable measures. *J. Mult. Anal.* 13, 194-201.

[234] Taqqu, M.S. (1970). Note on evaluations of R/S for fractional noises and geophysical records. *Water Resources Research*, 6, 349-350.

[235] Taqqu, M.S. (1975). Weak convergence to fractional Brownian motion and to the Rosenblatt process. *Z. Wahrscheinlichkeitstheorie verw. Geb.* 31, 287-302.

[236] Taqqu, M.S. (1977). Law of the iterated logarithm for sums of non-linear functions of the Gaussian variables that exhibit a long range dependence. *Z. Wahrscheinlichkeitstheorie verw. Geb.* 40, 203-238.

[237] Taqqu, M.S. (1978). A representation for self-similar processes. *Stochastic Proc. and their Applic.* 7, 55-64.

[238] Taqqu, M.S. (1979). Convergence of integrated processes of arbitrary Hermite rank. *Z. Wahrscheinlichkeitstheorie verw. Geb.* 50, 53-83.

[239] Taqqu, M.S. (1981). Self-similar processes and related ultraviolet and infrared catastrophes. In *Random Fields: Rigorous Results in Statistical Mechanics and Quantum Field Theory*. Colloquia Mathematica Societatis Janos Bolyai, Vol. 27, Book 2, pp. 1057-1096. North-Holland: Amsterdam.

[240] Taqqu, M.S. (1984). Sojourn in an elliptical domain. School of Operations Research and Industrial Engineering Technical Report No. 630, Cornell University. To appear in *Stochastic Processes and their Applications*.

[241] Taqqu, M.S. (1985). Self-similar processes. To appear in *Encyclopedia of Statistical Sciences*.

[242] Taqqu, M.S. and Czado, C. (1984). Reproducing kernel Hilbert space for some non-Gaussian processes. School of Operations Research and Industrial Engineering Technical Report No. 644, Cornell University.

[243] Taqqu, M.S. and Czado, C. (1985). A survey of functional laws of the iterated logarithm for self-similar processes. *Stochastic Models*, **1**(1), 77-115.

[244] Taqqu, M.S. and Goldberg, J. (1982). Regular multigraphs and their applications to the Monte Carlo evaluation of moments of non-linear functionals of Gaussian processes. *Stochastic Processes and Their Applications* 13, 121-138.

[245] Taqqu, M.S. and Levy, J. (1985). Using renewal processes to generate long-range dependence and high variability. School of Operations Research & Industrial Engineering Technical Report No. 566, Cornell University. Appears in this volume.

[246] Taqqu, M.S. and Wolpert, R. (1983). Infinite variance self-similar processes subordinate to a Poisson measure. *Z. Wahrscheinlichkeitstheorie verw. Geb.* 62, 53-72.

[247] Thu, N.V. (1984). Fractional calculus in probability. *Prob. and Math. Stat.* 3, 173-189.

[248] Todini, E. and O'Connell, P.E. (1979). Hydrological simulation of Lake Nasser: analysis and results. Vol. 1. Preprint: Institute of Hydrology, McLean Building, Crowmarsh Gifford, Wallingford, Oxfordshire 0X108BB, Great Britain.

[249] Van Harn, K. and Steutel, F.W. (1984). Integer-valued self-similar processes. Preprint.

[250] Vervaat, W. (1985). Sample path properties of self-similar processes with stationary increments. *Ann. Prob.* 13, 1-27.

[251] Vervaat, W. (1985). Stationary self-similar extremal processes and random semicontinuous functions. Appears in this volume.

[252] Weron, A. (1984). Stable processes and measures: a survey. Tech. Report 65, Center for Stochastic Processes. U. North Carolina.

[253] Wiener, N. (1958). *Nonlinear Problems in Random Theory*. Cambridge: MIT Press.

[254] Wolf, D. (1978), ed. *Noise in Physical Systems*, Proc. of the Fifth International Conf. on Noise, Bad Nauheim, March 1978-Springer Series in Electro-physics, Vol. 2, New York: Springer Verlag.

[255] Yaglom, A.M. (1958). Correlation theory of processes with random stationary nth increments. *Amer. Math. Soc. Transl.* (2), 8, 87-141.

[256] Zygmund, A. (1979). *Trigonometric Series*. Vols. 1 and 2. Cambridge: Cambridge Univ. Press.

Murad S. Taqqu
Department of Mathematics
Boston University
Boston, MA 02215

SECTION 2

MIXING CONDITIONS

BASIC PROPERTIES OF STRONG MIXING CONDITIONS

Richard C. Bradley

This is a survey of the basic properties of strong mixing conditions for sequences of random variables. The focus will be on the "structural" properties of these conditions, and not at all on limit theory. For a discussion of central limit theorems and related results under these conditions, the reader is referred to Peligrad [60] or Iosifescu [50]. This survey will be divided into eight sections, as follows:

1. Measures of dependence
2. Five strong mixing conditions
3. Mixing conditions for two or more sequences
4. Mixing conditions for Markov chains
5. Mixing conditions for Gaussian sequences
6. Some other special examples
7. The behavior of the dependence coefficients
8. Approximation of mixing sequences by other random sequences.

1. MEASURES OF DEPENDENCE

Suppose $(\Omega,\mathcal{F},P)$ is a probability space. For any two σ-fields $\mathcal{A}$ and $\mathcal{B} \subset \mathcal{F}$ define the following measures of dependence:

$$\alpha(\mathcal{A},\mathcal{B}) := \sup|P(A \cap B)-P(A)P(B)| \ , \ A \in \mathcal{A} \ , \ B \in \mathcal{B} \ . \tag{1.1}$$

$$\phi(\mathcal{A},\mathcal{B}) := \sup|P(B|A)-P(B)| \ , \ A \in \mathcal{A} \ , \ B \in \mathcal{B} \ , \ P(A) > 0 \ . \tag{1.2}$$

$$\phi_{rev}(\mathcal{A},\mathcal{B}) := \phi(\mathcal{B},\mathcal{A}) \quad (\text{"rev" stands for "reversed"}). \tag{1.3}$$

$$\psi(\mathcal{A},\mathcal{B}) := \sup \frac{|P(A\cap B)-P(A)P(B)|}{P(A)P(B)} \ , \ A \in \mathcal{A} \ , \ B \in \mathcal{B} \ . \tag{1.4}$$

$$\rho(\mathcal{A},\mathcal{B}) := \sup|\mathrm{Corr}(X,Y)| \ , \ X \in L_2(\mathcal{A}) \ , \ Y \in L_2(\mathcal{B}) \ ; \ X \ , \ Y \ \text{real}. \tag{1.5}$$

$$\beta(\mathcal{A},\mathcal{B}) := \sup \tfrac{1}{2} \Sigma_{i=1}^{I} \Sigma_{j=1}^{J} |P(A_i \cap B_j)-P(A_i)P(B_j)| \tag{1.6}$$

where this latter sup is taken over all pairs of partitions $\{A_1,\ldots,A_I\}$ and $\{B_1,\ldots,B_J\}$ of Ω such that $A_i \in \mathcal{A}$ for all i and $B_j \in \mathcal{B}$ for all j. In (1.4) and in the sequel, 0/0 is interpreted to be 0. These measures of dependence will be the basis for the mixing conditions that we shall study, starting with Section 2. Here in Section 1 we shall just study these measures of dependence.

The following inequalities hold:

$$2\alpha(\mathcal{A},\mathcal{B}) \le \beta(\mathcal{A},\mathcal{B}) \le \phi(\mathcal{A},\mathcal{B}) \le \tfrac{1}{2}\psi(\mathcal{A},\mathcal{B}) . \tag{1.7}$$

$$4\alpha(\mathcal{A},\mathcal{B}) \le \rho(\mathcal{A},\mathcal{B}) \le \psi(\mathcal{A},\mathcal{B}) . \tag{1.8}$$

$$\rho(\mathcal{A},\mathcal{B}) \le 2\phi^{\frac{1}{2}}(\mathcal{A},\mathcal{B}) \cdot \phi_{rev}^{\frac{1}{2}}(\mathcal{A},\mathcal{B}) . \tag{1.9}$$

$$\rho(\mathcal{A},\mathcal{B}) = \sup \|E(f|\mathcal{B})-Ef\|_2/\|f\|_2 , \ f \in L_2(\mathcal{A}) , \ f \text{ real} . \tag{1.10}$$

$$\alpha(\mathcal{A},\mathcal{B}) \le \tfrac{1}{4} , \ \beta(\mathcal{A},\mathcal{B}) \le 1 , \ \phi(\mathcal{A},\mathcal{B}) \le 1 , \ \rho(\mathcal{A},\mathcal{B}) \le 1 . \tag{1.11}$$

Eqn. (1.9), an improvement of the earlier well known inequality $\rho(\mathcal{A},\mathcal{B}) \le 2\phi^{\frac{1}{2}}(\mathcal{A},\mathcal{B})$, comes from Peligrad [59, p. 462, eqn. (4)]; independently the kindred inequality $\rho(\mathcal{A},\mathcal{B}) \le 2 \cdot \max\{\phi(\mathcal{A},\mathcal{B}) , \phi_{rev}(\mathcal{A},\mathcal{B})\}$ was given by Denker and Keller [34, p. 516, line -8]. In this last inequality as well as in (1.7), (1.8), and (1.9), equality is achieved in some simple cases such as when $\mathcal{A} = \mathcal{B} = \{\Omega, A, A^c, \phi\}$ where $P(A) = \frac{1}{2}$. Eqns. (1.7), (1.8), (1.10), and (1.11) are all either trivial or at least fairly easy to prove. Referring to eqn. (1.11), $\psi(\mathcal{A},\mathcal{B})$ can of course take on the value $+\infty$. Each of the measures of dependence in eqns. (1.1)-(1.6) takes the value 0 precisely when $\mathcal{A}$ and $\mathcal{B}$ are independent σ-fields.

The measures of dependence in eqns. (1.1)-(1.6) fit nicely into a more general framework using "norms" of the bilinear form "covariance". For any σ-field $\mathcal{A}$, let $S(\mathcal{A})$ denote the set of all complex-valued simple $\mathcal{A}$-measurable random variables. (The particularly nice form of eqn. (1.13) below depends on the use of complex-valued rather than just real-valued random variables; however, this is not of any special importance.) Define the following families of measures of dependence between pairs of σ-fields $\mathcal{A}$ and $\mathcal{B}$:

For $0 \le r , s \le 1$,

$$\alpha_{r,s}(\mathcal{A},\mathcal{B}) := \sup \frac{|P(A\cap B)-P(A)P(B)|}{[P(A)]^r[P(B)]^s} , \; A \in \mathcal{A} , \; B \in \mathcal{B} .$$

For $1 \le p , q \le \infty$,

$$R_{p,q}(\mathcal{A},\mathcal{B}) := \sup \frac{|EXY-EXEY|}{\|X\|_p\|Y\|_q} , \; X \in \mathcal{S}(\mathcal{A}) , \; Y \in \mathcal{S}(\mathcal{B}) .$$

(Note that $\alpha_{r,s}(\cdot,\cdot)$ is a variant of $R_{1/r,1/s}(\cdot,\cdot)$ using only indicator functions.) Obviously the measures of dependence in eqns. (1.1)-(1.5) are respectively $\alpha_{0,0}(\mathcal{A},\mathcal{B})$, $\alpha_{1,0}(\mathcal{A},\mathcal{B})$, $\alpha_{0,1}(\mathcal{A},\mathcal{B})$, $\alpha_{1,1}(\mathcal{A},\mathcal{B})$, and $R_{2,2}(\mathcal{A},\mathcal{B})$. (The equation $\rho(\mathcal{A},\mathcal{B}) = R_{2,2}(\mathcal{A},\mathcal{B})$ holds by [75, Theorem 1.1] and a simple calculation.) Also it is easy to show that if one modifies the definition of $R_{\infty,\infty}(\mathcal{A},\mathcal{B})$ in an appropriate way so as to allow the r.v.'s X and Y to take their values in the standard Banach sequence-spaces ℓ^1 and ℓ^∞ respectively, then one obtains a measure of dependence that is within a positive constant factor of $\beta(\mathcal{A},\mathcal{B})$ in eqn. (1.6) (see [22, Section 2.2]).

If $0 \le r_0 , r_1 , s_0 , s_1 \le 1$, $0 < \theta < 1$, $r := (1-\theta)r_0 + \theta r_1$, and $s := (1-\theta)s_0 + \theta s_1$, then for any two σ-fields $\mathcal{A}$ and $\mathcal{B}$,

$$\alpha_{r,s}(\mathcal{A},\mathcal{B}) \le [\alpha_{r_0,s_0}(\mathcal{A},\mathcal{B})]^{1-\theta} \cdot [\alpha_{r_1,s_1}(\mathcal{A},\mathcal{B})]^{\theta} , \tag{1.12}$$

$$R_{1/r,1/s}(\mathcal{A},\mathcal{B}) \le [R_{1/r_0,1/s_0}(\mathcal{A},\mathcal{B})]^{1-\theta} \cdot [R_{1/r_1,1/s_1}(\mathcal{A},\mathcal{B})]^{\theta} . \tag{1.13}$$

Eqn. (1.12) has a trivial one-line proof and is useful for comparing the various measures of dependence $\alpha_{r,s}$ (see [21, Theorems 3.1, 3.2, and 4.1 (i) (ii)]). Eqn. (1.13) is an application of Thorin's multilinear version of the Riesz-Thorin interpolation theorem (see e.g. [5, p. 18, Exercise 13]); eqn. (1.13) and variants of it are useful for comparing various measures of dependence and for studying the relations between them (see [67, Chapter 7] [56, Lemma 1] [21] [22]). For example, as a consequence of (1.13) one can show that if $1 \le p , q , t \le \infty$ and $1/p + 1/q + 1/t = 1$, then for any two σ-fields $\mathcal{A}$ and $\mathcal{B}$,

$$R_{p,q}(\mathcal{A},\mathcal{B}) \le (2\pi) \cdot [\alpha(\mathcal{A},\mathcal{B})]^{1/t} \cdot [\phi(\mathcal{A},\mathcal{B})]^{1/p} \cdot [\phi_{rev}(\mathcal{A},\mathcal{B})]^{1/q}$$

(see [21, Theorem 1.1(i)]). Except for a constant factor, this inequality covers some other previously known ones as special cases, including eqn. (1.9). In an obvious way, a "small" upper bound on $R_{p,q}(\mathcal{A},\mathcal{B})$ might lead to a "small" value of $\mathrm{Cov}(X,Y)$ if, say, X and Y are r.v.'s which are $\mathcal{A}$-measurable and $\mathcal{B}$-measurable respectively. Such bounds are often useful in the proofs of limit theorems for dependent random variables.

Further information about measures of dependence can be gained from the use of other methods and results in interpolation theory, such as the techniques in the Marcinkiewicz interpolation theorem and the Stein-Weiss [72] methods for handling indicator functions. That observation is due to W. Bryc. For example, using such techniques Bryc proved that if $1 < p, q < \infty$ and $1/p + 1/q = 1$, then for any two σ-fields $\mathcal{A}$ and $\mathcal{B}$,

$$R_{p,q}(\mathcal{A},\mathcal{B}) \le C \cdot \alpha_{1/p,1/q}(\mathcal{A},\mathcal{B}) \cdot [1 - \log \alpha_{1/p,1/q}(\mathcal{A},\mathcal{B})] \tag{1.14}$$

where C is a positive constant that depends only on p and q (see [21, Theorem 4.1(vi)]). For the case $p = q = 2$, (1.14) improved a similar but much weaker inequality established in [15] by different methods; see also [24].) For any choice of p and q meeting the given specifications, (1.14) is within a constant factor of being sharp (see [22, Section 1.1]).

Let us say that two measures of dependence are "equivalent" if each one becomes arbitrarily small as the other one becomes sufficiently small. Among the measures of dependence $\alpha_{r,s}$, $0 \le r, s \le 1$, and $R_{p,q}$, $1 \le p, q \le \infty$, there are only five equivalence classes which consist of more than one member:

(i) $\alpha_{r,s}$, $r + s < 1$, $R_{p,q}$, $1/p + 1/q < 1$;

(ii) $\alpha_{r,1-r}$, $0 < r < 1$, $R_{p,p'}$, $1 < p < \infty$ (where p' is defined by $1/p + 1/p' = 1$);

(iii) $\alpha_{1,0}$, $R_{1,\infty}$;

(iv) $\alpha_{0,1}$, $R_{\infty,1}$;

(v) $\alpha_{1,1}$, $R_{1,1}$ (actually $\alpha_{1,1} = R_{1,1}$ by a simple argument).

These five classes each contain one of the measures of dependence in eqns. (1.1)-(1.5). The measure β in (1.6) is not equivalent to any of the measures $\alpha_{r,s}$ or $R_{p,q}$. All of this is explained in [21, Remark 4.1] and [22, Sections 1.2 and 2.2].

2. FIVE STRONG MIXING CONDITIONS

Henceforth all random variables are real-valued. For any family $(Y_s\ ,\ s \in S)$ of random variables (where S is an index set), the notation $\sigma(Y_s\ ,\ s \in S)$ will mean the σ-field generated by this family, i.e. the smallest σ-field containing the events $\{Y_s < r\}$, $s \in S$, $r \in \mathbb{R}$.

Suppose $(X_k\ ,\ k \in \mathbb{Z})$ is a sequence of random variables. (No assumption of stationarity is made yet.) For $-\infty \le J \le L \le \infty$ define $\mathcal{F}_J^L := \sigma(X_k\ ,\ J \le k \le L)$. Referring to eqns. (1.1)-(1.6), for each $n = 1,2,3,\ldots$ define

$$\alpha(n) := \sup_{J\in\mathbb{Z}} \alpha(\mathcal{F}_{-\infty}^J, \mathcal{F}_{J+n}^\infty)\ ,$$

$$\phi(n) := \sup_{J\in\mathbb{Z}} \phi(\mathcal{F}_{-\infty}^J, \mathcal{F}_{J+n}^\infty)\ ,$$

$$\psi(n) := \sup_{J\in\mathbb{Z}} \psi(\mathcal{F}_{-\infty}^J, \mathcal{F}_{J+n}^\infty)\ ,$$

$$\rho(n) := \sup_{J\in\mathbb{Z}} \rho(\mathcal{F}_{-\infty}^J, \mathcal{F}_{J+n}^\infty)\ ,$$

$$\beta(n) := \sup_{J\in\mathbb{Z}} \beta(\mathcal{F}_{-\infty}^J, \mathcal{F}_{J+n}^\infty)\ .$$

The sequence (X_k) is said to be
strongly mixing [66] if $\lim_{n\to\infty}\alpha(n) = 0$,
ϕ-mixing [46] if $\lim_{n\to\infty}\phi(n) = 0$,
ψ-mixing ([8] essentially) if $\lim_{n\to\infty}\psi(n) = 0$,
ρ-mixing [52] if $\lim_{n\to\infty}\rho(n) = 0$,
absolutely regular [73] if $\lim_{n\to\infty}\beta(n) = 0$.
These are the five conditions on which we shall focus. The ψ-mixing condition actually evolved form the "*-mixing" condition, which was the condition studied in [8]: $\lim_{n\to\infty}\sup_J \psi(\mathcal{F}_{-\infty}^J, \mathcal{F}_{J+n}^{J+n}) = 0$. The maximal correlation coefficient $\rho(\mathcal{A},\mathcal{B})$ was studied in [44] [39], much earlier than the ρ-mixing condition. The absolute regularity condition was attributed in [73] to Kolmogorov; it also has the name "weak Bernoulli".

Several minor comments are in order:

(i) In defining the strong mixing condition $(\alpha(n) \to 0)$ for a "singly-infinite" sequence $(X_k\ ,\ k = 1,2,3,\ldots)$ one modifies the definition of $\alpha(n)$ as follows: $\alpha(n) := \sup_{J\geq 1}\alpha(\mathcal{F}_1^J, \mathcal{F}_{J+n}^\infty)$.

(ii) In defining the strong mixing condition for a *strictly stationary* doubly-infinite sequence $(X_k\ ,\ k \in \mathbb{Z})$ one can simply define $\alpha(n)$ by $\alpha(n) := \alpha(\mathcal{F}_{-\infty}^0, \mathcal{F}_n^\infty)$.

(iii) In whatever context one is dealing with, the sequence of numbers $\alpha(1)$, $\alpha(2)$, $\alpha(3),\ldots$ is obviously automatically non-increasing.

(iv) If a given random sequence $(X_k\ ,\ k \in \mathbb{Z})$ is strongly mixing, and for each $k \in \mathbb{Z}$, $f_k : \mathbb{R} \to \mathbb{R}$ is a Borel-measurable function, then the random sequence $(f_k(X_k)\ ,\ k \in \mathbb{Z})$ is also obviously strongly mixing, with the dependence coefficients $\alpha(n)$, $n = 1,2,\ldots$ for the sequence $(f_k(X_k))$ being no greater than the corresponding ones for (X_k) .

(v) If a *strictly stationary* strongly mixing singly-infinite sequence $(X_1,X_2,X_3,\ldots)$ is extended to a *strictly stationary* doubly-infinite sequence $(X_k\ ,\ k \in \mathbb{Z})$, then this new doubly-infinite sequence is also strongly mixing, with precisely the same dependence coefficients $\alpha(n)$, $n = 1,2,\ldots$. Conversely, if the doubly-infinite extension is strongly mixing, then the original singly-infinite sequence is strongly mixing too.

(vi) Comments (i)-(v) carry over verbatim to the other mixing conditions defined above (ϕ-mixing, ψ-mixing, ρ-mixing, and absolute regularity) and their dependence coefficients.

By eqns. (1.7), (1.8), (1.9), and (1.11) the following implications hold for a given random sequence:

(i) ρ-mixing $\Rightarrow$ strong mixing.

(ii) absolute regularity $\Rightarrow$ strong mixing.

(iii) ϕ-mixing $\Rightarrow$ ρ-mixing and absolute regularity.

(iv) ψ-mixing $\Rightarrow$ ϕ-mixing.

Among these five mixing conditions there are (aside from transitivity) no other general implications. (For special families of random sequences, however, e.g. Gaussian sequences, discrete Markov chains, etc., there are other implications; this will be seen in more detail in Sections 4 and 5 later on.) Since "strong mixing" is the weakest of these five conditions, these conditions — and others that imply strong mixing — are sometimes referred to collectively as "strong mixing conditions" (plural). The term "strong mixing condition" (singular) will refer to the condition $\alpha(n) \to 0$ as above. Of course all of these mixing conditions are satisfied by sequences of independent r.v.'s and also by m-dependent sequences. Other examples will be encountered in Sections 4, 5, 6, and 7 later on. Later in Section 2 here the strong mixing condition will be compared to standard conditions in ergodic theory.

For a given sequence $(X_k, k \in \mathbb{Z})$ the ϕ-mixing condition is not necessarily preserved if the direction of "time" is reversed. Referring to eqn. (1.3), define for each $n = 1,2,3,\ldots$, $\phi_{rev}(n) :=$ $\sup_{J\in\mathbb{Z}}\phi_{rev}(\mathcal{F}_{-\infty}^{J}, \mathcal{F}_{J+n}^{\infty}) = \sup_{J\in\mathbb{Z}}\phi(\mathcal{F}_{J+n}^{\infty}, \mathcal{F}_{-\infty}^{J})$. In [51, p. 414] there is an example of a strictly stationary countable-state Markov chain $(X_k, k \in \mathbb{Z})$ such that $\phi(n) \to 0$ as $n \to \infty$ and $\phi_{rev}(n) = 1$ for all $n \geq 1$. "Symmetric" versions of the ϕ-mixing condition, putting equal emphasis on $\phi(n)$ and $\phi_{rev}(n)$, have been useful in limit theory (see [34] [59]).

In the rest of Section 2, and also in most of the rest of this paper, we shall deal only with a strictly stationary doubly-infinite sequence $(X_k, k \in \mathbb{Z})$.

For measure-theoretic convenience, for the rest of Section 2 we shall assume that our probability space is $(\mathbb{R}^{\mathbb{Z}}, \mathcal{B}^{\mathbb{Z}}, P)$. (In a context such as this, the symbol $\mathcal{B}$ is intended to mean the standard Borel σ-field.) We shall of course assume that for each $\omega \in \mathbb{R}^{\mathbb{Z}}$ and each $k \in \mathbb{Z}$, $X_k(\omega) := \omega_k$. (The notation ω_k means the kth coordinate of ω.) We shall use a regular conditional distribution of $(X_1, X_2, X_3, \ldots)$ given $(X_0, X_{-1}, X_{-2}, \ldots)$. In this context a standard measure-theoretic argument will show that, under our assumption that (X_k) is strictly stationary, for each $n \geq 1$,

$$\phi(n) = \text{ess sup}[\sup|P(B|\mathcal{F}_{-\infty}^{0}) - P(B)|, B \in \mathcal{F}_{n}^{\infty}], \quad \text{and}$$

$$\beta(n) = E[\sup|P(B|\mathcal{F}_{-\infty}^{0}) - P(B)|, B \in \mathcal{F}_{n}^{\infty}].$$

There is another useful formulation of $\beta(n)$. Let Q denote the probability measure on $(\mathbb{R}^{\mathbb{Z}}, \mathcal{B}^{\mathbb{Z}})$ such that (i) under Q the σ-fields $\mathcal{F}_{-\infty}^{0}$ and $\mathcal{F}_{1}^{\infty}$ are independent and (ii) on each of these two σ-fields $\mathcal{F}_{-\infty}^{0}$ and $\mathcal{F}_{1}^{\infty}$ the measure Q is identical to P. Then (under our assumption of strict stationarity),

$$\beta(n) = [\sup|P(D) - Q(D)| \ , \ D \in F_{-\infty}^{0} \vee F_{n}^{\infty}] = \tfrac{1}{2}\|P_n - Q_n\| \tag{2.1}$$

where P_n (resp. Q_n) is the restriction of P (resp. Q) to $F_{-\infty}^{0} \vee F_{n}^{\infty}$ and $\|\cdot\|$ denotes total variation.

Let T denote the usual shift operator on $\mathbb{R}^{\mathbb{Z}}$; that is, for each $\omega \in \mathbb{R}^{\mathbb{Z}}$, $T\omega$ is defined by $(T\omega)_k := \omega_{k+1}$ for all $k \in \mathbb{Z}$. For any event $A \in F_{-\infty}^{\infty}$ $(= \mathcal{B}^{\mathbb{Z}})$ we use the notation $TA := \{\omega : T^{-1}\omega \in A\}$. Our (strictly stationary) sequence (X_k) is said to be "mixing", or "mixing in the ergodic-theoretic sense", if for all $A, B \in F_{-\infty}^{\infty}$, $\lim_{n\to\infty} P(A \cap T^nB) = P(A) \cdot P(B)$. (In ergodic theory this condition is sometimes referred to as "strong mixing", but we shall use the term "strong mixing" for the condition $\alpha(n) \to 0$ as before.) Our sequence (X_k) is said to be "regular" if its past tail σ-field $\bigcap_{n=1}^{\infty} F_{-\infty}^{-n}$ is trivial (i.e. contains only events of probability 0 or 1). It is well known that

(i) mixing (in the ergodic-theoretic sense) $\Rightarrow$ ergodic,

(ii) regular $\Rightarrow$ mixing (in the ergodic-theoretic sense), and

(iii) strong mixing $(\alpha(n) \to 0)$ $\Rightarrow$ regular.

Statements (ii) and (iii) are easy consequences of [47, p. 302, Theorem 17.1.1]. Naturally, in (ii) and (iii) one can replace "regular" by the condition that the future tail σ-field $\bigcap_{n=1}^{\infty} F_{n}^{\infty}$ be trivial. (In Example 6.2 in Section 6 we shall encounter a well known stationary regular sequence whose future tail σ-field fails to be trivial.)

If (X_k) is strictly stationary and absolutely regular, then its double tail σ-field $\bigcap_{n=1}^{\infty} (F_{-\infty}^{-n} \vee F_{n}^{\infty})$ is trivial (i.e. $P(D) = 0$ or 1 for every D in the double tail σ-field). This holds by (2.1) and an elementary measure-theoretic argument (one can use e.g. [74, Lemma 4.3]). In [19] a strictly stationary ρ-mixing sequence is constructed for which the double tail σ-field fails to be trivial.

Let us briefly give references for several other related mixing conditions, for strictly stationary sequences. The "information regularity" condition (see [65]) is like the strong mixing conditions defined above, using the "coefficient of information" as the basic measure of dependence. A "Cesaro" variant of strong mixing, known as "uniform ergodicity", was studied by Cogburn [25]; and Rosenblatt [68, Theorem 2] established a nice connection between this condition and the strong mixing condition itself. Another mixing condition weaker than strong mixing has played a nice role in extreme value theory (see e.g. [54]) as well as in convergence in distribution to non-normal stable laws (see [29]). A mixing condition based on characteristic functions was studied in [75]. Finally, by a theorem of Ornstein, a condition of weak dependence known as the "very weak Bernoulli" condition characterizes the strictly stationary finite-state sequences that are isomorphic to a Bernoulli shift. For more information on the very weak Bernoulli condition, including recent generalizations of it to stationary real sequences in connection with central limit theory, see [71] [36] [32] [17] and the references therein.

3. MIXING CONDITIONS FOR TWO OR MORE SEQUENCES

Suppose $(X_k , k \in \mathbb{Z})$ and $(Y_k , k \in \mathbb{Z})$ are strongly mixing sequences that are independent of each other. Then the sequence of random vectors $((X_k,Y_k) , k \in \mathbb{Z})$ is strongly mixing. Hence the sequence of sums $(X_k + Y_k , k \in \mathbb{Z})$ is also strongly mixing. The same comments apply to the other mixing conditions being discussed here. Pinsker [65, p. 73] noted this for absolute regularity. Under natural extra restrictions, such comments can be extended from two to countably many sequences that are independent of each other. Here we shall just present the basic propositions from which all of these comments can easily be deduced.

The first result is due to Csaki and Fischer [27, p. 40, Theorem 6.2]:

<u>Theorem 3.1 (Csaki and Fischer)</u>: Suppose $\mathcal{A}_n$ and $\mathcal{B}_n$, $n = 1,2,3,\dots$ are σ-fields and the σ-fields $(\mathcal{A}_n \vee \mathcal{B}_n)$, $n = 1,2,3,\dots$ are independent. Then

$$\rho\left(\bigvee_{n=1}^{\infty} \mathcal{A}_n , \bigvee_{n=1}^{\infty} \mathcal{B}_n\right) = \sup_{n\geq 1} \rho(\mathcal{A}_n , \mathcal{B}_n) .$$

For a short proof see Witsenhausen [76, Theorem 1]. In Example 4.4 in the next section an interesting application of Theorem 3.1 will be given. For the other dependence coefficients, slightly weaker statements hold:

Theorem 3.2: If the hypothesis of Theorem 3.1 is satisfied, then the following statements hold:

$$\text{(i)} \quad \alpha\Big(\bigvee_{n=1}^{\infty} A_n , \bigvee_{n=1}^{\infty} B_n\Big) \le \sum_{n=1}^{\infty} \alpha(A_n , B_n) .$$

$$\text{(ii)} \quad \beta\Big(\bigvee_{n=1}^{\infty} A_n , \bigvee_{n=1}^{\infty} B_n\Big) \le \sum_{n=1}^{\infty} \beta(A_n , B_n) .$$

$$\text{(iii)} \quad \phi\Big(\bigvee_{n=1}^{\infty} A_n , \bigvee_{n=1}^{\infty} B_n\Big) \le \sum_{n=1}^{\infty} \phi(A_n , B_n) .$$

$$\text{(iv)} \quad \psi\Big(\bigvee_{n=1}^{\infty} A_n , \bigvee_{n=1}^{\infty} B_n\Big) \le \Big(\prod_{n=1}^{\infty} [1 + \psi(A_n , B_n)]\Big) - 1 .$$

Statements (i)-(iii) can be found in [12, Lemma 8], [20, Lemma 2.2], and [11, Lemma 2.2]. Statements (ii)-(iii) can also be derived easily from [37, Lemma 1]. Statement (iv) is an elementary consequence of [14, Lemma 1].

4. MIXING CONDITIONS FOR MARKOV CHAINS

Here a brief discussion is given of Markov chains satisfying strong mixing conditions. For a more thorough discussion of this topic see Rosenblatt [67, Chapter 7].

The following theorem is fundamental to the study of mixing conditions on Markov chains:

Theorem 4.1. Suppose $(X_k , k \in \mathbb{Z})$ is a strictly stationary real Markov chain. Then for all $n \ge 1$ the following five statements hold:

(i) $\alpha(n) = \alpha(\sigma(X_0),\sigma(X_n))$.

(ii) $\rho(n) = \rho(\sigma(X_0),\sigma(X_n))$.

(iii) $\beta(n) = \beta(\sigma(X_0),\sigma(X_n))$.

(iv) $\phi(n) = \phi(\sigma(X_0),\sigma(X_n))$.

(v) $\psi(n) = \psi(\sigma(X_0),\sigma(X_n))$.

The proof is an elementary measure-theoretic exercise using the Markov property. For example, see [8, Lemma 8] for a proof of (v). (Thus for Markov chains, ψ-mixing is equivalent to the "*-mixing" condition studied in [8]). As a consequence of (iv), for Markov chains the

ϕ-mixing condition is equivalent to Doeblin's condition (see [67, p. 212, eqn. (18)]). (See [67, p. 209] for the definition of Doeblin's condition.)

For the next theorem we shall use the following terminology: A sequence of non-negative numbers $a_1, a_2, a_3, \ldots$ is said to "converge to 0 exponentially fast" if there exists a positive number r such that $a_n = O(e^{-rn})$ as $n \to \infty$.

<u>Theorem 4.2.</u> Suppose $(X_k, k \in \mathbb{Z})$ is a strictly stationary real Markov chain. Then the following three statements hold:

(i) If $\rho(n) \to 0$, then $\rho(n) \to 0$ exponentially fast.
(ii) If $\phi(n) \to 0$, then $\phi(n) \to 0$ exponentially fast.
(iii) If $\psi(n) \to 0$, then $\psi(n) \to 0$ exponentially fast.

Part (iii) was proved in [8, pp. 8-9, Theorem 5]. The arguments for parts (i) and (ii) are similar. (For part (i) a simple argument using (1.10) and the Markov property will show the well known inequality $\rho(m+n) \le \rho(m) \cdot \rho(n)$ for all positive integers m and n. For part (ii) see e.g. [67, p. 209, Lemma 3].) Theorem 4.2 does not extend to either $\alpha(n)$ or $\beta(n)$. As a consequence of the classic convergence theorem for transition probabilities, any strictly stationary countable-state irreducible aperiodic Markov chain is absolutely regular. (To put this another way, any stationary countable-state Markov chain which is mixing in the ergodic-theoretic sense is absolute regular.) Such Markov chains exist for which the rate of convergence of $\alpha(n)$ (and hence also the rate for $\beta(n)$) to 0 is slower than exponential (see e.g. [30, Examples 1 and 2] or [51, p. 414, Corollary 1]). (By Theorem 4.2 such Markov chains cannot be ρ-mixing.) Of course every stationary *finite*-state irreducible aperiodic Markov chain is ψ-mixing (with exponential mixing rate).

A strictly stationary real Markov chain (X_k) is said to be a "Harris chain" if it has the Harris recurrence property: There exists a regular version of the conditional distribution of $(X_1, X_2, X_3, \ldots)$ given X_0 such that for every $x \in \mathbb{R}$, for every Borel subset $B \subset \mathbb{R}$ such that $P(X_0 \in B) > 0$, one has that $P(X_n \in B$ for infinitely many positive integers $n | X_0 = x) = 1$. (Thus every stationary countable-state irreducible Markov chain is a stationary Harris chain.) It is well known that every stationary Harris chain has a well defined "period" $p \in \{1,2,3,\ldots\}$ (the chain is said to be "aperiodic" if

$p = 1$). This fact and the next theorem can be seen (with a little work) from Orey [57, p. 13, Theorem 3.1; p. 23, Theorem 5.1; and p. 25, lines 9-13].

Theorem 4.3 (i) Every strictly stationary real aperiodic Harris chain is absolutely regular. (ii) More generally, for any strictly stationary real Harris chain, $\lim_{n\to\infty}\beta(n) = 1 - 1/p$ where p is the period.

A sequence (X_k) is said to be an "instantaneous function" of a real Markov chain (Y_k) if there is a Borel-measurable function $f : \mathbb{R} \to \mathbb{R}$ such that for each $k \in \mathbb{Z}$, $X_k = f(Y_k)$. As a consequence of Theorem 4.3(i), any instantaneous function of a stationary real aperiodic Harris chain is a stationary absolutely regular sequence. In [4] there are some stationary ϕ-mixing sequences which *cannot* be represented (on any probability space) as an instantaneous function of a stationary real Harris chain (periodic or aperiodic). It is apparently unknown whether any such sequences exist which are ψ-mixing or even 1-dependent.

Example 4.4. Rosenblatt [67, p. 214, line -3 to p. 215, line 13] presents a class of stationary real Markov chains which are ρ-mixing but not absolutely regular. (Consequently, they are not Harris chains.) His construction is based on "random rotations". Here we shall construct one of those examples in a different way, as an application of Theorem 3.1.

As a preliminary step, consider a stationary Markov chain $(W_k , k \in \mathbb{Z})$ with two states $\{0,1\}$, with invariant probability vector $(\frac{1}{2},\frac{1}{2})$, and with one-step transition probability matrix (p_{ij}) given by $p_{00} = p_{11} = 3/4$, $p_{10} = p_{01} = \frac{1}{4}$. By an induction argument, for each $n \geq 1$ the n-step transition probability matrix $(p_{ij}^{(n)})$ is given by $p_{00}^{(n)} = p_{11}^{(n)} = (1+2^{-n})/2$, $p_{10}^{(n)} = p_{01}^{(n)} = (1-2^{-n})/2$. A simple argument will show that for each $n \geq 1$, (W_k) satisfies $\rho(n) = \rho(\sigma(W_0), \sigma(W_n)) = |\mathrm{Corr}(W_0, W_n)| = 2^{-n}$. Here the first equality comes from Theorem 4.1 and the second from the elementary fact that every function of a two-state r.v. is automatically an *affine* function.

Now let $(W_k^{(j)}, k \in \mathbb{Z})$, $j = 1,2,3,\ldots$ be independent Markov chains, each having the same distribution as the Markov chain $(W_k , k \in \mathbb{Z})$ above. Define the sequence $(X_k , k \in \mathbb{Z})$ by

$X_k := \Sigma_{j=1}^{\infty} 2^{-j} W_k^{(j)}$. Up to null sets, for each k , $\sigma(X_k) = \sigma(W_k^{(j)}$, $j = 1,2,\ldots)$ (i.e. from X_k one can calculate the value of $W_k^{(j)}$, $j = 1,2,3,\ldots$ a.s.). It follows that the sequence (X_k) is a Markov chain, and it is easy to see that it is strictly stationary. Further, (X_k) satisfies $\rho(n) = 2^{-n}$ for all $n \geq 1$, by Theorem 3.1 and the properties of (W_k) above; and hence (X_k) is ρ-mixing. In the framework of eqn. (2.1) one has that for each $n \geq 1$, by a simple calculation and the strong law of large numbers, $P(\lim_{J\to\infty} \frac{1}{J}\Sigma_{j=1}^{J} W_0^{(j)} W_n^{(j)} = (\frac{1}{4})(1+2^{-n})) = 1$ and $Q(\lim_{J\to\infty} \frac{1}{J}\Sigma_{j=1}^{J} W_0^{(j)} W_n^{(j)} = \frac{1}{4}) = 1$, and hence P and Q are mutually singular on $\sigma(X_0, X_n)$. Hence (X_k) satisfies $\beta(n) = 1$ for all $n \geq 1$, i.e. (X_k) fails to be absolutely regular. This completes Example 4.4

5. MIXING CONDITIONS FOR GAUSSIAN SEQUENCES

For stationary real Gaussian sequences, a thorough discussion of the various mixing conditions is given by Ibragimov and Rozanov [48] [49, Chapters 4-5]. Theorem 5.1 here essentially just lists a few basic results from that discussion.

Theorem 5.1. Suppose $(X_k$, $k \in \mathbb{Z})$ is a (non-degenerate) stationary real Gaussian sequence. Then the following four statements hold:

(1) The following two conditions are equivalent:

(a) (X_k) is regular.

(b) (X_k) has an absolutely continuous spectral distribution function, and its spectral density f (defined on $[-\pi,\pi]$) satisfies $\int_{-\pi}^{\pi} \log f(\lambda)d\lambda > -\infty$.

(2) The following three conditions are equivalent:

(a) (X_k) is strongly mixing.

(b) (X_k) is ρ-mixing.

(c) The spectral density f of (X_k) can be expressed in the form $f(\lambda) = |P(e^{i\lambda})|^2 \exp[u(e^{i\lambda}) + \tilde{v}(e^{i\lambda})]$ where P is a polynomial, u and v are continuous real functions on the unit circle (in the complex plane), and $\tilde{v}$ is the conjugate function of v .

(3) The following two conditions are equivalent:

(a) (X_k) is absolutely regular.

(b) The spectral density f of (X_k) can be expressed in the form $f(\lambda) = |P(e^{i\lambda})|^2 \exp[\Sigma_{j=-\infty}^{\infty} a_j e^{ij\lambda}]$ (the sum converging in $L_2[-\pi,\pi]$) where P is a polynomial whose roots (if there are any) lie on the unit circle and $\Sigma_{j=-\infty}^{\infty} |j| \cdot |a_j|^2 < \infty$.

(4) The following four conditions are equivalent:

(a) (X_k) is ϕ-mixing.

(b) (X_k) is ψ-mixing.

(c) (X_k) is m-dependent.

(d) The spectral density f of (X_k) can be expressed in the form $f(\lambda) = |P(e^{i\lambda})|^2$ where P is a polynomial.

A few comments are in order. If f is the spectral density of a stationary real (not complex) Gaussian sequence then of course f is symmetric about 0. In connection with statement (2), one has (for Gaussian sequences) that $\rho(n) \leq 2\pi \cdot \alpha(n)$ and that $\rho(n)$ is identical to the supremum of $|\mathrm{Corr}(Y,Z)|$ taken over all finite linear combinations $Y = a_0X_0 + a_{-1}X_{-1} + \ldots + a_{-M}X_{-M}$ and $Z = a_nX_n + a_{n+1}X_{n+1} + \ldots + a_{n+N}X_{n+N}$ (see [52, Theorems 1 and 2]). In (2) the equivalence of (c) with (b) comes from the formulation of the Helson-Sarason theorem given in [70, p. 62]. From (2) and (3) we see that in order to construct a stationary real Gaussian sequence which is ρ-mixing but not absolutely regular, one can simply choose a spectral density which is positive and continuous but very "jagged", such as $f(\lambda) = \exp[\Sigma_{j=1}^{\infty} 2^{-j}\cos(2^{2^j}\lambda)]$. For a stationary real ρ-mixing (or even absolutely regular) Gaussian sequence the spectral density need not be continuous or even bounded; consider an example with spectral density $f(\lambda) = \exp[\Sigma_{j=2}^{\infty} (j \log j)^{-1}\cos(j\lambda)]$ (which satisfies $\lim_{\lambda\to 0} f(\lambda) = +\infty$ by [77, p. 188, Theorem 2.15]). For more examples see [49, pp. 179-180]. For part (4) one can use the Wold decomposition theorem to show that any stationary real ϕ-mixing Gaussian sequence must be a moving average of i.i.d. Gaussian r.v.'s; one uses the fact that if Y and Z are jointly Gaussian r.v.'s with $\mathrm{Corr}(Y,Z) \neq 0$ then $\phi(\sigma(Y), \sigma(Z)) = 1$. To see this latter fact, say in the case where $\mathrm{Corr}(Y,Z) > 0$, one can first note that $P(Z > q)$ becomes arbitrarily small as $q > 0$ becomes sufficiently large, and then for

q fixed, $P(Z > q | Y > r)$ becomes arbitrarily close to 1 as $r > 0$ becomes sufficiently large.

Ibragimov and Rozanov [48] [49, p. 182, Lemma 17, and p. 190, Note 2] proved that every stationary Gaussian sequence satisfying $\Sigma_{n=1}^{\infty} \rho(2^n) < \infty$ has a continuous spectral density $f(\lambda)$; they derived for each $n \geq 1$ an upper bound on the "uniform error" of the "best" approximation of f by a trigonometric polynomial of degree $\leq n$. Their result introduced the (logarithmic) mixing rate $\Sigma_{n=1}^{\infty} \rho(2^n) < \infty$ into the literature, along with some of the techniques for handling this mixing rate. In central limit theory this mixing rate has turned out to be quite prominent for ρ-mixing (see e.g. [60] [58] [38] and the references therein).

6. SOME OTHER SPECIAL EXAMPLES

Here we shall briefly describe the strong mixing properties of a few stationary sequences that arise in other areas such as time series analysis, number theory, and interacting particle systems. (There will be a slight overlap with the Markov chains and Gaussian sequences studied in Sections 4 and 5.) In most of these examples we shall encounter strong mixing conditions with exponential mixing rates — a context in which the known limit theory under strong mixing conditions usually applies very nicely. However, in Example 6.2 below, we shall also look at a well known simple stationary AR(1) process (autoregressive process of order 1) which *fails* to be strongly mixing.

Example 6.1. Suppose $(Z_k, k \in \mathbb{Z})$ is an i.i.d. sequence and the distribution of Z_0 is absolutely continuous with a density which (to start with a few nice specific cases) is Gaussian, Cauchy, exponential, or uniform (on some interval). Suppose $a_1, a_2, a_3, \ldots$ is a sequence of real numbers with $|a_n| \to 0$ exponentially fast. Then the random sequence $(X_k, k \in \mathbb{Z})$ defined by

$$X_k := \sum_{j=0}^{\infty} a_j Z_{k-j} ,$$

is well defined, strictly stationary, and satisfies absolute regularity with exponential mixing rate. In particular this includes the cases where (X_k) is a stationary ARMA (autoregressive-mixed-moving-average) process based on the i.i.d. sequence (Z_k) given above. The conditions on (a_n) can be relaxed somewhat; if $a_n \to 0$ at a sufficiently fast polynomial rate, then (X_k) will still satisfy absolute regularity, but the mixing rate may be slower than exponential. Of course all these statements apply to a much broader class of density functions for Z_0 than just the ones given above. For details see [61] and the references therein. As the next example will show, the above results in general do not carry over to the case where the distribution of Z_0 is discrete.

Example 6.2. The following example is well known; see e.g. [69, p. 267]. Suppose $(Z_k\,,\ k \in \mathbb{Z})$ is i.i.d. with $P(Z_k = 0) = P(Z_k = 1) = \frac{1}{2}$. Define the sequence $(X_k\,,\ k \in \mathbb{Z})$ by

$$X_k := (\tfrac{1}{2})Z_k + (\tfrac{1}{4})Z_{k-1} + (\tfrac{1}{8})Z_{k-2} + (\tfrac{1}{16})Z_{k-3} + \dots$$

Then (X_k) is a strictly stationary AR(1) process; it can be represented by $X_k = (\frac{1}{2})X_{k-1} + (\frac{1}{2})Z_k$. For each k the r.v. X_k is uniformly distributed on the interval $[0,1]$. (Note that the digits in the binary expansion of X_k are $Z_k, Z_{k-1}, Z_{k-2}, \dots$.) For each k one also has that X_k is a (Borel-measurable) function of X_{k+1} (up to null sets): $X_k =$ (fractional part of $2X_{k+1}$) a.s. As a consequence of all this, for each $n \ge 1$, X_0 is a (Borel-measurable) function of X_n (by induction) and hence

$$\alpha(n) \ge \alpha(\sigma(X_0)\,,\ \sigma(X_n)) \ge \alpha(\sigma(X_0)\,,\ \sigma(X_0))$$

$$\ge P(X_0 < \tfrac{1}{2}) - [P(X_0 < \tfrac{1}{2})]^2 = \tfrac{1}{4}\,.$$

Hence (X_k) *fails* to be strongly mixing; in fact $\alpha(n) = \frac{1}{4}$ for all $n \ge 1$ by (1.11). Indeed, even though (X_k) is regular, its future tail σ-field $\bigcap_{n=1}^{\infty} F_n^{\infty}$ is non-trivial (it coincides with $F_{-\infty}^{\infty}$ up to null sets). Of course one can make this example symmetric about 0 by replacing Z_k by $Z_k - \frac{1}{2}$.

Example 6.3. This is another well known example, related to number theory. For every irrational number $x \in (0,1)$ there exists a unique sequence of positive integers $x_1, x_2, x_3, \dots$ such that the following "continued fraction" expansion holds:

$$x = \cfrac{1}{x_1 + \cfrac{1}{x_2 + \cfrac{1}{x_3 + \ddots}}}$$

Suppose we impose the Gauss probability measure on $[0,1]$, namely the measure which is absolutely continuous with density $f(x) := (\log 2)^{-1} \cdot (1 + x)^{-1}$. In this context the sequence $(x_1, x_2, x_3, \ldots)$ is a strictly stationary sequence of random variables, and it is ψ-mixing with exponential mixing rate (see [62, p. 450, Corollary 1]). For applications of limit theorems under strong mixing conditions to number theory (including this continued fraction expansion), see [62] [63].

Example 6.4. There has been a lot of research on the ergodic-theoretic properties of mappings $T : [0,1] \to [0,1]$. See e.g. [55] as well as the references given below. Example 6.3 above fits into this framework in a natural way (see e.g. [62, pp. 448-449]). Here we shall just describe one other simple example. Suppose $2^{\frac{1}{2}} < \lambda \le 2$. Consider the mapping $T : [0,1] \to [0,1]$ defined by

$$T(x) := \begin{cases} \lambda x + 2 - \lambda & \text{if} \quad 0 \le x \le 1 - 1/\lambda \\ \\ -\lambda x + \lambda & \text{if} \quad 1 - 1/\lambda \le x \le 1 \end{cases} .$$

The graph of this function looks somewhat like an inverted, chopped letter V with apex at the point $(1-1/\lambda, 1)$. By [53, Theorem 1] there exists on $[0,1]$ an absolutely continuous probability measure μ which is T-invariant. In [45, Theorem 1(vi)] a "canonical" method is described for defining μ. Under this measure μ the transformation T is "weak mixing" by [10, Theorem 2(ii)]. (We need not give the definition of "weak mixing" here.) Now let $I_1, I_2, \ldots, I_M$ be an arbitrary partition of [0,1] into finitely many intervals, and define on $([0,1], \mathcal{B}_{[0,1]}, \mu)$ the (strictly stationary) sequence $(X_k, k = 1,2,3,\ldots)$ by

$$X_k(x) := m \quad \text{if} \quad T^k(x) \in I_m , \; m = 1,2,\ldots,M .$$

By [45, p. 132, lines 4-6], (X_k) is absolutely regular with exponential mixing rate.

Example 6.5. Gibbs measures have sometimes been used in the study of interacting particle systems. We shall discuss Gibbs measures in just the simplest context. Suppose $\phi : \{0,1\}^{\mathbb{Z}} \to \mathbb{R}$ is a function with the following restriction on its "variation": $\exists a > 0$, $\exists C > 0$, such that for all $m = 1,2,3,\ldots$

$$[\sup\{|\phi(x)-\phi(y)| : x , y \in \{0,1\}^{\mathbb{Z}} \text{ such that } x_k = y_k \ \forall k = -m , -m+1,\ldots,m\}] \le Ce^{-am}$$

(where the kth coordinate of any $x \in \{0,1\}^{\mathbb{Z}}$ is denoted x_k). Then there exists a unique shift-invariant probability measure μ on $\{0,1\}^{\mathbb{Z}}$ with the following property: $\exists C_1 > 0$, $\exists C_2 > 0$, $\exists q \in \mathbb{R}$ such that for all $x \in \{0,1\}^{\mathbb{Z}}$, for all $m \ge 1$,

$$C_1 \le \frac{\mu\{y : y_k = x_k \text{ for all } k = 0,1,\ldots,m-1\}}{\exp[-qm + \Sigma_{0\le k\le m-1}\ \phi(T^k(x))]} \le C_2$$

where T is the usual shift operator on $\{0,1\}^{\mathbb{Z}}$. This measure μ is called the "Gibbs measure" based on the function ϕ . (For details see e.g. [9] [35].) On the probability space $(\{0,1\}^{\mathbb{Z}} , \mathcal{B}^{\mathbb{Z}}_{\{0,1\}} , \mu)$ define the (strictly stationary) random sequence $(X_k , k \in \mathbb{Z})$ by $X_k(\omega) := \omega_k$. One can interpret $X_k = 0$ resp. 1 to mean that the kth site is empty resp. occupied by a particle. This sequence (X_k) is ψ-mixing with exponential mixing rate (see [9, p. 24] or [35]).

7. THE BEHAVIOR OF THE DEPENDENCE COEFFICIENTS

First we shall examine the possible limit values of the dependence coefficients.

Theorem 7.1. If $(X_k , k \in \mathbb{Z})$ is strictly stationary and mixing in the ergodic-theoretic sense, then the following three statements hold:

(i) Either $\beta(n) \to 0$ as $n \to \infty$ or $\beta(n) = 1$ for all $n \ge 1$.

(ii) Either $\phi(n) \to 0$ as $n \to \infty$ or $\phi(n) = 1$ for all $n \ge 1$.

(iii) Either $\psi(n) \to 0$, $\psi(n) \to 1$, or $\psi(n) = \infty$ for all $n \ge 1$.

Statements (i) and (ii) can be found in [13, Theorem 1] and [11, Theorem 1], and statement (iii) is a trivial consequence of [14, Theorem 1]. Statement (i) is a slight extension of an earlier result of Volkonskii and Rozanov [74, Theorem 4.1]. Statement (ii) was previously known for stationary Markov chains (see [67, p. 209, Lemma 3]) and of course for stationary Gaussian sequences (see Section 5). Theorem 7.1 does not extend to either $\alpha(n)$ or $\rho(n)$. Instead, for stationary regular sequences, $\lim \alpha(n)$ can be any value in $[0,\frac{1}{4}]$ and $\lim \rho(n)$ can be any value in $[0,1]$. (See eqn. (1.11) and [12, Theorem 6]; regularity was not mentioned in this theorem, but is an elementary property of the construction given in the proof.) Berbee [2, Theorem 2.1] proved an analog of Theorem 7.1 for stationary sequences which are ergodic but not assumed to be mixing:

<u>Theorem 7.2 (Berbee).</u> If $(X_k, k \in \mathbb{Z})$ is strictly stationary and ergodic, then $\lim_{n\to\infty}\beta(n) = 1 - 1/p$ for some $p \in \{1,2,3,\ldots\} \cup \{\infty\}$. If this p satisfies $2 \le p < \infty$, then letting T denote the usual shift operator (on events in $\mathcal{F}_{-\infty}^{\infty}$), the invariant σ-field of T^p is identical to each tail σ-field (past, future, and double) of (X_k) (up to null sets) and is purely atomic with exactly p atoms, each having probability $1/p$ (the p atoms are A, TA, $T^2A,\ldots,T^{p-1}A$ where A is any one of the atoms), and conditional on any one of these atoms the sequence of random vectors $(Y_k, k \in \mathbb{Z})$ defined by $Y_k := (X_{(k-1)p+1}, X_{(k-1)p+2},\ldots,X_{kp})$ is strictly stationary and satisfies the absolute regularity condition.

Theorem 4.3(ii) is in essence a special case of Theorem 7.2. Also, as a simple corollary of Theorem 7.2 (after one applies Theorem 7.1(ii)(iii) to the sequence (Y_k) there if $2 \le p < \infty$) one has the following additional properties of strictly stationary ergodic sequences (X_k):

(i) $\phi(n) \to 1 - 1/p$ for some $p \in \{1,2,3,\ldots\} \cup \{\infty\}$,

(ii) $\psi(n) \to p - 1$ for some $p \in \{1,2,3,\ldots\} \cup \{\infty\}$.

(In particular, for example, if $\lim \beta(n) = 1 - 1/p$ for some (finite) positive integer p, then either $\lim \phi(n) = 1 - 1/p$ for the same p or else $\phi(n) = 1$ for all n.)

For strictly stationary sequences there is essentially no restriction on the mixing rates for the mixing conditions being discussed here. See e.g. [51, Theorems 2,3, and 4], [49, pp. 181-190], [11, Theorem 2],

[12, Theorem 6], and [14, Theorem 2]. Also, for most mixing rates used in the literature (in particular exponential, polynomial, or logarithmic), the mixing conditions being discussed here can all hold with essentially the same given rate; this is a consequence of the following theorem which (because of eqns. (1.7) and (1.8)) is just [18, Theorem 1]:

Theorem 7.3. Suppose $g : [0,\infty) \to (0,\infty)$ is a positive continuous non-increasing function such that $\lim_{x\to\infty} g(x) = 0$, $g(0) \le 1/24$, and $\log g$ is convex on $[0,\infty)$. Then there exists a strictly stationary sequence (X_k) such that for every $n \ge 1$,

$$(\tfrac{1}{4})g(n) \le \alpha(n)\ ,\ \beta(n)\ ,\ \rho(n)\ ,\ \phi(n)\ ,\ \psi(n) \le 8g(n)\ .$$

(The conditions on g here are of course somewhat redundant.) Under fewer restrictions on g, Theorem 7.3 was already shown for non-stationary sequences by Kesten and O'Brien [51, Theorem 1].

Theorems 7.1 and 7.2 do not extend to non-stationary sequences. In limit theory for non-stationary sequences, conditions such as "$\psi(n) < \infty$ for some n" or "$\phi(n) < 1$ for some n" are sometimes useful (see e.g. [26] [58]). Note that if $\psi(n) < \infty$ then (for the same n) $\phi(n) < 1$. (In fact it is not hard to show that $\phi(\mathcal{A},\mathcal{B}) < 1$ if either $\sup P(A \cap B)/[P(A)P(B)] < \infty$ or $\inf P(A \cap B)/[P(A)P(B)] > 0$, the sup and inf being taken over all $A \in \mathcal{A}$ and $B \in \mathcal{B}$ with $P(A) > 0$ and $P(B) > 0$.) For non-stationary sequences, conditions such as $\psi(n) < \infty$ or $\rho(n) < 1$ do not impose any essential restrictions on the moments of the r.v.'s X_k or on the rate at which $\alpha(n)$ or $\beta(n)$ or $\rho(n)$ might perhaps converge to 0, nor do they preclude *second-order* stationarity. Defining $\phi_{rev}(n) := \sup_{J\in\mathbb{Z}} \phi_{rev}(F_{-\infty}^{J}, F_{J+n}^{\infty})$ for each $n = 1,2,\dots$, one has the following theorem:

Theorem 7.4: Suppose $d_1, d_2, d_3, \dots$ is a non-increasing sequence of positive numbers such that $\lim_{n\to\infty} d_n = 0$. Suppose $0 < c \le 1$. Then there exists a (not strictly stationary) sequence $(X_k\ ,\ k \in \mathbb{Z})$ with the following eight properties:

(i) For all $k \in \mathbb{Z}$, $EX_k = 0$ and $EX_k^2 = 1$.

(ii) For all $k, \ell \in \mathbb{Z}$ with $k \ne \ell$, $EX_k X_\ell = 0$.

(iii) For all $k \in \mathbb{Z}$, $|X_k| \le 2$ a.s.

(iv) For all $n \geq 1$, $\psi(n) = c$.

(v) For all $n \geq 1$, $\phi(n) = \phi_{rev}(n) = c/2$.

(vi) $\rho(n) \overset{\cup}{\cap} d_n^{\frac{1}{2}}$ as $n \to \infty$.

(vii) $\beta(n) \overset{\cup}{\cap} d_n$ as $n \to \infty$.

(viii) $\alpha(n) \overset{\cup}{\cap} d_n$ as $n \to \infty$.

Here the notation $a \overset{\cup}{\cap} b$ means that $a = O(b)$ and $b = O(a)$. This theorem apparently has not been published anywhere; its proof will be given here, based on an argument used by Kesten and O'Brien [51, Theorem 1].

First some preliminary calculations are needed. Suppose $0 < c \leq 1$ and $0 < d \leq \frac{1}{4}$. Suppose U and V are r.v.'s, each with state space $\{-2, -1, 1, 2\}$ and marginal probability vector $[d, \frac{1}{2} - d, \frac{1}{2} - d, d]$, and with joint probability function (f_{ij}), where $f_{ij} := P(U = i, V = j)$, given by the matrix

$$\begin{bmatrix} f_{-2,-2}, & f_{-2,-1}, & f_{-2,1}, & f_{-2,2} \\ f_{-1,-2}, & f_{-1,-1}, & f_{-1,1}, & f_{-1,2} \\ f_{1,-2}, & f_{1,-1}, & f_{1,1}, & f_{1,2} \\ f_{2,-2}, & f_{2,-1}, & f_{2,1}, & f_{2,2} \end{bmatrix} :=$$

$$:= \begin{bmatrix} d^2 & d(\frac{1}{2}-d)(1+c) & d(\frac{1}{2}-d)(1-c) & d^2 \\ d(\frac{1}{2}-d)(1-c) & (\frac{1}{2}-d)^2 & (\frac{1}{2}-d)^2 & d(\frac{1}{2}-d)(1+c) \\ d(\frac{1}{2}-d)(1+c) & (\frac{1}{2}-d)^2 & (\frac{1}{2}-d)^2 & d(\frac{1}{2}-d)(1-c) \\ d^2 & d(\frac{1}{2}-d)(1-c) & d(\frac{1}{2}-d)(1+c) & d^2 \end{bmatrix}.$$

By elementary calculations one can show that $\psi(\sigma(U), \sigma(V)) = c$, $\phi(\sigma(U), \sigma(V)) = \phi_{rev}(\sigma(U), \sigma(V)) = c(\frac{1}{2}-d)$, $\beta(\sigma(U), \sigma(V)) = 4cd(\frac{1}{2}-d)$, and $\alpha(\sigma(U), \sigma(V)) \geq P(U = 1 \text{ or } 2, V = -2 \text{ or } 1) - P(U = 1 \text{ or } 2)P(V = -2 \text{ or } 1) = 2cd(\frac{1}{2}-d)$. By the first equation in (1.7), in fact $\alpha(\sigma(U), \sigma(V)) = 2cd(\frac{1}{2}-d)$. Also, by eqn. (1.12), $\alpha_{\frac{1}{2},\frac{1}{2}}(A,B) \leq [\alpha(A,B) \cdot \psi(A,B)]^{\frac{1}{2}} \leq cd^{\frac{1}{2}}$. Since U and V each have only four states, $\rho(\sigma(U), \sigma(V)) \leq 16\alpha_{\frac{1}{2},\frac{1}{2}}(A,B) \leq 16cd^{\frac{1}{2}}$. Also,

$\rho(\sigma(U), \sigma(V)) \geq \mathrm{Corr}(I_{\{U=2\}}, I_{\{V=1\}}) \geq (c/2)d^{1/2}$ (using the fact that $d \leq 1/4$). Also, $EU = EV = EUV = 0$, $EU^2 = EV^2 > 1$, and $\|U\|_\infty = \|V\|_\infty = 2$.

Now we mimic the construction in [51, Theorem 1]. For each $m = 1,2,3,\ldots$ let $(X_k, k = m^2, m^2 + m)$ be a random vector with the same distribution as the random vector (U,V) above, in the case where the parameters are c and d_m from the hypothesis of Theorem 7.4. We assume without loss of generality that $d_m \leq 1/4$ for all m. For all integers k that are *not* of the form m^2 or $m^2 + m$, $m = 1,2,\ldots$, let X_k be a r.v. such that $P(X_k = 1) = P(X_k = -1) = 1/2$. We impose the additional restriction that (X_1,X_2), (X_4,X_6), (X_9,X_{12}), $(X_{16},X_{20}),\ldots$ and $\ldots X_{-2}$, X_{-1}, X_0, X_3, X_5, X_7, X_8, X_{10}, X_{11}, $X_{13},\ldots$ are independent. Following the argument in [51, Theorem 1] we have for each $n = 1,2,3,\ldots$

$$\alpha(n) = \sup_{m\geq n}\alpha(\sigma(X_{m^2}), \sigma(X_{m^2+m}))$$

and the analogous statement holds for the other dependence coefficients. Using the calculations in the preceding paragraph, one can show that the sequence (X_k) constructed here satisfies all properties (i)-(viii) in Theorem 7.4, except that $EX_k^2 > 1$ for $k = m^2$, $m^2 + m$, $m = 1,2,\ldots$. Dividing each X_k by its standard deviation, we obtain a new sequence (X_k) satisfying all properties in Theorem 7.4. This completes the proof.

8. APPROXIMATION OF MIXING SEQUENCES BY OTHER RANDOM SEQUENCES

In proving limit theorems for a given sequence of random variables satisfying strong mixing conditions, it is often useful to approximate the sequence by another random sequence with certain properties. Here we shall give some pertinent references on such techniques.

The technique of directly approximating a mixing sequence by a sequence of martingale differences was introduced by Gordin [40]. It has been exploited by many people (see e.g. [64] [42] as well as the exposition of it in [50, pp. 127-131]). A long-standing, previously unnoticed error in one particular application of Gordin's technique was pointed out by Herrndorf [43] (see [20] for further details).

The technique of directly approximating a mixing sequence by a sequence of independent random variables was introduced by Berkes and Philipp [6] [7]. This technique is a versatile one which often allows one to handle very slow (logarithmic) mixing rates (see e.g. [7] [28] [16]), other types of conditions of weak dependence besides the strong mixing conditions being discussed here (see e.g. [36]), and also random variables taking their values in general Banach spaces (see e.g. [33] and the references therein). This technique works particularly nicely with the absolute regularity condition (see [1, Corollary 4.2.5] [3] [23] [33]). Dehling [31] exposed a weakness in this technique under just the strong mixing condition (or even, by the same argument, the ρ-mixing condition) when the random variables are taking their values in general Banach spaces.

For proving theorems in renewal theory, "coupling" techniques are often useful. Berbee [1, p. 104, Theorem 4.4.7] characterized the strictly stationary absolutely regular sequences in terms of a coupling property:

Theorem 8.1 (Berbee). A strictly stationary sequence $(X_k\ ,\ k \in \mathbb{Z})$ of random variables is absolutely regular if and only if there exists a probability space with sequences $(X'_k\ ,\ k \in \mathbb{Z})$ and $(X''_k\ ,\ k \in \mathbb{Z})$, each having the same distribution as $(X_k\ ,\ k \in \mathbb{Z})$, such that

(i) $(X'_k\ ,\ k \le 0)$ and $(X''_k\ ,\ k \in \mathbb{Z})$ are independent and

(ii) $P(\exists n \ge 1$ such that $X'_k = X''_k$ for all $k \ge n) = 1$.

Berbee [1, p. 106] also explains that if (X_k) is absolutely regular (i.e. satisfies $\beta(n) \to 0$), then in the context of Theorem 8.1 the inequality

$$P(\exists k \ge n \text{ such that } X'_k \ne X''_k) \ge \beta(n) \tag{8.1}$$

automatically holds for all $n \ge 1$, and that there exists an "optimal" coupling in which equality in eqn. (8.1) is achieved simultaneously for all $n \ge 1$. This is analogous to Griffeath's [41] well known "maximal coupling" result for Markov chains.

Acknowledgements: This paper arose from conversations with a lot of people. Special thanks are due to M. Denker (who suggested Examples 6.3, 6.4, and 6.5), M. Peligrad, and M. Rosenblatt for their comments.

REFERENCES

[1] H. Berbee. Random walks with stationary increments and renewal theory. Mathematical Centre, Amsterdam, 1979.

[2] H. Berbee. Weak Bernoulli processes, periodicity and renewal theory on the integers. Report no. 167. Mathematical Centre, Amsterdam, 1981.

[3] H. Berbee. Convergence rates in the strong law for bounded mixing sequences. Report MS-R8412. Mathematical Centre, Amsterdam, 1984.

[4] H. Berbee and R.C. Bradley. A limitation of Markov representation for stationary processes. Stochastic Process. Appl. 18 (1984) 33-45.

[5] J. Bergh and J. Löfström. Interpolation spaces. Springer-Verlag, Berlin, 1976.

[6] I. Berkes and W. Philipp. An almost sure invariance principle for the empirical distribution function of mixing random variables. Z. Wahrsch. Verw. Gebiete 41 (1977) 115-137.

[7] I. Berkes and W. Philipp. Approximation theorems for independent and weakly dependent random vectors. Ann. Probab. 7 (1979)29-54.

[8] J.R. Blum, D.L. Hanson, and L.H. Koopmans. On the strong law of large numbers for a class of stochastic processes. Z. Wahrsch. Verw. Gebiete 2 (1963) 1-11.

[9] R. Bowen. Equilibrium states and the ergodic theory of Anosov diffeomorphisms. Lecture Notes in Mathematics 470. Springer-Verlag, Berlin, 1975.

[10] R. Bowen. Bernoulli maps of the interval. Israel J. Math. 28 (1977) 161-168.

[11] R.C. Bradley. On the ϕ-mixing condition for stationary random sequences. Duke Math. J. 47 (1980) 421-433.

[12] R.C. Bradley. Central limit theorems under weak dependence. J. Multivariate Anal. 11 (1981) 1-16.

[13] R.C. Bradley. Absolute regularity and functions of Markov chains. Stochastic Process. Appl. 14 (1983) 67-77.

[14] R.C. Bradley. On the ψ-mixing condition for stationary random sequences. Trans. Amer. Math. Soc. 276 (1983) 55-66.

[15] R.C. Bradley. Equivalent measures of dependence. J. Multivariate Anal. 13 (1983) 167-176.

[16] R.C. Bradley. Approximation theorems for strongly mixing random variables. Michigan Math. J. 30 (1983) 69-81.

[17] R.C. Bradley. On a very weak Bernoulli condition. Stochastics 13 (1984) 61-81.

[18] R.C. Bradley. Identical mixing rates. Preprint, 1984. (Submitted for publication.)

[19] R.C. Bradley. A bilaterally deterministic ρ-mixing stationary random sequence. Trans. Amer. Math. Soc. (1986), to appear.

[20] R.C. Bradley. On the central limit question under absolute regularity. Ann. Probab. (1985), to appear.

[21] R.C. Bradley and W. Bryc. Multilinear forms and measures of dependence between random variables. J. Multivariate Anal. 16 (1985) 335-367.

[22] R.C. Bradley, W. Bryc, and S. Janson. Some remarks on the foundations of measures of dependence. Tech. Rept. No. 105, Center for Stochastic Processes, University of North Carolina, Chapel Hill, 1985.

[23] W. Bryc. On the approximation theorem of I. Berkes and W. Philipp. Demonstratio Math. 15 (1982) 807-816.

[24] A.V. Bulinskii. On measures of dependence close to the maximal correlation coefficient. Soviet Math. Dokl. 30 (1984) 249-252.

[25] R. Cogburn. Asymptotic properties of stationary sequences. Univ. Calif. Publ. Statist. 3 (1960) 99-146.

[26] H. Cohn. On a class of dependent random variables. Rev. Rounanie Math. Pures Appl. 10 (1965) 1593-1606.

[27] P. Csaki and J. Fischer. On the general notion of maximal correlation. Magyar Tud. Akad. Mat. Kutato Int. Kozl. 8 (1963) 27-51.

[28] A.R. Dabrowski. A note on a theorem of Berkes and Philipp for dependent sequences. Statist. Probab. Letters 1 (1982) 53-55.

[29] R.A. Davis. Stable limits for partial sums of dependent random variables. Ann. Probab. 11 (1983) 262-269.

[30] Y.A. Davydov. Mixing conditions for Markov chains. Theory Probab. Appl. 18 (1973) 312-328.

[31] H. Dehling. A note on a theorem of Berkes and Philipp. Z. Wahrsch. Verw. Gebiete 62 (1983) 39-42.

[32] H. Dehling, M. Denker, and W. Philipp. Versik processes and very weak Bernoulli processes with summable rates are independent. Proc. Amer. Math. Soc. 91 (1984) 618-624.

[33] H. Dehling and W. Philipp. Almost sure invariance principles for weakly dependent vector-valued random variables. Ann. Probab. 10 (1982) 689-701.

[34] M. Denker and G. Keller. On U-statistics and V. Mises' statistics for weakly dependent processes. Z. Wahrsch. Verw. Gebiete 64 (1983) 505-522.

[35] M. Denker and W. Philipp. Approximation by Brownian motion for Gibbs measures and flows under a function. Ergodic Theory Dynamical Systems 4 (1984) 541-552.

[36] E. Eberlein. Strong approximation of very weak Bernoulli processes, Z. Wahrsch. Verw. Gebiete 62 (1983) 17-37.

[37] E. Eberlein. Weak convergence of partial sums of absolutely regular sequences. Statist. Probab. Letters 2 (1984) 291-293.

[38] M. Falk. On the convergence of spectral densities of arrays of weakly stationary processes. Ann. Probab. 12 (1984) 918-921.

[39] H. Gebelein. Das Statistische Problem der Korrelation als Variations- und Eigenwertproblem and sein Zusammenhang mit der Ausgleichungsrechnung. Z. Angew. Math. Mech. 21 (1941) 364-379.

[40] M.I. Gordin. The central limit theorem for stationary processes. Soviet Math. Dokl. 10 (1969) 1174-1176.

[41] D. Griffeath. A maximal coupling for Markov chains. Z. Wahrsch. Verw. Gebiete 31 (1975) 95-106.

[42] P. Hall and C.C. Heyde. Martingale Limit Theory and Its Application. Academic Press, New York, 1980.

[43] N. Herrndorf. Stationary strongly mixing sequences not satisfying the central limit theorem. Ann. Probab. 11 (1983) 809-813.

[44] H.O. Hirschfeld. A connection between correlation and contingency. Proc. Camb. Phil. Soc. 31 (1935) 520-524.

[45] F. Hofbauer and G. Keller. Ergodic properties in invariant measures for piecewise monotonic transformations. Math. Z. 180 (1982) 119-140.

[46] I.A. Ibragimov. Some limit theorems for strictly stationary stochastic processes. Dokl. Akad. Nauk SSSR 125 (1959) 711-714. (In Russian).

[47] I.A. Ibragimov and Y.V. Linnik. Independent and Stationary Sequences of Random Variables. Wolters-Noordhoff, Groningen, 1971.

[48] I.A. Ibragimov and Y.A. Rozanov. Gaussian Random Processes. Nauka, Moscow, 1970. (In Russian).

[49] I.A. Ibragimov and Y.A. Rozanov. Gaussian Random Processes. Springer-Verlag, Berlin, 1978.

[50] M. Iosifescu. Recent advances in mixing sequences of random variables. In: Third international summer school on probability theory and mathematical statistics, Varna 1978, pp. 111-138. Publishing House of the Bulgarian Academy of Sciences, Sofia, 1980.

[51] H. Kesten and G.L. O'Brien. Examples of mixing sequences. Duke Math. J. 43 (1976) 405-415.

[52] A.N. Kolmogorov and Y.A. Rozanov. On strong mixing conditions for stationary Gaussian processes. Theory Probab. Appl. 5 (1960) 204-208.

[53] A. Lasota and J.A. Yorke. On the existence of invariant measures for piecewise monotonic transformations. Trans. Amer. Math. Soc. 186 (1973) 481-488.

[54] M.R. Leadbetter, G. Lindgren, and H. Rootzen. Extremes and Related Properties of Random Sequences and Processes. Springer-Verlag, Berlin, 1983.

[55] F. Ledrappier. Some properties of absolute continuous invariant measures on an interval. Ergodic Theory Dynamical Systems 1 (1981) 77-93.

[56] B.A. Lifshits. Invariance principle for weakly dependent variables. Theory Probab. Appl. 29 (1984) 33-40.

[57] S. Orey. Lecture notes on limit theorems for Markov chain transition probabilities. Mathematical Studies 34. Van Nostrand Reinhold, New York, 1971.

[58] M. Peligrad. Invariance principles for mixing sequences of random variables. Ann. Probab. 10 (1982) 968-981.

[59] M. Peligrad. A note on two measures of dependence and mixing sequences. Adv. in Appl. Probab. 15 (1983) 461-464.

[60] M. Peligrad. Recent advances in the central limit theorem and its invariance principle for mixing sequences of random variables. This Birkhäuser volume.

[61] T.D. Pham and L.T. Tran. Some mixing properties of time series models. Stochastic Process. Appl. 19 (1985) 297-303.

[62] W. Philipp. Some metrical theorems in number theory II. Duke Math. J. 37 (1970) 447-458.

[63] W. Philipp. Mixing sequences of random variables and probabilistic number theory. Mem. Amer. Math. Soc. No. 114, 1971.

[64] W. Philipp and W. Stout. Almost sure invariance principles for partial sums of weakly dependent random variables. Mem. Amer. Math. Soc. No. 161, 1975.

[65] M.S. Pinsker. Information and Information Stability of Random Variables and Processes. Holden-Day, San Francisco, 1964.

[66] M. Rosenblatt. A central limit theorem and a strong mixing condition. Proc. Nat. Acad. Sci. U.S.A. 42 (1956) 43-47.

[67] M. Rosenblatt. Markov Processes, Structure and Asymptotic Behavior. Springer-Verlag, Berlin, 1971.

[68] M. Rosenblatt. Uniform ergodicity and strong mixing. Z. Wahrsch. Verw. Gebiete 24 (1972) 79-84.

[69] M. Rosenblatt. Linear processes and bispectra. J. Appl. Probab. 17 (1980) 265-270.

[70] D. Sarason. An addendum to 'Past and Future'. Math. Scand. 30 (1972) 62-64.

[71] P. Shields. The Theory of Bernoulli Shifts. Univ. of Chicago Press, Chicago, 1973.

[72] E.M. Stein and G. Weiss. An extension of a theorem of Marcinkiewicz and some of its applications. J. Math. Mech. 8 (1959) 263-284.

[73] V.A. Volkonskii and Y.A. Rozanov. Some limit theorems for random functions I. Theory Probab. Appl. 4 (1959) 178-197.

[74] V.A. Volkonskii and Y.A. Rozanov. Some limit theorems for random functions II. Theory Probab. Appl. 6 (1961) 186-198.

[75] C.S. Withers. Central limit theorems for dependent random variables I. Z. Wahrsch. Verw. Gebiete 57 (1981) 509-534.

[76] H.S. Witsenhausen. On sequences of pairs of dependent random variables. SIAM J. Appl. Math. 28 (1975) 100-113.

[77] A. Zygmund. Trigonometric Series, Vol. I. Cambridge Univ. Press, Cambridge, 1959.

Department of Mathematics
Indiana University
Bloomington
Indiana 47405
USA

This paper was written at the Center for Stochastic Processes, University of North Carolina, Chapel Hill, and was supported in part by AFOSR Grant No. F49620 82 C 0009 and by NSF Grant No. DMS 84-01021.

RECENT ADVANCES IN THE CENTRAL LIMIT THEOREM AND ITS WEAK INVARIANCE PRINCIPLE FOR MIXING SEQUENCES OF RANDOM VARIABLES (A Survey)

by

Magda Peligrad*

ABSTRACT. The purpose of this paper is to describe the progress that has recently been made in the study of the central limit theorem and its weak invariance principle for mixing sequences of random variables and to point out some open problems in this subject.

§0. Introduction.

Mixing sequences of random variables are sequences for which past and distant future are asymptotically independent. The definitions of the mixing coefficients use measures of dependence between two σ-algebras and are based either on pairs of events or on correlations. Recent work of Bradley and Bryc (1985) allows a unified presentation of some of the principal mixing coefficients.

Let $\{X_n\}$ be a sequence of random variables on a probability space (Ω,K,P) and $F_n^m := \sigma(X_k; n \le k \le m)$. Let us suppose there are two positive reals p and r and a positive sequence $\beta_n(p,r) \to 0$ as $n \to \infty$ such that

(0.1) $$|EXY - EXEY| \le \beta_n(p,r) \|X\|_p \|Y\|_r$$

for every $X \in L_p(F_1^m)$, $Y \in L_r(F_{n+m}^\infty)$ and for every m.

---

*The work was support in part by a NSF grant DMS-8503016 and by a Taft Grant in aid for travel from the Charles Phelps Taft Memorial Fund, University of Cincinnati.

1. If $\frac{1}{p} + \frac{1}{r} < 1$ the sequence is called strong mixing, $\beta_n(p,r) \to 0$ being equivalent to $\alpha_n \to 0$, where

$\alpha_n := \sup\{|P(A \cap B)-P(A)P(B)| : A \in F_1^k,\ B \in F_{k+n}^\infty,\ k > 1\}$

(Rosenblatt (1956)).

2. If $\frac{1}{p} + \frac{1}{r} = 1$ and $p \neq 1$, and $r \neq 1$, the sequence is called ρ-mixing (Ibragimov (1975)) and (0.1) is equivalent to $\lambda_n \to 0$, where

$$\lambda_n := \sup\{\frac{|P(A \cap B)-P(A)P(B)|}{[P(A)P(B)]^{1/2}},\ A \in F_1^k,\ B \in F_{k+n}^\infty,\ P(A)P(B) \neq 0,\ k > 1\}$$

(Bradley (1982)). This class was apparently first studied by Kolmogorov and Rozanov (1960). The traditional dependence coefficient that generates this class is based on the maximal correlation ρ_n and it is defined by

$\rho_n := \beta_n(1/2,1/2) = \sup\{\mathrm{corr}(X,Y),\ X \in L_2(F_1^k),\ Y \in L_2(F_{k+n}^\infty), k>1\}$.

3. If $p = 1$ and $r = \infty$ we obtain an equivalent definition of ϕ-mixing sequences, $\phi_n \to 0$ where

$\phi_n := \sup\{|P(B|A)-P(B)|; A \in F_1^k,\ B \in F_{k+n}^\infty,\ P(A) \neq 0,\ k > 1\}$

(Ibragimov (1962)).

4. If $p = \infty$, $r = 1$ we obtain an equivalent of ϕ mixing for the reversed sequence namely $\phi_n^r \to 0$, where

$\phi_n^r := \sup\{|P(A|B) - P(A)|;\ A \in F_1^k,\ B \in F_{k+n}^\infty,\ P(B) \neq 0,\ k > 1\}$.

5. If $p = 1$ and $r = 1$ $\beta_n(1,1) \to 0$ is equivalent to $\psi_n \to 0$, where

$$\psi_n := \sup\{\frac{|P(A \cap B)-P(A)P(B)|}{P(A)P(B)},\ A \in F_1^k,\ B \in F_{k+n}^\infty,\ P(A)P(B) \neq 0,\ k > 1\}.$$

It is easy to see that every ρ-mixing sequence is strong mixing, every ϕ-mixing is ρ-mixing and every ψ-mixing is ϕ-mixing sequence.

We denote by $S_n = \sum_{i=1}^{n} X_i$ and $\sigma_n^2 = \mathrm{Var} S_n$.

Definition (0.1). The partial sum S_n is said to be attracted (respectively, partially attracted) to a distribution D as $n \to \infty$ if there are sequences of real numbers $\{a_n\}$ and $\{b_n\}$ with $a_n > 0$, $a_n \to \infty$ such that $\frac{S_n - b_n}{a_n} \to D$ in distribution as $n \to \infty$ (respectively, along a subsequence of positive integers).

In connection with the limiting behavior of S_n, if $\{X_n\}$ is a strictly stationary strong mixing sequence (not necessarily having finite first or second moment) then by a result in Ibragimov (1962) S_n can be attracted only to a stable law and if this law has exponent α then $a_n = n^{1/\alpha} h(n)$ with $h(n)$ a slowly varying function. By a result of Cogburn (1960) S_n can be partially attracted only to infinitely divisible laws.

Definition (0.2). We say that $\{X_n\}$ satisfies the C.L.T. (central limit theorem) if $\{\frac{S_n - b_n}{a_n}\}$ is attracted to the standard normal distribution.

For each $t \in [0,1]$ let us put $W_n(t) = \frac{S_{[nt]}}{\sigma_n}$ where $[x]$ denotes the greatest integer function. The function $\omega \to W_n(\cdot,\omega)$ is a measurable map from (Ω,K) into (D,B) where D is the set of all functions on the interval $[0,1]$ which have left hand limits and are continuous from the right at every point and B, the Borel σ-algebra on D induced by the Skorohod topology.

Definition (0.3). We say that $\{X_n\}$ satisfies the functional form of the C.L.T. (or the weak invariance principle) if W_n is weakly convergent to the standard Brownian motion process on [0,1], denoted by W in the sequel.

It should be noted that a great amount of work has been done on the central limit theorem (C.L.T.) and its invariance principle for mixing sequences of random variables. The basic results about C.L.T. can be found in Iosifescu-Theodorescu (1969), Ibragimov-Linnik (1971) and they are synthesized in Iosifescu's (1980) survey paper.

This paper presents the progress that was done on C.L.T. and its invariance principle the last decade, and discusses some open problems.

The mixing type of dependence was considered in connection with other topics such as: the strong laws of large numbers, almost sure invariance principles, empirical processes, U-statistics, Banach space valued random vectors, set indexed sums of random variables. These topics alone deserve a separate review so we shall not discuss them here.

For the basic properties and examples of mixing sequences the reader is referred to Bradley's (1985) survey paper that will appear in this volume.

The way we shall present the results will be from general to particular. First we shall discuss the C.L.T. and its weak invariance principle for strong mixing sequences, this being the largest class we consider here. Then we shall consider other classes namely ρ-mixing, which is contained in strong mixing, and ϕ-mixing, which is included in ρ-mixing. Because these classes are more and more restrictive, it is natural that their special characteristics will allow us to find weaker conditions than in the strong mixing case in order that the C.L.T. and its weak invariance principle still hold.

We begin by giving necessary and sufficient conditions for the C.L.T. under mixing assumptions. In general these necessary and sufficient conditions are related to the behavior of the moments of the partial sums S_n, while sufficient conditions consists of moment restrictions imposed on the individual summands together with the existence of a certain rate of decay to zero of the mixing coefficients. These sufficient conditions are shown to be minimal in the sense that if they are violated by a small amount then counterexamples to the theorems are known.

Finally some open problems will be discussed. Whenever a partial answer is available this will be mentioned. These open problems are mostly related with the problem of replacing the sufficient conditions for the C.L.T. involving certain rates of convergence to zero of the strong mixing coefficients with other sufficient conditions which could be easier to verify in certain practical situations. These types of conditions will involve in one way or another the behavior of two or more mixing coefficients at the same time.

About the techniques that solve the C.L.T. for mixing sequences, it should be said that at the beginning two kinds of techniques were developed. First was Bernstein's method, which consists of approximating mixing sequences with independent ones. Gordin's method uses a martingale approximation of a mixing sequence. While variations of Bernstein's method are still successfully used in mixing topics, the Gordin method leads to some classes that generalise martingales and helps in studying the asymptotics of those classes. See McLeish (1975, 1977), Peligrad (1981), Eberlein (1984).

Recently, new techniques were developed which are specific to each class of mixing sequences and lead to progress in the field. We expect that the open problems will be solved not using Bernstein's or Gordin's approximations but rather by working with the mixing properties directly.

In this survey the notation is the usual one. We use I_A for the indicator function of A, $\|\cdot\|_p$ for the norm in L_p, => for the weak convergence, $\to$ for convergence when n approaches ∞, << instead of $O(\cdot)$.

§1. Strong (α) Mixing Sequences.

The first characterization of the C.L.T. for strong mixing was obtained by using Bernstein's method known also by the name of "the big small block technique". The idea is to represent S_n as a sum of "almost" independent random variables (big blocks) alternating with other terms (small blocks) whose sum is negligible under suitable norming. The following fundamental auxiliary result can be obtained by induction, and it is stated in Iosifescu (1980).

Lemma (1.1). Let $m_1 < n_1 < \dots < m_k < n_k$, and denote $\tau = \min_{r<k} (m_{r+1}-n_r)$. Consider complex valued random variables Y_r such that $|Y_r| \le 1$ and Y_r is $F_{m_r}^{n_r}$ measurable, $r \le k$. Then:

$$|E(Y_1 \cdot \ldots \cdot Y_k) - EY_1 \ldots EY_k| \le 2\pi(k-1)\, \alpha_\tau$$

By alternating blocks of size p(n) that are approximated by a sum of independent random variables, with small blocks of size q(n) having a negligible sum (in L_2), the following result was obtained:

Theorem (1.1). Assume $\{X_n\}$ is a centered strictly stationary strong mixing sequence with finite second moments. In order for the C.L.T. to hold for $\{S_n/\sigma_n\}$ and $\lim_{n\to\infty} \sigma_n = \infty$ it is necessary that:

i) $\sigma_n^2 = nh(n)$, where h is a slowly varying function defined on the positive axis;

ii) for all systems of natural valued functions $p = p(n)$ and $q = q(n)$ satisfying the conditions

a) $p \to \infty$, $q \to \infty$, $q = o(p)$, $p = o(n)$;

b) $\lim_{n\to\infty} n^{1-\beta} q^{1+\beta}/p^2 = 0$ for all $\beta > 0$;

c) $\lim_{n\to\infty} n\, \alpha_q/p = 0$;

we have:

(1.1) $$\lim_{n\to\infty} \frac{n}{p\sigma_n^2} E\, S_p^2\, I_{\{|S_p|>\varepsilon\sigma_n\}} = 0 \qquad \text{for all } \varepsilon > 0 .$$

Conversely, if condition (i) is fulfilled and (1.1) holds for a system of functions p and q as in (ii) then C.L.T. holds for $\{S_n/\sigma_n\}$.

For the proof of this theorem see Ibragimov-Linnik (1971) Theorem 18.4.1.

Now if σ_n^2 has the representation

(1.2) $$\sigma_n^2 = n\sigma^2(1+o(1)) \quad \text{for a } \sigma > 0 \text{ as } n \to \infty,$$

and in addition $\{S_n^2/\sigma_n^2\}$ is uniformly integrable it is easy to see that (1.1) holds for every choice of $p(n) = o(n)$. Thus, under (1.2) and strong mixing, the C.L.T. holds for $\{S_n/\sigma_n\}$ if and only if $\{S_n^2/\sigma_n^2\}$ is uniformly integrable. (Volkonskii and Rozanov (1959). See also Ibragimov-Linnik (1971) Theorem 18.4.2). The condition (1.2) can be removed. Withers (1981), Theorem (2.2) and its corrigendum (1983) made some steps in this direction. He replaced the asymptotic linearity in n of σ_n^2 (1.2) by the condition that σ_n^2/n is bounded above and below by strictly positive constants. Denker (1978) removed the condition of asymptotic linearity in n of σ_n^2. The publication of his manuscript was delayed (see Denker (1978) (1979)) but a proof of his result using Theorem (1.1) will appear in these proceedings. The result is:

Theorem (1.2). Let $\{X_n\}$ be a strictly stationary centered strong mixing sequence having finite second moments and $\sigma_n^2 \to \infty$. Then the C.L.T. holds for $\{S_n/\sigma_n\}$ if and only if $\{S_n^2/\sigma_n^2\}$ is uniformly integrable.

The proof is based on the following remark from Ibragimov-Linnik, (1971), page 330, which can be stated as:

Theorem (1.3). Let $\{X_n\}$ be a strictly stationary strong mixing sequence, centered, with $EX_1^2 < \infty$ and $\sigma_n^2 \to \infty$. If $\{S_n^2/\sigma_n^2\}$ is uniformly integrable, then σ_n^2 has the representation $\sigma_n^2 = nh(n)$ with h a slowly varying function on R^+.

Nahapetian (1984) independently announced the conclusion of Theorem (1.2), but he imposed the condition $\sigma_n^2 = nh(n)$, with h(n) a slowly varying function.

In order to derive the functional form of the C.L.T. it should be noted that under strong mixing W_n has asymptotically independent increments (Lemma (1.1) could be used in order to see this). So the C.L.T. will imply that the finite dimensional distributions of W_n will converge to the corresponding ones of the Brownian motion. By Prohorov's theorem (see Billingsley (1968) Theorem (6.1)) the functional form of the C.L.T. will result by proving the tightness of W_n. In order to verify the tightness some criteria are available (see Billingsley (1968), (8.12), Theorem (8.4) and page 73). So, we can get the following extension of Theorem (1.2):

Theorem (1.4). Let $\{X_n\}$ be strictly stationary strong mixing, centered, $EX_1^2 < \infty$ and $\sigma_n^2 \to \infty$. In order that $W_n \Rightarrow W$ it is necessary and sufficient that the family $\{S_n^2/\sigma_n^2\}$ be uniformly integrable and for each positive ε there exists $\lambda > 1$ such that

(1.3) $$P(\max_{i \leq n} |S_i| > \lambda \sigma_n) < \varepsilon/\lambda^2$$

Also Theorem (1.1) can be extended to a weak invariance principle. The following theorem is a combination of Theorem (1.1) and Theorem (4.1) in Billingsley (1968) (see also Herrndorf (1983 c)).

Theorem (1.5). Assume $\{X_n\}$ is a strictly stationary strong mixing sequence, centered, having finite second moments and $\sigma_n^2 = n\, h(n)$ where h is a slowly varying function. Suppose that for a system of functions satisfying (a), (b), (c) of Theorem (1.1) the condition (1.1) holds. Suppose also that:

$$np^{-1} P(\max_{i \leq p} |S_i| > \varepsilon \sigma_n) \to 0 \quad \text{as } n \to \infty .$$

for every $\varepsilon > 0$.

Then $W_n \Rightarrow W$.

A variant of this theorem was stated in Herrndorf (1983 c).

The conditions in Theorems (1.4) or (1.5) are implied for instance by the boundedness of $\{\|S_n\|_{2+\delta}/\sigma_n\}$, where $\delta > 0$. So it is clear that moment conditions imposed on $|S_n|$ are important for the C.L.T. problem, when mixing dependence is assumed. A recent paper of Dehling, Denker, Philipp (1984) came to stress this characteristic. Their results connect the C.L.T. for strong mixing sequences with the boundedness of $\{\sigma_n/E|S_n|\}$. The proof is also based on the "big-small block" type of argument with a very fine choice of the sequences p(n) and q(n). One of their results is:

Theorem (1.6). Suppose that $\{X_n\}$ is strictly stationary strong mixing, centered, $EX_1^2 < \infty$ and $\sigma_n^2 = n\,h(n)$ with $h(n)$ a slowly varying function. Then the C.L.T. holds for $\{S_n/\sigma_n\}$ if and only if

$$\limsup_n \sigma_n/E|S_n| < (\pi/2)^{1/2}$$

In general for mixing sequences, the conditions imposed on the moments of the partial sums $|S_n|$ are not easy to check. There are instead sufficient conditions expressed in terms of the moments of the individual summands $\{X_n\}$ and the rate of convergence to zero of the mixing coefficients. These conditions are sometimes strong enough to assure more than the C.L.T.. Many times the weak invariance principle is available at no extra cost.

The C.L.T. included in the following theorem is the classic C.L.T. of Ibragimov (1962) (see Ibragimov-Linnik (1971) Theorems 18.5.3 and 18.5.4). By improving a result of Davydov (1968), Oodaira and Yoshihara (1972) established the functional form of the C.L.T.

Theorem (1.7). Suppose $\{X_n\}$ is strong mixing, strictly stationary, centered, and one of the following two conditions hold:

i) for some $\delta > 0$, $E|X_1|^{2+\delta} < \infty$ and $\sum_{n=1}^{\infty} \alpha_n^{\delta/(2+\delta)} < \infty$

or

ii) for some $C < \infty$, $|X_1| < C$ a. s., and $\sum_{n=1}^{\infty} \alpha_n < \infty$

Then $\sigma^2 = EX_1^2 + 2\sum_{k=2}^{\infty} E(X_1X_k)$ exists. If in addition $\sigma^2 > 0$, then W_n is weakly convergent to W.

Herndorff (1983 c) extended this result by replacing (i) with a more flexible moment assumption. Let F be the class of all positive functions $\phi(x)$ defined on the positive real axis that are convex, vanishing at 0 and with $\phi(x)/x^2$ increasing to $+\infty$. Let $\phi^{inv}(x)$ be the inverse of $\phi(x)$. Denote $f_\phi(x) = (\phi^{inv}(1/x))^2\, x$ for $x > 0$ and $f_\phi(0) = 0$. With this notation, Theorem (1.7) still holds with (i) replaced by

i') for some $\phi \in F$, $E\phi(|X_1|) < \infty$ and $\sum_{n=1}^{\infty} f_\phi(\alpha_n) < \infty$

Of course (i) is contained in (i'). In addition some other combinations of moment assumptions and mixing rates can be considered.

For example if for some $a > 0$

$$EX_1^2(\log(1+|X_1|))^a < \infty \text{ and } \sum_{n=1}^{\infty} |\log \alpha_n|^{-a} < \infty \tag{1.5}$$

then the conclusion of theorem (1.7) holds. As a consequence, if

$$\text{for some } \varepsilon > 0 \quad EX_1(\log(1+|X_1|))^{1+\varepsilon} < \infty \tag{1.6}$$

and the mixing rate decays to zero exponentially fast, then the functional form of C.L.T. holds.

The proof of Theorem (1.7) was obtained by using a blocking procedure and then by applying a variant of Theorem (1.5). Alternatively this theorem could be proved using Theorem (5.1) of Dvoretzky (1972) or Lemma (3.1) of Withers (1981).

Comment (1.1). The first question that appeared with respect to Theorem (1.7) was if the mixing rate condition in this theorem could be improved or removed. Davydov (1969), (1973) constructed some counterexamples to the C.L.T. He made the observation that in Theorem (1.7) the assumptions are very nearly as minimal as they can be. His examples are strictly stationary, centered, countable state Markov chains, satisfying one of the following:

i) For some $\delta > 0$ $E|X_1|^{2+\delta} < \infty$ and $\alpha_n = o(n^{-(1-\varepsilon)(1+2/\delta)})$, as $n \to \infty$

or

ii) For some $c > 0$, $|X_1| < c$ a.s. and $\alpha_n = o(n^{-(1-\varepsilon)})$ as $n \to \infty$.

In his example $\sigma_n^2 \approx n^d$ for some $1 < d < 2$ and S_n is attracted to a symmetric law with exponent $\alpha = 2/d$ satisfying $1 < \alpha < 2$.

These examples show in our context that, if the mixing rate condition is violated by a small amount it is possible that the C.L.T. fails. The conditions considered by Herrndorf (1983 c), (1.6), are also minimal. Two constructions, one by Herrndorf (1983 c) in the nonstationary case and the other of Bradley (1984 a) in the stationary case show that, under an exponential mixing rate the moment assumption $EX_1^2(\log(1+|X_1|))^{\varepsilon} < \infty$ with $\varepsilon > 1$ is almost minimal in order to have the C.L.T. For instance a moment assumption of the type $EX_1^2(\log(1+|X_1|))^{\varepsilon} < \infty$ with $\varepsilon < 1$ could not be enough for C.L.T.

Comment (1.2). Another question that could be raised with respect to Theorem (1.7) is whether it is possible to find a strong mixing rate, which together with $EX_1^2 < \infty$ assures the C.L.T.. Two examples of Bradley (1983 a) and Herrndorf (1983 b) answered this question negatively. Bradley (1983 a) constructed a strong mixing sequence, strictly stationary, having only finite second order moments, an arbitrary fast mixing rate, σ_n^2/n^2 converges to 0 arbitrarily slowly and S_n is attracted to all infinitely divisible laws. Herrndorf's example (1983 b) is a strong mixing sequences with an arbitrary mixing rate, which is orthogonal, such that the family of distributions of

S_n is tight, and S_n cannot be asymptotically normal under any kind of normalisation. The novelty is that Herrndorf's example satisfies $\sigma_n^2 = nEX_1^2$. This example uses moving averages of i.i.d. random variables.

Comment (1.3). Gordin claimed that, if $\{\sigma_n^2/n\}$ converges to a positive limit $\sigma^2 > 0$, then the mixing rate in Theorem (1.7) could be weakened to $\sum_{n=1}^{\infty} \alpha_n^{(1+\delta)/(2+\delta)} < \infty$. This is stated in Ibragimov Linnik (1971) page 420 and given again in Hall and Hyde (1980) page 139.

Gordin's claim however wasn't true. For the case $\delta = 0$ Herrndorf's (1983 b) above mentioned example provides a counterexample to Gordin's statement. For $\delta > 0$, Bradley (1984 a) constructed a series of counterexamples using moving averages of dependent random variables. The conclusion is that even with the assumption $\sigma_n^2/n \to \sigma^2 > 0$ as $n \to \infty$, the rate imposed on the strong mixing coefficients in Theorem (1.7) is still essentially sharp.

Comment (1.4). One of the conclusions of Theorem (1.7) is that $\{\sigma_n^2/n\}$ is a convergent sequence. But the fact that $\lim_n \sigma_n^2/n$ is a strictly positive constant is a condition in this theorem. In Ibragimov-Linnik's book it is asked (problem 4, page 393) what is the slowest mixing rate for $\alpha(n)$ that will insure the C.L.T. without the additional assumption $\lim \sigma_n^2/n > 0$.

Bradley's paper (1984 a) gives an answer to this problem. His observation that if $\sum_{n=1}^{\infty} n|\text{cov}(X_1,X_n)| < \infty$ then either $\sup \text{Var}\, S_n < \infty$ or $\text{Var}\, S_n$ is asymptotically linear in n, together with the Theorem (1.7) yield:

Theorem (1.8). Let $\{X_n\}$ be a strictly stationary mixing sequence, centered, $\sigma_n^2 \to \infty$ and one of the following two conditions hold:

i) for some $\delta > 0$, $E|X_1|^{2+\delta} < \infty$ and $\sum_n n\alpha_n^{\delta/(2+\delta)} < \infty$

or

ii) for some $c > 0$, $|X_1| < c$ and $\sum_n n\, \alpha_n < \infty$

Then W_n is weakly convergent to W.

In the same paper Bradley showed that these mixing rates are almost minimal. The examples he constructed satisfy

i) $E|X_1|^{2+\delta} < \infty$ for some $\delta > 0$ and $\alpha_n \ll n^{-2(2+\delta)/\delta}(\log n)^{3(2+\delta)/\delta}$

or

ii) $|X_1| < c$ for some $c > 0$ and $\alpha_n \ll n^{-2}(\log n)^3$

and the sequence cannot be asymptotically normal under any kind of normalisation (they satisfy $\inf_{n \geq 1} (P(S_n = 0) > 0)$. The mixing rates used in these examples still leave a slight gap, essentially a certain power of log n, between the mixing rates in Theorem (1.8) and the rates that allow the construction of counterexamples.

The sufficient conditions we have discussed up to now all involved an estimation of the rate of convergence to 0 of the strong mixing coefficients. In many situations of interest it is possible to establish that a certain system is strong mixing but it appears hard to get good estimates for the strong mixing coefficients. Therefore it is of interest to replace the strong mixing rate, by other sufficient conditions that might be in some situations easier to verify. Bradley (1981) and Peligrad (1982), (1984 b) had the idea to replace the strong mixing rate by other sufficient conditions, involving bounds for other mixing coefficients, to insure the C.L.T. A bound for $\rho^* = \lim_n \rho_n$ can be found in Bradley (1981 b) and it was improved in Peligrad (1984 b). The proof of the following result is based on the estimate of $E|S_n|^{2+\delta}$ in terms of ρ^* and can be found in Peligrad (1984 b).

Theorem (1.9). Suppose $\{X_n\}$ is strong mixing sequence strictly stationary, centered, $\sigma_n^2 \to \infty$ and $E|X_1|^{2+\delta} < \infty$ for some $0 < \delta < 1$. If ρ^* satisfies:

$$(1.7) \quad 2(1 + \frac{\delta(1+\delta)}{2} \rho^{*2\delta/(2+\delta)} + (1+\delta)\rho^{*2/(2+\delta)}) < [2(1-\rho^*)]^{(2+\delta)/2}$$

then:

a) $\sigma_n^2 = n\, h(n)$ where $h(n)$ is a slowly varying function on R.

b) W_n is weakly convergent to W.

c) For every $\ell < 2+\delta$, $\{E|S_n/\sigma_n|^{\ell}\}$ converges to the absolute moment of order ℓ of the normal distribution.

A computer printout made by J. Deddens (1985) shows the relationship between δ and ρ^* that satisfy (1.7). (This is available at the University of Cincinnati). For instance the pair of conditions

$$(1.8) \qquad \delta = 1 \quad \text{and} \quad \rho^* < 0.037$$

verify (1.7). Theorem (1.9) can be extended also for values of $\delta > 1$. However this seems to be interesting only for the convergence of the moments.

Another idea for replacing the strong mixing rate by other sufficient conditions came from ergodic theory (the theory of Bernoulli

Shifts). This consists in the following: for each n to find a subset A_n of the basic space Ω, with $P(A_n) \underset{n}{\to} 0$ such that on $\Omega - A_n$, the sequence $\{X_n\}$ generates ϕ_n or ρ_n mixing coefficients that vanish at ∞. If outside A_n, ϕ_n is estimated to converge to 0 we get (according to Shields (1973), Chapter 6) the class of absolutely regular sequences (Volkonskii and Rozanov (1959)) that was studied in many papers for deriving limit theorems for dependent random variables; see e.g. Gastwirth and Rubin (1975), Yoshihara (1978), Berbee (1979) and Dehling and Philipp (1982).

In a paper of Bradley and Peligrad (1985) there are established limit theorems for the situation when outside A_n, with $P(A_n) \underset{n}{\to} 0$, ρ_n is estimated to converge to 0. The idea to look at this type of condition appeared from discussions with E. Presutti (see Boldrighini, DeMassi, Noguina, Presutti (1985)), concerning Bernoulli flows. Our study led to an extension of the class of strong mixing sequences that satisfies the C.L.T. A certain rate of converging to zero is imposed only to $a_n = P(A_n)$ while outside A_n , ρ_n tends to zero arbitrarily slowly.

Another aspect of the asymptotic behavior of mixing sequences is treated in Dehling Denker Philipp's (1984) paper. Their problem is to study what sufficient conditions insure that S_n is in the domain of partial attraction of the normal distribution and to get the C.L.T. by using normalising constants different from σ_n. The normalising constants they use are $a_n = \sqrt{\frac{\pi}{2}}\, E|S_n|$ and one of their results states that if $\{\sigma_n/a_n\}$ does not approach 0, then $\{S_n/a_n\}$ is in the domain of partial attraction of the normal law. Some other sufficient conditions can be added to the previous one in order for $\{S_n/a_n\}$ to satisfy the C.L.T. For instance if a_n has the representation $a_n = n^{1/2}h_1(n)$ with $h_1(n)$ a slowly varying function, then $\{S_n/a_n\}$ is asymptotically normally distributed.

§2. ρ-Mixing Sequences.

This class has a different C.L.T. behavior from that of strong mixing sequences. This is partially due to the special behavior of σ_n^2 under ρ-mixing.

The following is Theorem (2.1) of Ibragimov (1975).

Theorem (2.1). If the sequence $\{X_n\}$ is second order stationary, centered, ρ-mixing and $\sigma_n^2 \to \infty$, then $\sigma_n^2 = n\, h(n)$ where h is a slowly varying function on R^+, in the sense of Karamata.

Remark (2.1). Sometimes it is useful to apply the observation that this theorem still holds under an assumption less restrictive than ρ-mixing namely:

(2.1) $|\mathrm{corr}(S_n, S_{m+n} - S_n)| < \varepsilon$ for all n and m sufficiently large. (Ibragimov-Linnik (1971), page 330).

Theorem (2.2). Let $\{X_n\}$ be a second order stationary, centered sequence and:

(2.2) $$\sum_n \rho_n/n < \infty \quad \text{(equivalently } \sum_i \rho\,(2^i) < \infty).$$

Then $\{X_n\}$ has a continuous spectral density $f(\lambda)$. If in addition $f(0) \neq 0$ then

$$\sigma_n^2 = 2\pi f(0)n(1+o(1)).$$

The proof of this theorem can be found in Ibragimov Rozanov (1978) page 190. At pages 179-180 it is showed that the ρ-mixing rate used in

Theorem (2.2) is somehow as slow as permissible. Using a collection of stationary Gaussian sequences they construct examples of ρ-mixing sequences with $\rho(n) = O((\log n)^{-1})$. Under such a rate the spectral density may fail to be continuous, the quantity σ^2 defined to be $\lim_{n\to\infty} n^{-1}\sigma_n^2$ may fail to exist, or σ^2 may be 0.

Remark (2.2). In Theorem (2.2) the condition $\sigma_n^2 \to \infty$ can replace the condition $\sigma^2 > 0$. (See Bradley (1981 a) or Peligrad (1982).)

Another characteristic of this class is that the moments higher than 2 of $|S_n|$ can be estimated in terms of σ_n. In general the proofs of the C.L.T. for ρ-mixing sequences are based (either directly or after a truncation argument) on the boundedness of $E(|S_n|/\sigma_n)^{2+\delta}$. This fact assures on one hand uniform integrability of $\{S_n^2/\sigma_n^2\}$ and on the other hand makes easier the verification of the tightness condition contained in Theorem (1.3).

The following theorem combines a result in Ibragimov (1975) with one of Peligrad (1982).

Theorem (2.3) a) Let $\{X_n\}$ be a ρ-mixing sequence, second order stationary, centered, $\sigma_n^2 \to \infty$ and $\sum \rho(2^i) < \infty$. Then $\{S_n/\sigma_n\}$ satisfies C.L.T.

b) If in addition $\sum \rho^{1/2}(2^i) < \infty$ (or $\lim \phi_n < 1$, or $\lim \psi_n < \infty$) then W_n is weakly convergent to W.

The conditions (a) of this theorem are once again as weak as they are permitted to be. Two papers of Bradley (1980) and (1984 b) show that in general the ρ-mixing rate cannot be removed or weakened. The constructions are based on Gaussian sequences with $\sigma_n^2/n \to 0$. The example is a ρ-mixing sequence, strictly stationary, having only second order moments, centered, with $\sigma_n^2 \to \infty$ and $\rho_n = O(\log n)^{-1}$, and it does not satisfy the C.L.T. The mixing rate in Theorem (2.1) can be removed if the moments higher than 2 are supposed to exist.

The following is a corollary of Theorem (1.9). Part (a) was proved in Ibragimov (1975) and part (b) is an extension of part (c) of Theorem (1.9).

Theorem (2.4). Let $\{X_n\}$ be a strictly stationary ρ-mixing sequence, centered, $\sigma_n \to \infty$ and $E|X_1|^{\ell} < \infty$ for some real $\ell > 2$. Then:

a) W_n is weakly convergent to W.

b) For every $1 \leq j \leq \ell$ the moments $\{E|S_n/\sigma_n|^j\}$ converge to the absolute moments of order j of the normal distribution.

Comment (2.1). One might ask whether in this theorem the existence of the moments of order $2+\delta$ with $\delta > 0$ is too strong a condition for the C.L.T. A weaker moment assumption of the type $Eg|X_1| < \infty$ where g is a certain function satisfying $x^2 < g(x) < x^{2+\delta}$ for $x > 0$, might be sufficient for the C.L.T. without any restriction on the ρ-mixing rate. Bradley (1984 b) raised the problem of finding the slowest permissible ρ-mixing rates to assure the C.L.T. under more flexible moment assumptions. His paper solved the opposite problem of constructing as efficiently as possible counterexamples of ρ-mixing sequences satisfying a moment assumption of the type $Eg|X| < \infty$ with g in a certain class of functions, and not satisfying the C.L.T.. He constructed a centered ρ-mixing sequence with $\sigma_n^2 \to \infty$, $\rho_n \ll \tau_n$ where τ_n is an arbitrary given sequence satisfying the pair of conditions:

$$Eg(|X_1|) < \infty \quad \text{for some } g(x) > x^2$$

and for some positive number d,

$$g([n \exp(-d \sum_{k=1}^{\infty} k^{-1}\tau_k)]^{1/2}) = o(n)$$

and not satisfying the C.L.T. Instead the sequence $\{S_n/\sigma_n\}$ is partially attracted to a poisson mixture of normal distributions (including the point mass at 0). The direct question suggested by this very sharp counterexample is the following (Bradley (1984 b)).

Conjecture (2.1). Let $\{X_n\}$ be a strictly stationary sequence, ρ-mixing, centered and $Eg|X_1| < \infty$ for a function $x^2 < g(x) < x^{2+\delta}$ with $\delta > 0$. If

$$g([n \exp(-d \sum_{k=1}^{n} k^{-1}\rho_k)]^{1/2}) >> n \quad \text{as } n \to \infty \tag{2.3}$$

for every $d > 0$, then $\{S_n/\sigma_n\}$ is assymptotically normal distributed.

§3. ϕ-Mixing Sequences.

A special property of the ϕ-mixing class of random variables is a nice extension of the Ottaviani inequality. The following lemma seems to be a powerful tool in proving the C.L.T. (see Peligrad (1984a)).

Lemma (3.1). Let $\{X_n\}$ be a sequence of random variables. Then:

(a)
$$[1-\phi_m - \max_{m<i<n} P(|S_n-S_i| > \frac{ba}{2})]\, P(\max_{i<n} |S_i| > (1+b)a)$$
$$< P(|S_n| > a) + P(\max_{i<n} |X_i| > \frac{ba}{2(m-1)})$$

and

(b)
$$P(|S_n| > (1+2b)a) < [\phi_m + \max_{i<n} P(|S_n-S_i| > \frac{ba}{2})]\, P(\max_{i<n} |S_i| > (1+b)a)$$
$$+ P(\max_{i<n} |X_i| > \frac{ba}{2m})$$

Part (a) of this Lemma was established for $\phi_1 < 1$ by Iosifescu (see Iosifescu and Theodorescu (1969) Lemma 1.1.6), and in the general case by Peligrad (1981b).

By combining parts (a) and (b) it is possible to prove that, if $\lim_n \phi_n < 1/4$ and $\{X_n\}$ is centered then $\{S_n^2/\sigma_n^2\}$ is uniformly integrable if and only if $\{\max_{i \leq n} X_i^2/\sigma_n^2\}$ is uniformly integrable. This is the key step in proving the following (Peligrad (1984)):

<u>Theorem (3.1).</u> (a) Let $\{X_n\}$ be a ϕ-mixing sequence, second order stationary, centered, $\sigma_n^2 \to \infty$ and satisfying the Lindeberg condition

(L) $$\lim_{n\to\infty} \frac{1}{\sigma_n^2} \sum_{i=1}^{n} EX_i^2 \, I_{\{|X_i| > \varepsilon\sigma_n\}} = 0 \quad \text{for every } \varepsilon > 0.$$

Then W_n is weakly convergent to W.

(b) If in addition $\phi_1 < 1$ the converse is true: $W_n \Rightarrow W$ implies (L).

<u>Comment (3.1).</u> This theorem is still true if one replaces ϕ-mixing by the pair of conditions: strong mixing and $\lim \phi_n < 1/2$.

<u>Comment (3.2).</u> The condition (L) is the usual Lindeberg Feller condition used in the C.L.T. for independent random variables. However under ϕ-mixing the behavior of σ_n^2 is not yet known. Of course this class, being included in the class of ρ-mixing variables, we know only that if $\sigma_n^2 \to \infty$, σ_n^2 is almost linear in n (Theorem (2.1)).

If the sequence is strictly stationary, (L) becomes

(L') $$\lim_{n\to\infty} \frac{n}{\sigma_n^2} \, EX_1^2 \, I_{\{|X_1| > \varepsilon\sigma_n\}} = 0 \text{ for every } \varepsilon > 0$$

and this is true for instance if

$$\liminf_n \frac{\sigma_n^2}{n} > 0$$

On the other hand Herrndorf (1983 a), showed that if there exists a strictly stationary ϕ-mixing sequence with $\sigma_n^2 \to \infty$ and $\liminf \frac{\sigma_n^2}{n} = 0$ a counterexample to the invariance principle can be constructed in the following way: if the sequence does not satisfy the C.L.T. it does not satisfy invariance principle and if the sequence $\{X_n\}$ satisfies the C.L.T. and $\sigma_n^2/n \to 0$, there is an iid sequence $\{Q_n\}$ such that $Y_n = X_n + Q_n - Q_{n-1}$ does not satisfy weak invariance principle. Theorem (3.1) is related to the Ibragimov-Iosifescu's conjecture:

<u>Conjecture (3.1).</u> If $\{X_n\}$ is ϕ-mixing centered strictly stationary, having only finite second moments and $\sigma_n^2 \to \infty$. Then the C.L.T. holds. The functional form of the C.L.T. holds.

The answer to this conjecture depends now on the answer to the following:

<u>Conjecture (3.2).</u> Every ϕ-mixing sequence which is, centered, strictly stationary with $EX_1^2 < \infty$, $\sigma_n^2 \to \infty$ satisfies $\liminf_n \sigma_n^2/n > 0$.

<u>Comment (3.3).</u> The Lindeberg condition (L) from Theorem (3.1) is satisfied in many situations of interest. This fact allows us to obtain sufficient conditions for the weak invariance principle. For instance if g(x) is a function such that $g(x)/x^2$ in nondecreasing it is easy to see that

$$\frac{n}{\sigma_n^2} EX_1^2 I_{\{|X_1| > \varepsilon\sigma_n\}} < \frac{n}{g(\sigma_n)} E\, g(|X_1|)\, I_{\{|X_1| > \varepsilon\sigma_n\}} \tag{3.1}$$

On the other hand, one can deduce from Bradley (1981a) or from relations (4.4) and (4.15) of Peligrad (1982), that for every integer $p > 1$ one can find a constant C_p such that for all n sufficiently big:

$$\sigma_n^2 > C_p\, n \prod_{i=p}^{\log n-1} (1 - \rho([2^{i/3}])) \tag{3.2}$$

On the other hand, if $\rho^* = \lim_n \rho(n) < 1$, one can find a constant c_1, that depends on ρ^* such that for all m sufficiently big, $(1-\rho(m)) > \exp(-d_1 \rho(m))$. By this observation and by (3.2) we can find a constant d_2 such that $\sigma_n^2 > C_p\, n \exp(-d_2 \sum_{i=1}^{n} \rho_i/i)$. It is easy to see now that there is a positive constant d such that

$$\sigma_n^2 > n \exp\left(-d \sum_{i=1}^{n} \rho_i/i\right) \tag{3.3}$$

(We can take for instance $d = d_2 + |\ln C_p|/\rho_1$). Because g(x) is a nondecreasing function it follows from (3.1) and (3.3) that there is a constant $d > 0$ such that:

$$\frac{n}{\sigma_n^2} EX_1^2 I\{|X_1| > \epsilon\sigma_n\} < \frac{n}{g([n \exp(-d \sum_{k=1}^{n} k^{-1}\rho_k)]^{1/2})}\, o(1)$$

From this we deduce:

<u>Corollary (3.1)</u>. Let $\{X_n\}$ be a strictly stationary sequence, ϕ-mixing centered with $\sigma_n^2 \to \infty$. If for a function g(x) with $g(x)/x^2$ nondecreasing, $Eg|X| < \infty$ and (2.3) is satisfied, then $W_n \Rightarrow W$.

This is a partial answer to the Conjecture (2.1). This shows that if one replaces ρ-mixing with ϕ-mixing the conjecture is true.

If $g(x) = x^2$ (2.3) is satisfied if $\sum_{n=1}^{\infty} n^{-1} \rho_n < \infty$. Because $\rho_n < 2(\phi_n \phi_n^r)^{1/2}$ we have the following result (Peligrad (1983).

Corollary (3.2). Let $\{X_n\}$ be a ϕ-mixing sequence, centered second order stationary, $EX_1^2 < \infty$ and $\sigma_n^2 \to \infty$. If

$$\sum_n n^{-1}(\phi_n \phi_n^r)^{1/2} < \infty$$

then W_n is weakly convergent to W.

The last year some other interesting results were obtained on the limiting behavior of ϕ-mixing sequences. I would like to mention here Samur's (1984) paper that deals with the asymptotic behavior of ϕ-mixing triangular arrays of random vectors, Bryc's (1984) preprint containing related results with those presented here, Jokubowski's (1985) result about the necessary and sufficient conditions that assure the C-tightness of W_n for ϕ-mixing sequences and Hahn and Kuelbs' paper on asymptotic normality of trimmed sums of ϕ-mixing sequences.

§4. Other General Comments and Open Problems.

In general all the theorems discussed in this survey were obtained under a weaker form of stationarity than that presented here. We have chosen this presentation for the sake of simplicity. In general, if strict stationarity is not assumed some other conditions that replace the stationarity are imposed on the second moments. For instance the family $\{X_n^2\}$ is taken to be uniformly integrable and

$\{\sup_k \frac{E(S_{k+n}-S_k)^2}{\sigma_n^2}\}$ to be bounded.

The lack of strict stationarity allows also discussion of some problems that have no sense otherwise. For instance one of the questions can be what is the maximum rate of increase of $E|X_n|^\ell$ for a given $\ell > 2$, that assure the C.L.T. The answer to this question is contained in papers of Bradley (1981b), Herrndorf (1984a) and Peligrad (1984b).

The maximal condition that assures the truth of Theorems (1.9) and (2.4) is

$$E|X_n|^{2+\delta} = 0\,(n^{\frac{\delta}{2}-\varepsilon}) \quad \text{for some } \varepsilon > 0 . \tag{4.1}$$

In (4.1), ε cannot be 0 as an example in Herrndorf (1984 a) shows. The Lindeberg condition (L) in Theorem (3.1) holds for instance if:

$$\max_{i \leq n} E|X_i|^{2+\delta} = o(\sigma_n^{2+\delta}/n) \tag{4.2}$$

We shall list now a few open problems that appear to be the next steps in solving some problems we have discussed.

1. A first problem is to narrow the gap between the strong mixing rate used in Theorem (1.3) and the rates that allow the construction of counterexamples to this theorem. This could be done by constructing counterexamples to C.L.T. for strong mixing sequences with $\alpha_n < \tau_n$ where $\{\tau_n\}$ satisfies:

$$\sum_n n\tau_n^{\delta/(2+\delta)} = \infty \quad \text{when} \quad E|X_1|^{2+\delta} < \infty \quad \text{for some } \delta < 0 ,$$

or

$$\sum_n n\tau_n = \infty \quad \text{when } |X_1| < c \quad \text{for some } c > 0 \,.$$

2. Another problem is to find the best size $0 < c < 1$ such that the condition $\rho^* < c$ could replace the condition (1.7) of Theorem (1.9) and such that if $\rho^* > c$ a counterexample to Theorem (1.9) could be found.

3. For Theorem (2.3) part (b) the problem is to provide a proof for tightness of W_n using only the rate $\sum_i \rho(2^i) < \infty$ instead of $\sum_i \rho^{1/2}(2^i) < \infty$. In other words the problem is to prove the weak invariance principle under the same sufficient conditions that assume the C.L.T.. This could be done by establishing a Ottaviani like inequality for ρ-mixing sequences.

4. Another problem is to prove an invariance principle for ρ-mixing sequences, centered, strictly stationary with $\sigma_n^2 \to \infty$ and with $Eg|X_1| < \infty$ where $g(x)/x^2$ is nondecreasing and satisfying (2.3).

5. To prove that there is a ϕ-mixing sequence strictly stationary, centered having only finite second moments, with $\sigma_n^2 \to \infty$ and with

$$\liminf_n \sigma_n^2/n = 0 \tag{4.3}$$

or to prove that for such sequences (4.3) is not possible . □

I would like to thank R. Bradley, W. Bryc, M. Denker, and D. Surgailis for providing me with some preprints, and to M. Iosifescu, J. Mitro and the referee for valuable comments.

BIBLIOGRAPHY

1. Berbee, H. C. P., (1979); Random Walks With Stationary Increments and Renewal Theory. Mathematical Centre, Amsterdam.

2. Boldrighini, C., De Massi, A., Nogueira, A., Presutti, E., (1985); The Dynamics of a particle interacting with a semi-infinite ideal gas in a Bernoulli flow. Statistical Physics and Dynamical Systems (Progress in Physics) Birkhäuser, p. 153.

3. Bradley, R. C. (1983), Equivalent Measures of Dependence, J. Multivariate Analysis 13, 167-176.

4. Bradley, R. C. (1980), A Remark on the Central Limit Question for Dependent Random Variances, J. Appl. Prob. 17, 94-101.

5. Bradley, R. C. (1981 a), A Sufficient Condition for Linear Growth of Variables in a Stationary Random Sequence, Proc. Amer. Math. Soc. 83, 586-589.

6. Bradley, R. C. (1981 b), Central Limit Theorems under Weak Dependence, J. Multivariate Analysis, 11, 1-16.

7. Bradley, R. C. (1984 a), On the Central Limit Question under Absolute Regularity (to appear in Ann. of Prob.).

8. Bradley, R. C. (1984 b), The Central Limit Question under ρ-mixing (to appear in Rocky Mtn. J. Math.).

9. Bradley, R. C. (1983 a), Information Regularity and Central Limit Question, Rocky Mtn. J. Math., 13, 77-97.

10. Bradley, R. C., Bryc, W. (1985), Multilinear Forms and Measures of Dependence Between Random Variables, J. Multivariate Analysis, 16, 335-367.

11. Bradley, R. C. (1982), Counterexamples to the Central Limit Theorem Under Strong Mixing Conditions, Colloquia Mathematica Soc. János Bolyai. 36 Limit Theorems in Probability and Statistics, Veszprém (Hungary).

12. Bradley, R. C. (1985), The Basic Properties of Strong Mixing Conditions, (to appear in Dependence in Probability and Statistics, Birkhäuser).

13. Bradley, R. C., Peligrad, M. (1985), Invariance Principles Under A Two-Part Mixing Assumption. Preprint.

14. Bryc, W. (1985), Central Limit Theorem for ϕ-mixing Random Sequences Under Lindeberg-type Condition, (preprint).

15. Cogburn, R. (1960), Asymptotic Properties of Stationary Sequences, Univ. Calif. Publ. Statist. 3, 99-146.

16. Davydov, Y. A. (1968), Convergence of Distributions Generated by Stationary Stochastic Processes., Teor. Probability Appl. 13, 691-696.

17. Davydov, Y. A. (1969), On the Strong Mixing Property for Markov Chains with a Countable Number of States, Soviet Math. Dokl. 10, 825-827.

18. Davydov, Y. A. (1973), Mixing Conditions for Markov Chains, Theory Probab. Appl. 18, 312-328.

19. Dehling, H., Denker, M., Philipp, W., (1984), Central Limit Theorems for Mixing Sequences of Random Variables under Minimal Conditions (to appear in Ann. Probability).

20. Dehling, H. and Philipp, W. (1982), Almost Sure Invariance Principle for Weakly Dependent Vector-Valued Random Variables, Ann. Prob. 10, 689-701.

21. Denker, M. (1978), A Note on the Central Limit Theorem for Strongly Mixing Random Variables. Preprint.

22. Denker, M. (1979), A Limit Theorem for Mixing Stationary Processes and its Applications. Preprint.

23. Denker, M. (1985), Uniform Integrability and the Central Limit Theorem (to appear in Dependence in Probability and Statistics. Birkhäuser).

24. Dvoretzky, A. (1972), Asymptotic Normality for Sums of Dependent Random Variables, Proc. Sixth Berkeley Sympos. Math. Stat. Prob., Vol. 2, 515-535.

25. Eberlein, E. (1984), On Strong Invariance Principle Under Dependence Assumptions, (to appear in Ann. Probability).

26. Gastwirth, J. L., Rubin, H. (1975), The Asymptotic Distribution Theory of the Empiric c.d.f. for Mixing Stochastic Processes, Ann. Statist., 3, 809-824.

27. Hall, P., Heyde, C. C. (1980), Martingale Limit Theory and Its Applications, Academic Press, New York.

28. Hahn, M. G., Kuelbs, J. (1985), Asymptotic Normality of Trimmed Sums of ϕ Mixing Random Variables. Preprint.

29. Herrndorf, N. (1983 a), The Invariance Principle for ϕ-mixing Sequences, Z. Wahrsch. Verw. Gebiete 63, 1, 97-109.

30. Herrndorf, N., (1983 b), Stationary Strongly Mixing Sequences Not Satisfying the Central Limit Theorem, Annals Prob. 11, 809-813.

31. Herrndorf, N., (1983 c), A Functional Central Limit Theorem for Strongly Mixing Sequences of Random Variables, (Z. Wahrsch. verw. Gebiete, submitted).

32. Herrndorf, N., (1984 a), A Functional Central Limit Theorem for ρ-Mixing Sequences, J. of Multiv. Anal. 15, 141-146.

33. Herrndorf, N., (1984 b), A Functional Central Limit Theorem for Weakly Dependent Sequences of Random Variables, Ann. of Prob., 12, 141-153.

34. Hoffman-Jorgensen, J., (1974), Sums of Independent Banach Space Valued Random Variables, Studia Math. 52, 159-186.

35. Ibragimov, I. A., (1962), Some Limit Theorem for Stationary Processes, Theory Prob. Appl. 7, 349-382.

36. Ibragimov, I. A., (1975), A Note on the Central Limit Theorem for Dependent Random Variables, Theory Prob. Appl. 20, 135-141.

37. Ibragimov, I. A., Linnik, Y. V. (1971), Independent and Stationary Sequences of Random Variables. Wolters-Noordhoff, Groningen.

38. Ibragimov, I. A., Rozanov, Y. A., (1978), Gaussian Random Processes, Springer-Verlag, Berlin.

39. Iosifescu, M. (1980), Recent Advances in Mixing Sequences of Random Variables. Third International Summer School on Probability Theory and Mathematical Statistics, Varna. 1978.

40. Iosifescu, M., Teodorescu, R. (1969), Random Processes and Learning, Springer-Verlag, New York.

41. Jakubowski, A. (1985), A Note on the Invariance Principle for Stationary ϕ-Mixing Sequences. Tightness via Stopping Times. Preprint.

42. Kesten, H., O'Brien, G. L. (1976), Examples of Mixing Sequences, Duke Math. J. 43, 405-415.

43. Kolmogorov, A. N., Rozanov, Y. A. (1960), On Strong Mixing Conditions for Stationary Gaussian Processes. Theory Probab. Appl. 5, 204-208.

44. Lai, T. L. (1977), Convergence Rates and r-Quick Version of the Strong Law for Stationary Mixing Sequences, Ann. Prob. 5, 693-706.

45. Lai, T. L., Robbins, H. (1978), A Class of Dependent Random Variables and Their Maxima, Z. Wahrsch. Verw. Gebiete 42, 89-111.

46. McLeish, D. L. (1975), Invariance Principles for Dependent Variables, Z. Wahrsch. verw. Gebiete 32, 165-178.

47. McLeish, D. L. (1977), On the Invariance Principles for Nonstationary Mixingales, Ann. Prob. 5, 616-621.

48. Moricz, F. (1982), A General Moment Inequality for the Maximum of Partial Sums of Single Series, Acta Sci. Math. 44, 67-75.

49. Moricz, F., Serfling, R. J. and Stout, W. (1982), Moment and Probability Bounds with Quasi-Superadditive Structure for the Maximum Partial Sum, Ann. Prob. 10, 1032-1040.

50. Nahapetian, B. S. (1984), On Limit Theorem for Dependent Random Variables. The Sixth International Symposium on Information Theory. September 18-22, 1984, Tashkent.

51. Oodaira, H. and Yoshihara, K. (1971), The Law of the iterated Logarithm for Stationary Processes Satisfying Mixing Conditions, Kodai Math. Sem. Rep. 23, 311-334.

52. Oodaira, H. and Yoshihara, K. (1972), Functional Central Limit Theorems for Strictly Stationary Processes Satisfying the Strong Mixing Condition, Kodai Math. Sem. Rep. 24, 259-269.

53. Peligrad, M. (1981 a), An Invariance Principle for Dependent Random Variables, Z. Wahrsch. verw. Gebiete 57, 495-507.

54. Peligrad, M. (1981 b), A Maximal Inequality for ϕ-Mixing Sequences, Center of Math. Statistics Notes, Bucharest.

55. Peligrad, M. (1982), Invariance Principles for Mixing Sequences of Random Variables, The Ann. of Prob. 10, 4, 968-981.

56. Peligrad, M. (1983), A Note on Two Measures of Dependence and Mixing Sequences, Adv. Appl. Prob. 15, 461-464.

57. Peligrad, M. (1984 a), An Invariance Principle for ϕ-Mixing Sequences, (to appear in Ann. of Prob.).

58. Peligrad, M. (1984 b), Invariance Principles under Weak Dependence (to appear in J. of Multivariate Analysis).

59. Philipp, W. (1979), An Almost Sure Invariance Principle for Sums of B-Valued Random Variables, Lecture Notes in Math. 709, 171-193, Springer-Verlag, New York.

60. Philipp, W. (1980), Weak and L^p-Invariance Principle for Sums of B-Valued Random Variables, The Ann. of Prob. 8, 1, 68-82.

61. Philipp, W., Stout, W. F. (1975), Almost Sure Invariance Principles for Sums of Weakly Dependent Random Variables, Amer. Math. Soc. Memoir, No. 161.

62. Philipp, W., Webb, G. R. (1973), An Invariance Principle for Mixing Sequences of Random Variables, Z. Warhsch. verw. Gebiete 25, 223-237.

63. Rosenblatt, M. (1956), A Central Limit Theorem and a Strong Mixing Condition, Proc. Nat. Acad. Sci. USA 42, 43-47.

64. Rosenblatt, M. (1971), Markov Processes, Structure and Asymptotic Behavior, Springer-Verlag, Berlin.

65. Samur, J. D. (1984), Convergence of Sums of Mixing Triangular Arrays of Random Vectors with Stationary Rows, Ann. Prob. 12, 390-426.

66. Serfling, R. J. (1968), Contributions to Central Limit Theory for Dependent Variables, Ann. Math. Stat. 39, 1158-1175.

67. Shields, P., (1973), The Theory of Bernoulli Shifts. University of Chicago Press, Chicago.

68. Simons, G., Stout, W. (1978), A Weak Invariance Principle with Applications to Domains of Attraction, Ann. Prob. 6, 294-315.

69. Volkonskii, V. A., Rozanov, Y. A. (1959), Some Limit Theorems for Random Functions I. Theory Prob. Appl. 4, 178-197.

70. Yokoyama, R. (1980), Moment Bounds for Stationary Mixing Sequences, Z. Wahrsch. verw. Gebiete 52, 45-57.

71. Yoshihara, K. (1978), Probability Inequalities for Sums of Absolutely Regular Processes and Their Applications, Z. Wahrsch. verw. Gebiete 43, 319-329.

72. Withers, C. L. (1981), Central Limit Theorems for Dependent Variables I, Z. Wahrsch. verw. Gebiete 57, 509-534.

73. Withers, C. L. (1983), Corringendum to Central Limit Theorems for Dependent Random Variables, Z. Wahrsch. verw. Gebiete, 63, 555 .

Department of Mathematical Sciences
University of Cincinnati
Cincinnati, Ohio 45221-0025

Added in Proof: After the survey was written, the author managed to solve the conjecture (2.1) listed in 2.

Peligrad, M (1985): A central limit theorem for ρ-mixing sequences of random variables (preprint).

INVARIANCE PRINCIPLES FOR INDEPENDENT AND WEAKLY DEPENDENT RANDOM VARIABLES

WALTER PHILIPP

1. Introduction

During the last three decades the concept "invariance principle" has undergone several subtle changes. In the early 1950's an invariance principle was a result that nowadays often would be called a functional central limit theorem (FCLT). At present the term "invariance principle" generally stands as a synonym for an approximation theorem: A given process, such as a partial sum process, an empirical process, an extremal process, a U-statistic, etc. is approximated in distribution, in probability, in L^p or almost surely by a canonical process, such as a Brownian motion, a Kiefer process, a special extremal process or in case of a U-statistic by a multiple stochastic integral.

In this section I will attempt to illustrate the content of the above paragraph by means of the partial sum process and the empirical process. Moreover, I will try to demonstrate the utility and power of these approximation theorems by the great many consequences they have for limit theorems in probability theory.

Let $C = C[0,1]$ denote the space of real-valued continuous functions equipped with the supremum norm and let $\mathcal{C}$ denote the σ-field generated by the open sets in C. Let $\{\xi_j, j \geq 1\}$ be a sequence of i.i.d.r.v.'s with mean zero, variance 1 and partial sums S_k. Then

$$\begin{aligned} f_n(t) &= n^{-\frac{1}{2}} S_k \qquad \text{if } t = \frac{k}{n}, \ 0 \leq t \leq 1 \\ &= \text{linear on } [\tfrac{k-1}{n}, \tfrac{k}{n}], \ 1 \leq k \leq n \end{aligned} \tag{1.1}$$

defines a sequence of r.v.'s with values in C. Let p_n denote the distribution of f_n, i.e. p_n is defined as $p_n(A) = P\{f_n(\cdot) \in A\}$, $A \in \mathcal{C}$. Let W denote Wiener measure on $(C[0,1],\mathcal{C})$, i.e. the distribution of standard Brownian motion $\{B(t), 0 \leq t \leq 1\}$, considered as a r.v. with values in $C[0,1]$. Then by the classical Donsker (1951) theorem

$$P_n \to W \quad (n \to \infty) \tag{1.2}$$

in the sense of weak convergence of probability measures on $(C[0,1],\mathcal{C})$, i.e. for any bounded continuous functional $h: C \Rightarrow \mathbb{R}$

$$h(f_n) \to h(B) \quad (n \to \infty) \tag{1.3}$$

hence the term FCLT. But results of this type are also called <u>invariance principles in distribution</u> (IPID). The term invariance principle was coined by Erdös and Kac (1946, 1947) who proved (1.3) for four different functionals h observing that the limit distribution $\mathcal{L}(h(B))$ did not depend on the common distribution function F of $\{\xi_j, j \geq 1\}$, i.e. remained invariant under a change of F as long as

$$\int x dF(x) = 0 \quad \text{and} \quad \int x^2 dF(x) = 1 . \tag{1.4}$$

If, for instance h is defined by $h(x(\cdot)) = x(1)$, $x \in C$ we obtain the classical CLT, or if $h(x(\cdot)) = \max(x(t) : 0 \leq t \leq 1)$, $x \in C$ we obtain for $\alpha \geq 0$

$$P\{n^{-\frac{1}{2}} \max_{k \leq n} S_k \leq \alpha\} \to P\{\sup_{0 \leq t \leq 1} B(t) \leq \alpha\}$$
$$= (2/\pi)^{\frac{1}{2}} \int_0^{\alpha} e^{-\frac{1}{2}u^2} du .$$

A conceptually simpler yet stronger form of Donsker's theorem is the <u>invariance principle in probability</u> (IPIP). Let F be a distribution function satisfying (1.4). Then there are two sequences $\{\xi_j, j \geq 1\}$ and $\{\eta_j, j \geq 1\}$ of i.i.d.r.v. defined on some probability space with common laws $\mathcal{L}(\xi_1) = F$ and $\mathcal{L}(\eta_1) = \mathcal{N}(0,1)$ respectively such that

$$n^{-\frac{1}{2}} \max_{k \leq n} |S_k - T_k| \to 0 \text{ in prob.} \quad (n \to \infty) . \tag{1.5}$$

Here $T_k := \Sigma_{j \leq k} \eta_j$. (Major (1976a)).

Since $\{n^{-\frac{1}{2}} \max_{k \leq n} S_k^2, n \geq 1\}$ is uniformly integrable the convergence in (1.5) is also in L^2, an example of an <u>L^2-invariance principle</u> (Philipp (1980)).

Finally we shall consider almost sure approximation and the resulting theorems are called <u>almost sure</u> or <u>strong invariance principles</u> (ASIP) or strong approximation theorems. The classical result in this direction is Strassen's (1964) theorem. If F satisfies (1.4) then there are two sequences $\{\xi_j, j \geq 1\}$ and $\{\eta_j, j \geq 1\}$ of i.i.d.r.v. defined on some probability space and with common laws $\mathcal{L}(\xi_1) = F$ and $\mathcal{L}(\eta_1) = \mathcal{N}(0,1)$ respectively such that

$$S_n - T_n = o((nLLn)^{\frac{1}{2}}) \quad \text{a.s.} \tag{1.6}$$

Here and throughout we write $Lx = \log(x \vee e)$.

There are several equivalent formulations. All of them apply in a possibly modified form to (1.5) and to (1.9) below. First, S_n is often replaced by $S_{[t]}$, T_n by $B(t)$ and n by t. For the purpose of proving limit theorems this does not contain more generality. On the contrary proofs are only more time consuming since one needs to estimate $\sup_{m \leq n} \max_{0 \leq t \leq 1} |B(m+t)|$. Second, some formulations start out with a given sequence $\{\xi_j, j \geq 1\}$ of i.i.d.r.v. whose common distribution function F satisfies (1.4). Then without changing its distribution one can redefine the sequence $\{\xi_j, j \geq 1\}$ on a richer probability space on which (or phrased more concisely, WLOG in the sense of Strassen) there exists a sequence $\{\eta_j, j \geq 1\}$ of i.i.d. $\mathcal{N}(0,1)$-r.v.'s such that (1.6) holds. The problem of course is that the given probability space might not be rich enough to support $\mathcal{N}(0,1)$ random variables. Thus in some formulations it is stipulated that the underlying probability space supporting the given sequence $\{\xi_j, j \geq 1\}$ is rich enough, so that one does not have to go through the process of redefining $\{\xi_j, j \geq 1\}$ without changing its law. As was shown in Dudley and Philipp (1983) taking the product of the given probability space and a copy of the unit interval with Lebesgue measure makes the product space rich enough.

Three more remarks all of which apply in a possibly modified form to (1.5) and to (1.9) below. First, the sequences $\{\xi_j, j \geq 1\}$ and $\{\eta_j, j \geq 1\}$ cannot be independent since then (1.6) would violate the law of the iterated logarithm (LIL) applied to $\{\xi_j - \eta_j, j \geq 1\}$. Second, (1.6) with o replaced by O is trivially true for any sequence $\{\xi_j, j \geq 1\}$ satisfying a bounded LIL. Third the sequence $\{\eta_j, j \geq 1\}$ in (1.6) cannot be obtained from a well-known theorem of Skorohod applied to (1.2). Skorohod's theorem only guarantees that the laws $\mathcal{L}(f_n(\cdot))$, $n \geq 1$ remain unchanged, but usually the (joint) law of the sequence $\{f_n, n \geq 1\}$ will be destroyed. To see what is happening apply Skorohod's theorem to the text-book example of a sequence of real-valued functions converging to zero in measure, but not converging at a single point.

Moreover, this would yield (1.6) with error term $o(n^{\frac{1}{2}})$ which is wrong. Major (1976a) showed that the error term in (1.6) is best possible under the assumption (1.4). Hence (1.6) does not imply the CLT, but (1.2) and (1.5) do. So both have earned their rightful places.

Among the many consequences of (1.6) I mention the Hartman-Wintner LIL. This follows immediately from the LIL for the sequence $\{\eta_j, j \geq 1\}$ which is much easier to obtain. Another consequence of (1.6) is the FLIL, also due to Strassen (1964). (Again the FLIL is much easier to derive for the sequence $\{\eta_j, j \geq 1\}$.) Define $g_n := (LLn)^{-\frac{1}{2}} f_n$. Then $\{g_n, n \geq 1\}$ is relatively compact in $(C[0,1], \|\cdot\|)$ and has as its set of limits the class $K := \{g: g(0) = 0,\ g \text{ absolutely continuous and } \int_0^1 (g'(t))^2 dt \leq 1\}$. It is easy to see that for every bounded continuous functional $h: C \Rightarrow \mathbb{R}$ the sequence $\{h(g_n), n \geq 1\}$ is also relatively compact in $C[0,1]$ and has as its set of limits $h(K) := \{h(g) : g \in K\}$. From this fact a host of corollaries can be derived (see Strassen (1964)). One of these is as follows:

$$\limsup_{n \to \infty} n^{-3/2} (2LLn)^{-\frac{1}{2}} \Sigma_{k \leq n} |S_k| = 3^{-\frac{1}{2}} \quad \text{a.s.}$$

The significance of relations of the form (1.6) is then clear. Any strong limit theorem valid for sums of independent $\mathcal{N}(0,1)$ random variables continues to hold for sums S_n of random variables (independent or dependent) provided that the error term in (1.6) is small enough. For example, if the error term in (1.6) is $O((n/LLn)^{\frac{1}{2}})$ (this holds under (1.4) and $\int |x|^{2+\delta} dF(x) < \infty$, $\delta > 0$) then the upper and lower class statements for the LIL, due to Kolmogorov, Erdös, Petrowski, continue to hold for such sums of random variables. Hence for any non-decreasing functions ψ

$$P(S_n > n^{\frac{1}{2}}\psi(n) \text{ i.o.}) = 0 \text{ or } 1 \tag{1.7}$$

according as

$$\int_1^\infty \frac{\psi(t)}{t} \exp(-\tfrac{1}{2}\psi^2(t))dt < \infty \text{ or } = \infty . \tag{1.8}$$

More applications can be found in Philipp and Stout (1975), Section 1 and in Csörgő and Révész (1981), Section 7.2.

In studying empirical processes it is advantageous to adopt a more general point of view. The classical result in this direction is due to Kiefer (1972). Let $\epsilon > 0$. There exist a sequence $\{x_j, j \geq 1\}$ of i.i.d. random variables with uniform distribution on $[0,1]$ and a sequence of $\{B_j(s), 0 \leq s \leq 1; j \geq 1\}$ of independent standard Brownian bridges (considered as $C[0,1]$-valued random variables) defined on some probability space such that with probability 1

$$\sup_{0 \leq s \leq 1} |\Sigma_{j \leq n}(1\{x_j \leq s\} - s) - \Sigma_{j \leq n} B_j(s)| \ll n^{\frac{1}{3}+\epsilon}. \quad (1.9)$$

(Here and throughout the paper $f(x) \ll g(x)$ means $f(x) = O(g(x)$.) Recall that the empirical distribution function $F_n(s)$ of a sample of size n is defined as $n^{-1}\Sigma_{j \leq n} 1\{x_j \leq s\}$. If we write $\xi_j(s) = 1\{x_j \leq s\} - s$ then $\{\xi_j, j \geq 1\}$ is a sequence of independent $D[0,1]$-valued random elements. We give $D[0,1]$ the topology defined by the supremum norm $\|\cdot\|$. Rewriting (1.9) in the form

$$\|\Sigma_{j \leq n} \xi_j - \Sigma_{j \leq n} B_j\| \ll n^{\frac{1}{3}+\epsilon} \quad \text{a.s.}$$

we can reinterpret (1.9) as an almost sure invariance principle for sums of independent $(D[0,1],\|\cdot\|)$-valued random elements. (1.9) not only improves upon Donsker's (1952) FCLT for empirical distribution functions but also avoids the difficulties of measurability and topology caused by the fact that $(D[0,1],\|\cdot\|)$ is non-separable (see Billingsley (1968), p. 153). Because of these measurability problems we called the ξ_j's random elements and not random variables. They are independent in the sense that they are functions $h(x_j)$ of independent r.v. x_j with $h: [0,1] \to D[0,1]$. But (1.9) also provides a quick proof of the Chung-Smirnov LIL for empirical distribution functions since it is easier to prove an LIL for independent Gaussian C-valued r.v.

For these reasons one might be tempted to replace invariance principles in distribution by invariance principles in probability or almost sure invariance principles. We shall do this whenever possible. However, in many instances invariance principles in distribution are the only ones available in the literature. Many of them will be quoted in the following sections. But the main goal of this article is to survey pathwise approximation theorems, i.e. invariance principles in probability, L^p and almost sure invariance principles for sums of random elements. In the most general setting these random elements will be assumed to satisfy a weak dependence relation, they need not be measurable and they will be allowed to assume values in a Banach space which need not be separable. But we will also treat invariance principles for extrema of random variables, U-statistics, etc.

There are many excellent books and surveys on certain areas of probability theory which devote space to invariance principles. They are quoted throughout this paper. For this reason I did not feel the need to try to make this survey complete. Just listing all the relevant papers would more than fill my allowed quota of pages. Instead, I will try to explain the methods to prove invariance principles and to quote the key papers in order to facilitate the interested reader his entry into an important branch of probability theory.

2. Weakly Dependent Random Variables

Let us call a sequence $\{\xi_j, j \geq 1\}$ of random variables weakly dependent if for all $k \geq 1$

$$E\left|E(\xi_{k+n}|\xi_1,\ldots,\xi_k)\right| \to 0 , \quad n \to \infty . \tag{2.1}$$

Unfortunately this definition does not encompass all sequences of random variables which have been called weakly dependent in the literature, but for a start this definition serves its purpose. Of course, martingale differences satisfy (2.1) and so do mixingale difference sequences. Gaussian sequences with pairwise correlation coefficients tending to zero, lacunary random variables, and many others all satisfy (2.1) either directly or in a slightly modified form. In this section I list several dependence relations for easy reference.

2.1. Mixing conditions. In contrast to (2.1) most of these do not presuppose the existence of a finite expectation. Let $\mathcal{M}_a^b$ denote the σ-field generated by the random variables $\xi_a,\xi_{a+1},\ldots,\xi_b$.

Definition. The sequence $\{\xi_j, j \geq 1\}$ is called strongly mixing if

$$\alpha(n) := \sup\{|P(A \cap B) - P(A)P(B)| : A \in \mathcal{M}_1^k,\ B \in \mathcal{M}_{k+n}^\infty,\ k \geq 1\} \to 0 \tag{2.2}$$

or equivalently

$$\alpha'(n) : = \sup\{E|P(B|\mathcal{M}_1^k) - P(B)| : B \in \mathcal{M}_{k+n}^\infty, k \geq 1\} \to 0 . \quad (2.3)$$

The sequence $\{\xi_j, j \geq 1\}$ is called <u>absolutely regular</u> if

$$\beta(n) : = E \sup\{|P(B|\mathcal{M}_1^k) - P(B)| : B \in \mathcal{M}_{k+n}^\infty, k \geq 1\} \to 0 . \quad (2.4)$$

(The supremum is measurable since in a Polish space it is sufficient to extend the supremum only over countably many sets B.) The sequence is called <u>ρ-mixing</u> if the maximal correlation coefficient

$$\rho(n) : = \sup\{\frac{|Exy|}{\|x\|_2\|y\|_2} : x \in L^2(\mathcal{M}_1^k), y \in L^2(\mathcal{M}_{k+n}^\infty),$$

$$Ex = Ey = 0, k \geq 1\} \to 0 . \quad (2.5)$$

The sequence $\{\xi_j, j \geq 1\}$ is called <u>φ-mixing</u> or <u>uniformly mixing</u> if

$$\phi(n) : = \sup\{|P(B|\mathcal{M}_1^k) - P(B)| : B \in \mathcal{M}_{k+n}^\infty, k \geq 1\} \to 0 \quad (2.6)$$

with probability 1 or equivalently

$$\phi(n) = \sup\{\frac{|P(A \cap B) - P(A)P(B)|}{P(A)} : A \in \mathcal{M}_1^k,$$

$$B \in \mathcal{M}_{k+n}^\infty, k \geq 1\} \to 0 . \quad (2.7)$$

It is called <u>ψ-mixing</u> if

$$\psi(n) : = \sup\{\frac{|P(A \cap B) - P(A)P(B)|}{P(A)P(B)} : A \in \mathcal{M}_1^k,$$

$$B \in \mathcal{M}_{k+n}^\infty, k \geq 1\} \to 0 \quad (2.8)$$

and it is called <u>*-mixing</u> if in the last supremum $\mathcal{M}_{k+n}^\infty$ is replaced by $\mathcal{M}_{k+n}^{k+n}$.

The concepts of ψ-mixing and *-mixing are of lesser importance. Well-known lemmas imply that under the appropriate moment conditions $\frac{1}{4}\alpha(n) \leq \rho(n) \leq 2\phi^{\frac{1}{2}}(n)$ and $\alpha(n) \leq \alpha'(n) \leq 2\alpha(n)$ and it is obvious that $\alpha(n) \leq \beta(n) \leq \phi(n) \leq \psi(n)$ so that a ψ-mixing sequence is φ-mixing which in turn implies absolutely regularity and this in turn implies strong mixing. There are well-known counterexamples to the reverse implications. For a survey on mixing conditions, see Bradley (1985b).

Other related dependence structures occurring in the literature are the following. Let $\{\xi_j, j \geq 1\}$ be a sequence of square-integrable random variables and let $\{\mathcal{M}_n, -\infty < n < \infty\}$ be a non-decreasing sequence of sub σ-fields. Then $\{\xi_n, \mathcal{M}_n, n \geq 1\}$ is called a <u>mixingale (difference) sequence</u> if there exist sequences c_n and $\psi_m \to 0$ of non-negative constants such that for all $n \geq 1$, $m \geq 0$

$$\begin{cases} \|E(\xi_n|\mathcal{M}_{n-m})\|_2 \leq \psi_m c_n \\ \text{and} \\ \|\xi_n - E(\xi_n|\mathcal{M}_{n+m})\|_2 \leq \psi_{m+1} c_n . \end{cases} \tag{2.10}$$

The sequence $\{\xi_j, j \geq 1\}$ is said to satisfy a <u>retarded strong mixing</u> condition if there exist an $\alpha^*(m) \to 0$ such that for all $n, k \geq 1$

$$\sup\{|P(A \cap B) - P(A)P(B)|: A \in \mathcal{M}_1^k, B \in \mathcal{M}_{k+n}^\infty\} \leq \alpha^*(nk^{-\tau}) \tag{2.11}$$

for some constant $\tau \geq 0$. (Note that $\tau = 0$ gives the strong mixing condition.) It is said to be a <u>retarded asymptotic martingale difference sequence</u> if there exists $\psi^*(m) \to 0$ such that for all n, $k \geq 1$

$$\|E(\xi_{k+n}|\mathcal{M}_1^k)\|_p \leq \psi^*(nk^{-\tau}) \tag{2.12}$$

for some $p \geq 1$ and $\tau \geq 0$. Thus mixingales are retarded asymptotic martingales.

The strong mixing condition was introduced by Rosenblatt (1956), the ϕ-mixing condition by Ibragimov (1962), the *-mixing condition by Blum, Hanson and Koopmans (1963). Mixingales were first defined by McLeish (1975) and the retarded strong mixing condition and retarded asymptotic martingales by Philipp and Stout (1975).

<u>2.2. Veršik and very weak Bernoulli sequences</u>. These have their roots in ergodic theory. For discrete random variables these processes have been introduced by Veršik and by Ornstein (1974) respectively. The definitions have been generalized by Dehling, Denker and Philipp (1984), Eberlein (1983) and Strittmatter (1982) to cover random variables not necessarily discrete. For a precise statement of these definitions (which are somewhat lengthy) the reader is referred to these papers. There is a close connection between Veršik and very weak Bernoulli processes on the one hand and mixing sequences on the

other (Dehling, Denker and Philipp (1984)). But there are very weak Bernoulli sequences which are not strongly mixing in the sense (2.2) (see Bradley (1984)).

2.3. Lacunary sequences of random variables. Let $\{n_k, k \geq 1\}$ be a sequence of positive numbers satisfying

$$n_{k+1}/n_k \geq q > 1 \ , \quad k \geq 1 \ . \tag{2.13}$$

Let ω be a random variable uniformly distributed over $[0,1]$ and let $\langle x \rangle := x - [x]$ denote the fractional part of a real number x. The sequence $\{\langle n_k \omega \rangle, k \geq 1\}$ is called a lacunary sequence of random variables. We call such a sequence quasilacunary if the corresponding sequence $\{n_k, k \geq 1\}$ satisfies

$$n_{k+1}/n_k \geq 1 + \phi(k)k^{-\frac{1}{2}} \ , \quad k \geq 1 \tag{2.14}$$

for some non-decreasing sequence $\phi(k) = o(k^{\frac{1}{2}})$, but tending to infinity.

2.4. Multiplicative systems. A sequence $\{\xi_k, k \geq 1\}$ centered at expectations is called a multiplicative system if for all $n \geq 1$ and $r_j = 0,1$ $(1 \leq j \leq n)$

$$E\xi_1^{r_1} \ldots \xi_n^{r_n} = E\xi_1^{r_1} \ldots E\xi_n^{r_n} \ . \tag{2.15}$$

It is called strongly multiplicative if $E\xi_k^2 = 1$ and if (2.15) holds for all $r_j = 0,1,2$ $(1 \leq j \leq n)$. In other words the left side of (2.15) is assumed to exist and it equals 1 if all $r_j = 0,2$ and it vanishes if $r_j = 1$ for some $1 \leq j \leq n$.

2.5. Associated sequences. A sequence $\{\xi_j, j \geq 1\}$ is called associated if for all $j,k \geq 1$ and all $x,y \in \mathbb{R}$

$$P(\xi_j > x, \xi_k > y) \geq P(\xi_j > x)P(\xi_k > y) \ .$$

3. Methods of Proof

3.1. Convergence of finite-dimension distributions and tightness. These are the magic words in the proof of invariance principles in distribution. They apply to any sequence of random vectors, but only for the proof of invariance principles in distribution. For the partial sum process considered in (1.1) the proof of convergence of the finite-dimensional distributions is usually obtained by a multi-dimensional CLT and the proof of tightness usually involves a maximal inequality. The method is well-known and is treated in many standard texts (see e.g. Billingsley (1968)).

3.2. Skorohod embedding. The Skorohod embedding theorem is extremely useful for the proof of all four types of invariance principles for martingales and any partial sum process which can be approximated by martingales. Unfortunately, the applicability of this method is, in essence, restricted to real-valued random variables. The following improved version of the Skorohod embedding theorem is due to Scott and Huggins (1983).

Theorem 3.1. Let $\{\xi_n, \mathcal{L}_n, n \geq 1\}$ be a square integrable martingale difference sequence. Then (on a possibly enlarged version of the underlying probability space) there exists a Brownian motion $\{B(t), \mathcal{L}_t^*, t \geq 0\}$ and a non-decreasing sequence $\{T_n, n \geq 1\}$ of stopping times such that $\mathcal{L}_n \subset \mathcal{L}^*_{T_n}$ and

$$B(T_n) = \Sigma_{j \leq n} \xi_j , \quad n \geq 1 \quad \text{a.s.} \tag{3.1}$$

Furthermore, there exists an increasing sequence $\{\mathcal{B}_n, n \geq 1\}$ of σ-fields such that τ_n is $\mathcal{B}_n$-measurable and that

$$E(\tau_n | \mathcal{B}_{n-1}) = E(\xi_n^2 | \mathcal{L}_{n-1}) \quad \text{a.s.} \tag{3.2}$$

and for $p > 1$

$$M_p E(\tau_n^p | \mathcal{B}_{n-1}) \leq E(|\xi_n|^{2p} | \mathcal{L}_{n-1}) \leq N_p E(\tau_n^p | \mathcal{B}_{n-1}) \quad \text{a.s.} \tag{3.3}$$

for some positive constants M_p and N_p, depending only on p. Here we set $\tau_n := T_n - T_{n-1}$.

Using the Skorohod embedding theorem Strassen (1967) proved his now classical almost sure invariance principle for martingales. Perhaps the hitherto sharpest version is due to Philipp and Stout (1985).

Theorem 3.2. Let f be a non-increasing function satisfying some mild regularity conditions and let $\{\xi_n, \mathcal{L}_n, n \geq 1\}$ be a square integrable martingale difference sequence such that with probability one

$$V_n := \Sigma_{j \leq n} E(\xi_j^2 | \mathcal{L}_{j-1}) \to \infty \tag{3.4}$$

and

$$|\xi_n| \leq f(V_n) V_n (LLV_n)^{-\frac{1}{2}} . \tag{3.5}$$

For $t > 0$ we define

$$S(t) := \Sigma_{j \leq n} \xi_j \quad \text{if} \quad V_n \leq t < V_{n+1} . \tag{3.6}$$

Then we can redefine $\{\xi_n, \mathcal{F}_n, n \geq 1\}$ without changing its distribution on a possibly richer probability space on which there exists a standard Brownian motion $\{B(t), t \geq 0\}$ and with probability one a t_0 such that

$$|S(t) - B(t)| \leq 10^3 (f(t) tLLt)^{\frac{1}{2}} \text{ for all } t \geq t_0 . \tag{3.7}$$

Remark. Of course, for independent random variables, (3.5) is the Kolmogorov condition for the LIL, usually expressed as $\xi_n = o(V_n (LLV_n)^{-\frac{1}{2}})$ a.s. It is easy to derive variants of this theorem for unbounded martingale differences by using a truncation argument.

The basic idea in the proofs of invariance principles via the Skorohod embedding theorem is to first prove a strong law of large numbers in the form

$$\frac{T_n}{V_n} = 1 + O(f(V_n)) \quad \text{a.s.} \tag{3.8}$$

This is obtained from the moment inequalities (3.3). The second step consists of feeding (3.8) into the Brownian motion. For this purpose let $\{t_k, k \geq 1\}$ be a sequence tending to infinity at a suitable rate depending, in general, on f, so that

$$V_{t_{k-1}} = V_{t_k} (1 + O(f(V_{t_k})) \quad \text{a.s.} \tag{3.9}$$

Then by (3.1) and (3.9)

$$\sup\{|S(t) - B(t)| : V_{t_{k-1}} \leq t \leq V_{t_k}\} \leq \sup|B(u) - B(v)|$$

where the last supremum is extended over all u and v in an interval with center $V_{t_k}^2$ and length $O(V_{t_k}^2 f(V_{t_k}))$. But this last supremum is easily estimated by Lévy's maximal inequality for Brownian motion and the Borel Cantelli lemma. (3.7) follows then in a straightforward fashion.

3.3. Embedding in a continuous parameter martingale. Another approach based on embedding in Brownian motion of a given process was recently discovered by Scott and Huggins (1983). Their method rests on an observation by Heath (1977) that every discrete parameter martingale can be embedded in a continuous parameter martingale with continuous sample paths so that these two martingales agree for integral values of the parameters. By a theorem of Knight (1971) (which extends a result of Kunita and Watanabe (1967)) this continuous parameter martingale equals a time changed Brownian motion. A theorem of Millar (1968) says that the p-th moments of the quadratic variation up to time t of a continuous parameter martingale is bounded above and below by constant multiples of the 2p-th moments of the martingale at time t. Using these three facts Scott and Huggins (1983) obtain Theorem 3.1 above, embedding theorems and a FLIL for reverse martingales, etc.

3.4. The quantile transform method. A totally different approach was developed by M. Csörgő and Révész (1974a) and later refined by Komlós, Major and Tusnády (1975). The main advantage of this method, now known as the quantile transform method, is that in the almost sure invariance principle for sums of i.i.d. random variables having finite p-th moment or a moment generating functions it yields sharp error terms. However, it is restricted to independent random variables and has only recently been extended by Berger (1982) to cover the case of i.i.d. random vectors with values in $\mathbb{R}^d$.

I will now describe the method in its simplest setting. Let $\{\xi_j, j \geq 1\}$ be a sequence of i.i.d. random variables with continuous distribution function F and $E\xi_1 = 0$, $E\xi_1^2 = 1$ and $E|\xi_1|^{2+\delta} < \infty$ for some $\delta > 0$. Let $n_k := k^\alpha$, for some large suitable chosen integer α. Put

$$t_k := \Sigma_{j < k}\, n_j \quad (\sim \frac{1}{\alpha+1} k^{\alpha+1}) \tag{3.10}$$

and

$$X_k := n_k^{-\frac{1}{2}} \Sigma_{j=t_k+1}^{t_{k+1}} \xi_j \,. \tag{3.11}$$

Note that the CLT implies $\mathcal{L}(X_k) \sim \mathcal{N}(0,1)$ with an error term $O(n_k^{-\frac{1}{2}\delta})$, if $\delta \leq 1$. Let F_k denote the distribution function of X_k. Observe that F_k is continuous since F is, and that $F_k(X_k)$ has uniform distribution on [0,1]. Define

$$Y_k := \phi^{-1}(F_k(X_k)) \tag{3.12}$$

where ϕ^{-1} is the inverse of the Gaussian distribution function ϕ. Since

$$P(Y_k \le y) = P(F_k(X_k) \le \phi(y)) = \phi(y)$$

we conclude that $\{Y_k, k \ge 1\}$ is a sequence of i.i.d. $\mathcal{N}(0,1)$-random variables by (3.10) and (3.11). Since $F_k \sim \phi$ it is not difficult to show that

$$Y_k - X_k = O(n_k^{-\beta}) \quad \text{a.s.} \tag{3.13}$$

for some $\beta = \beta(\delta) > 0$. Hence

$$n_k^{\frac{1}{2}} Y_k - n_k^{\frac{1}{2}} X_k = O(n_k^{\frac{1}{2}-\beta}) \quad \text{a.s.} \tag{3.14}$$

Now $\mathcal{L}(n_k^{\frac{1}{2}} Y_k) = \mathcal{N}(0, n_k)$. Hence by a simple measure theoretic lemma (see Lemma A1 of Berkes and Philipp (1979a)) we can write

$$n_k^{\frac{1}{2}} Y_k = \Sigma_{j=t_k+1}^{t_{k+1}} \eta_j \tag{3.15}$$

where $\{\eta_j, j \ge 1\}$ is a sequence of i.i.d. $\mathcal{N}(0,1)$-random variables. Thus summing (3.13) over $1 \le k < M$ we obtain by (3.10), (3.11) and (3.14)

$$\Sigma_{j \le t_M} \xi_j - \Sigma_{j \le t_M} \eta_j = O(\Sigma_{k < M} n_k^{\frac{1}{2}-\beta})$$

$$= O(\Sigma_{k < M} k^{\alpha(\frac{1}{2}-\beta)}) = O(M^{\frac{1}{2}(\alpha+1)-(\alpha\beta-\frac{1}{2})}) = O(t_M^{\frac{1}{2}-\lambda}) \text{ a.s.}$$

where $\lambda = (\alpha\beta-\frac{1}{2})/(\alpha+1) > 0$ if $\alpha > 1/(2\beta)$. Hence (1.6) holds with an error term $O(n^{\frac{1}{2}-\lambda})$, at least along the subsequence $\{t_M, M \ge 1\}$. From Ottaviani's maximal inequality and well-known estimates for the $(2+\delta)$-th moment of a sum of i.i.d. random variables we obtain by (3.10)

$$P(\max_{t_M < n \le t_{M+1}} |\Sigma_{j=t_M+1}^{n} \xi_j| \ge t_M^{\frac{1}{2}-\lambda})$$

$$= O(n_M^{1+\frac{1}{2}\delta} t_M^{-(\frac{1}{2}-\lambda)(1+\frac{1}{2}\delta)}) = O(M^{-2})$$

if λ is sufficiently small. Of course the same estimate holds with ξ_j replaced by η_j. Hence the Borel Cantelli lemma applies and we conclude that the fluctuation of the partial sums between t_M and t_{M+1} is negligible. Hence we obtain (1.6) with an error term $O(n^{\frac{1}{2}-\lambda})$ along the whole sequence of integers n.

<u>3.5. Approximation theorems based on estimates of the Prohorov distance and the Strassen-Dudley theorem.</u> The quantile transform method ostensibly only works for independent random variables ξ_j because of the way the random variables Y_k are constructed. The construction of a sequence $\{Y_k, k \geq 1\}$ of independent $\mathcal{N}(0,1)$ random variables satisfying (3.13) is really the crux of the method! There are now several so-called approximation theorems in the literature yielding constructions different from (3.12) of such desirable sequences $\{Y_k, k \geq 1\}$. Most of these approximation theorems are based on estimates of the Prohorov-distance and the Strassen-Dudley theorem (Strassen (1965), Dudley (1968)). The most versatile of these approximation theorems, due to Berkes and Philipp (1979a), is as follows.

<u>Theorem 3.3.</u> Let $\{X_k, k \geq 1\}$ be a sequence of random variables with values in $\mathbb{R}^{d_k}$, $d_k \geq 1$ and let $\{\mathfrak{F}_k, k \geq 1\}$ be a non-decreasing sequence of σ-fields such that X_k is $\mathfrak{F}_k$-measurable. Let $\{G_k, k \geq 1\}$ be a sequence of probability distributions on $\mathbb{R}^{d_k}$ with characteristic functions $g_k(u)$, $u \in \mathbb{R}^{d_k}$, respectively. Suppose that for some non-negative numbers λ_k, δ_k and $T_k \geq 10^8 d_k$

$$E\left|E\{\exp(i\langle u,X_k\rangle) \mid \mathfrak{F}_{k-1}\} - g_k(u)\right| \leq \lambda_k \tag{3.16}$$

for all u with $|u| \leq T_k$ and

$$G_k\{u: |u| > \tfrac{1}{4}T_k\} \leq \delta_k . \tag{3.17}$$

Then without changing its distribution we can redefine the sequence $\{X_k, k \geq 1\}$ on a richer probability space together with a sequence $\{Y_k, k \geq 1\}$ of independent random variables such that Y_k has distribution G_k and

$$P\{|X_k - Y_k| \geq \alpha_k\} \leq \alpha_k , \quad k = 1,2,\ldots \tag{3.18}$$

where $\alpha_1 = 1$ and

$$\alpha_k = 16 d_k T_k^{-1} \log T_k + 4\lambda_k^{\frac{1}{2}} T_k^{d_k} + \delta_k , \quad k \geq 2 . \tag{3.19}$$

This theorem applies directly to most of the dependence structures mentioned in Section 2 without first approximating by martingales. But it also works for vector-valued martingales whereas the Skorohod embedding method does not.

To illustrate the use of Theorem 3.3 I will sketch the proof of Theorem 3.2 but with (3.5) strengthened to

$$\xi_n \ll V_n^{1-\delta} \quad \text{a.s.} \tag{3.20}$$

for some $\delta > 0$. A detailed proof for $\mathbb{R}^d$-valued martingales is given in Morrow and Philipp (1982). We define n_k and t_k as in (3.10) above and put $\tau(0) = 0$ and

$$\tau(k) = \min\{\tau \geq \tau(k-1) : V_\tau^2 \geq t_k\} . \tag{3.21}$$

Then $\tau(k) - 1$ is a stopping time relative to $\{\mathcal{L}_n, n \geq 1\}$. We define

$$X_k := n_k^{-\frac{1}{2}} \Sigma_{j \geq 1} \xi_j 1\{\tau(k-1) \leq j < \tau(k)\} . \tag{3.22}$$

The sum in (3.22) is finite a.s. because of (3.4) and (3.21). Moreover, by (3.21) and (3.10) the sum of the conditional variances of the terms is about 1. We apply Theorem 3.3 with $\mathfrak{F}_k = \mathcal{L}_{\tau(k)-1}$. Then X_k is $\mathfrak{F}_k$ measurable and by the CLT for martingales

$$\sup_{|u| \leq k^2} E\left|E\{\exp(iuX_k)|\mathfrak{F}_{k-1}\} - e^{-\frac{1}{2}u^2}\right| \ll n_k^{-\gamma}$$

for some $\gamma > 0$. As a matter of fact this last estimate is not easy to obtain. In Morrow and Philipp (1982) a method of Brown (1971) came in handy. Hence with $T_k = k^2$, $\delta_k \ll e^{-k}$ and $\lambda_k = k^{-\alpha\gamma}$ we obtain from (3.19) $\alpha_k \ll k^{-\frac{1}{2}\alpha\gamma+2}$. Since for α sufficiently large $\Sigma\,\alpha_k < \infty$ we obtain from (3.10), (3.18) and the Borel Cantelli lemma relation (3.13) with a suitable $\beta > 0$. We now can complete the proof along the same lines as in Section 3.2, but using Doob's martingale inequalities to "break into the blocks".

As mentioned before, Theorem 3.3 also works for strongly mixing $\mathbb{R}^d$-valued random variables. But, as Dehling (1983a) showed by a counterexample, it is impossible to obtain an approximation theorem with a conclusion of the form (3.18) for infinite-dimensional random variables satisfying a strong mixing condition.

The following result, due to Berger (1982b), is the latest step in a series of improvements of another approximation theorem of Berkes and Philipp (1979a) (see Philipp (1979) and Dehling and Philipp (1982)).

Theorem 3.4. Let $\{B_k, m_k, k \geq 1\}$ be a sequence of Polish spaces. Let $\mathcal{B}_k$ denote the Borel field of B_k, let $\{X_k, k \geq 1\}$ be a sequence of random variables with values in B_k and let $\{\mathfrak{F}_k, k \geq 1\}$ be a sequence of non-decreasing σ-fields such that X_k is $\mathfrak{F}_k$-measurable. Suppose that for some sequence $\{\beta_k, k \geq 1\}$ of non-negative numbers

$$E \sup_{A \in \mathcal{B}_k} |P(X_k \in A | \mathfrak{F}_{k-1}) - P(X_k \in A)| \leq \beta_k$$

for all $k \geq 1$. Denote by F_k the distribution of X_k and let $\{G_k, k \geq 1\}$ be a sequence of distributions on $(B_k, \mathcal{B}_k)$ such that

$$F_k(A) \leq G_k(A^{\rho_k]}) + \sigma_k \quad \text{for all } A \in \mathcal{B}_k . \tag{3.23}$$

Here ρ_k and σ_k are non-negative numbers and $A^{\epsilon]} = \bigcup_{x \in A}\{y: m_k(x,y) \leq \epsilon\}$. Then without changing its distribution we can redefine the sequence $\{X_k, k \geq 1\}$ on a richer probability space on which there exists a sequence $\{Y_k, k \geq 1\}$ of independent random variables with distribution G_k such that for all $k \geq 1$

$$P\{m_k(X_k, Y_k) > \rho_k\} \leq \beta_k + \sigma_k .$$

Here is a sketch of the proof. It is a modification of the proof of Theorem 5 of Dehling and Philipp (1982), pp. 699-700 who obtained the bounds $2(\beta_k^{1/2} + \rho_k)$ and $2(\beta_k^{1/2} + \sigma_k)$ instead of ρ_k and $\beta_k + \sigma_k$. We follow the proof until the last line of p. 699 and rewrite (7.1) as

$$\Sigma_{D \in \mathcal{A}_{k-1}} P(D)\epsilon_k(D) \leq \beta_k \tag{3.24}$$

where

$$\epsilon_k(D) := \sup_{A \in \mathcal{B}_k} \left|\frac{P(X_k \in A, D)}{P(D)} - P(X_k \in A)\right| . \tag{3.25}$$

If we define P_1 and P_2 as in (2.1.2) of Berkes and Philipp (1979a) we obtain from (3.25) $|P_1(A) - P_2(A)| \leq \varepsilon_k(D)$. Thus by the proof of Theorem 2 of Berkes and Philipp (1979a) and by using Theorem 2 of Dudley (1968) instead of Lemma 2.1 in Berkes and Philipp (1979a) we obtain a sequence of random variables Y_k having the same distribution as X_k and

$$P(X_k \neq Y_k | D) \leq \varepsilon_k(D) .$$

Thus by (3.24)

$$P(X_k \neq Y_k) \leq \beta_k .$$

This proves the theorem when all X_k are discrete and $G_k = F_k$, $k \geq 1$. For the general case we more or less copy the proof of Theorem 3 of Philipp (1979) to obtain for each $\delta_j > 0$ a sequence $\{Y_k^{(j)}, k \geq 1\}$ such that

$$P(m_k(X_k, Y_k^{(j)}) > \rho_k + \delta_j) \leq \beta_k + \sigma_k + \delta_j .$$

Letting $\delta_j \to 0$ and using the completeness of B_k we finally arrive at a limit sequence $\{Y_k, k \geq 1\}$ with the desired properties.

Remarks. For $\rho_k = \sigma_k = 0$, i.e. $F_k = G_k$, $k \geq 1$, Theorem 3.4 was also discovered recently by Berbee (1984) using a different method. Instead of the Strassen-Dudley theorem Berbee (1984) uses results of Schwarz (1980) and his Ph.D. thesis (Berbee (1979)).

For obtaining sharp results it is sometimes important to have $\rho_k \to \infty$ and $\sigma_k \to 0$ (see Philipp (1979) and Einmahl (1984)).

Other approximation theorems based on Theorem 3.3 or its proof are Theorem 5 of Berkes and Philipp (1979a) and Theorem 1 of Eberlein (1983). Approximation theorems based on different ideas can be found in Bradley (1983) and Bryc (1982).

3.6. The blocking technique. There are several ways to prove invariance principles such as (1.6) for weakly dependent random variables. Most of them rely on the so-called blocking technique. To fix the ideas we shall consider a stationary sequence $\{\xi_j, j \geq 1\}$ of random variables satisfying a strong mixing condition (2.2) with $\alpha(n) \ll n^{-4}$, $E\xi_1 = 0$, and $E|\xi_1|^4 < \infty$. The following lemma, due to Davydov (1970), is useful.

Lemma 3.1. Let X and Y be random variables $\mathcal{M}_1^k$ and $\mathcal{M}_{k+n}^{\infty}$-measurable respectively and $E|X|^r < \infty$, $E|Y|^s < \infty$ where $1/r + 1/s < 1$. Then

$$|EXY - EXEY| \leq 15\|X\|_r \|Y\|_s (\alpha(n))^{1/t}, \; \frac{1}{r} + \frac{1}{s} + \frac{1}{t} = 1 \tag{3.26}$$

and

$$\|E(Y|\mathcal{M}_1^k) - EY\|_p \leq 15\|Y\|_s (\alpha(n))^{1/t}, \; \frac{1}{s} + \frac{1}{t} = \frac{1}{p}, \; p \geq 1 \; . \tag{3.27}$$

The result continues to hold with $r = \infty$ or $s = \infty$.

From this lemma it is easy to see that there exists a constant $\sigma \geq 0$ such that

$$ES_n^2 = \sigma^2 n + O(n^{3/4}) \; . \tag{3.28}$$

$\sigma = 0$ can happen, for instance, if $\xi_j = \eta_j - \eta_{j-1}$ with η_j i.i.d. We assume $\sigma = 1$ from now on since $\sigma = 0$ does not lead to interesting results.

The blocking technique can be traced back to a paper of Bernstein in 1927. There are several variations to this technique and the one we are going to use is as follows. Similar to Section 3.4 we put

$$n_k = k^\alpha \quad \text{and} \quad p_k = k^{2\alpha/3} \tag{3.29}$$

where α is a large suitably chosen integer, divisible by 3. Define

$$t_k = \Sigma_{j < k} (n_j + p_i) \; (\sim \frac{1}{\alpha+1} k^{\alpha+1}) \; . \tag{3.30}$$

The sums

$$x_k := \Sigma_{j=t_k+1}^{t_k+n_k} \xi_j \quad \text{and} \quad z_k := \Sigma_{j=t_k+n_k+1}^{t_{k+1}} \xi_j \;, \quad k \geq 0$$

are called the long blocks and the short blocks respectively. It is an easy exercise involving (3.28), Chebyshev's inequality and the Borel Cantelli lemma to show that as $M \to \infty$

$$\Sigma_{k \leq M} z_k \ll t_M^{\frac{1}{2}-\lambda} \quad \text{a.s.}$$

for some $\lambda > 0$ and this in essence is the reason why the short blocks can be discarded without affecting the conclusion of the final result (1.6). So why have they been introduced in the first place? They have been introduced in order to make the long blocks almost independent. For instance, by (3.27) with $p = 1$, $s = t = 2$ and setting

$\mathfrak{F}_k = \mathcal{M}_1^{t_{k-1}+n_{k-1}}$ we obtain

$$\|E(x_k|\mathfrak{F}_{k-1})\|_1 \ll \|x_k\|_2 \alpha(p_k)^{\frac{1}{2}} \ll n_k^{\frac{1}{2}}/p_k^2 \ll k^{\frac{1}{2}\alpha-4\alpha/3} = o(1) \quad (3.31)$$

by (3.28) and (3.29). Similarly, by (3.27) with $Y = x_k^2$, $p = 1$, $s = 4$, $t = \frac{4}{3}$ and (3.28) we obtain

$$\|E(x_k^2|\mathfrak{F}_{k-1}) - n_k\|_1 \ll \|x_k\|_4^2 p_k^{-3} \ll k^{2\alpha-3\beta} \ll 1 . \quad (3.32)$$

From this point on there are several options to proceed. The first one is to put

$$y_k = x_k - E(x_k|\mathfrak{F}_{k-1}) . \quad (3.33)$$

Then $\{y_k, \mathfrak{F}_k, k \geq 1\}$ is a martingale difference sequence. By (3.31), Markov's inequality and the Borel Cantelli lemma $y_k \sim x_k$. By (3.31)-(3.33) $E(y_k^2|\mathfrak{F}_{k-1}) \sim n_k$ a.s. Hence one can apply one of the many variants of Theorem 3.2 mentioned in the remark following it to obtain (1.6) with an error term $O(n^{\frac{1}{2}-\lambda})$ along the subsequence t_M. By an inequality for mixing random variables similar to Ottaviano's maximal inequality one can break into the blocks and obtain the desired result.

The second option is to apply Theorem 3.3 to $X_k = n_k^{-\frac{1}{2}} x_k$. By (3.27) with $r = s = \infty$ and $t = 1$ and $Y = \exp(iuX_k)$ one obtains

$$E|E(\exp(iuX_k)|\mathfrak{F}_{k-1}) - E\exp(iuX_k)| \ll \alpha(p_k) .$$

Now from Ibragimov's CLT for stationary strongly mixing sequences $E\exp(iuX_k) \sim e^{-\frac{1}{2}u^2}$ and one can proceed as in Section 3.5. (Compare this with (3.23).)

Another possibility is to apply the Skorohod embedding theorem directly to the martingale difference sequence $\{y_k, \mathfrak{F}_k, k \geq 1\}$. This approach has been sketched in Section 2 of Philipp and Stout (1975). In that paper it was shown that many dependent sequences, such as mixing, lacunary, Gaussian, asymptotic martingales, etc. satisfy conditions similar to (3.31) and (3.32).

That these two conditions are all that are really needed in the proof of almost sure invariance principles for arbitrary sequences of random variables such that the variances of the partial sums grow asymptotically linearly has recently been noticed by Eberlein (1985). For brevity we consider here only the case of real-valued random variables. Eberlein treats $\mathbb{R}^d$-valued random variables. His theorem with the obvious modifications continues to hold for Hilbert space valued random variables, but with an error term $o((nLLn)^{\frac{1}{2}})$.

Theorem 3.5. Let $\{\xi_j, j \geq 1\}$ be a sequence of random variables with finite $(2+\delta)$-th moments uniformly bounded for some $\delta > 0$. Put

$$S_n(m) := \Sigma_{j=m+1}^{m+n} \xi_j \ .$$

Suppose there exists a constant $\theta > 0$ such that uniformly in m

$$\|E(S_n(m)|\mathcal{M}_1^m)\|_1 \ll n^{\frac{1}{2}-\theta} \tag{3.34}$$

and

$$\|E(S_n^2(m)|\mathcal{M}_1^m) - n\|_1 \ll n^{1-\theta} \ . \tag{3.35}$$

Then (1.6) holds with an error term $O(n^{\frac{1}{2}-\lambda})$.

Of course, (3.34) and (3.35) imply that Var $S_n \sim n$. For a brief sketch of the proof we define n_k and t_k by (3.10) with $\alpha = 1 + 2/\theta$ and $x_k := S_{t_k}(n_k)$. We define y_k by (3.33) and proceed from there on in a similar fashion.

4. Selected Results

4.1. Partial sums

4.1.1. I.I.D. random variables

4.1.1.1. IPIP and L^p-invariance principles

For real-valued random variables Donsker's theorem (see Section 1) was first stated in the form (1.5) by Major (1976a). A proof of (1.5), but without the explicit statement, can be found in Freedman (1971), (130). Freedman's goal was a proof of Donsker's theorem via the Skorohod embedding theorem. For random variables with values in a separable Banach space Philipp (1980 & correction) proved that if a sequence is in the domain of attraction to a stable law of index $\alpha \leq 2$ then an IPIP and an L^p-invariance principle with $p < \alpha$ similar to (1.5) holds. In particular, this gives a new proof of Donsker's theorem for $\mathbb{R}^d$-valued random variables. The first IPIP for $\alpha < 2$ and real-valued random variables was proved by Simon and Stout (1979). For random elements not necessarily measurable and assuming values in a not necessarily separable Banach space Dudley and Philipp (1983) proved similar results pertaining to the Gaussian case and showed that the theory of empirical processes readily fits into the framework of probability on Banach spaces. Quantitative improvements for $\mathbb{R}^d$-valued random vectors having moments of order r with $2 < r < 4$ are due to Einmahl (1984).

For triangular arrays of Banach space valued r.v.'s deAcosta (1982) proved a corresponding general theorem: If the row sums converge to a(n infinitely divisible) limit law then an IPIP holds. This improves on a theorem of Skorohod (1957). For his proof deAcosta uses the Kantorovich-Rubinstein theorem. Subsequently, Dabrowski, Dehling and Philipp (1984) gave a simpler proof of deAcosta's theorem and observed that in this case convergence in probability can even be replaced by almost sure convergence (and hence is equivalent to it). See also Csörgő and Hall (1982).

4.1.1.2. ASIP

For real-valued random variables having finite second moments Strassen (1964) proved (1.6) using the Skorohod embedding theorem. Major (1976a) proved that this result is best possible (see also Section 1). Strassen's theorem was extended to $\mathbb{R}^d$-valued random variables by Philipp (1979) who used an early version of Theorem 3.4 above with $\rho_k \to \infty$ and $\Sigma\, \sigma_k < \infty$. He also gave for B-valued random variables necessary and sufficient conditions for the ASIP to hold. For similar results for non-measurable random elements with applications to empirical processes see Dudley and Philipp (1983). Here also a version of Theorem 3.4 is used to construct the limit process.

For random variables having moments of order r with $2 < r \leq 3$ relation (1.6) with an error term $o(n^{1/r})$ is due to Major (1976b). For $r > 3$ this result was proved by Komlós, Major and Tusnády (1975). They also proved (1.6) with an error term $O(\log n)$ if ξ_1 has a moment generating function. Their proofs use the conditional quantile transform, a substantial refinement of the method described in Section 3.4. In the same papers the authors proved that these error terms are best possible. For $\mathbb{R}^d$-valued random variables the conditional quantile transform method and the results pertaining to $r > 2$ have been extended by Berger (1982). In the range $2 < r < 4$ Einmahl (1984) has given new proofs and for $r \sim 2$ he has given refinements of these results. He uses Theorem 3.4 with $\rho_k \to \infty$ and $\Sigma\, \sigma_k < \infty$, but instead of applying Yurinskii's theorem for the estimates of the Prohorov distance (which was good enough to successfully handle the case $r = 2$) Einmahl develops the estimates (3.23) directly.

For real-valued random variables in the domain of normal attraction to a stable law the first ASIP has been proved by Stout (1979). Stout's results have subsequently been generalized somewhat by Berkes, Dabrowski, Dehling and Philipp (1985) who extended them to $\mathbb{R}^d$-valued random variables. For real-valued random variables in the domain of attraction to a Gaussian law (but of course with $E\xi_1^2 = \infty$) the ASIP is due to Mijnher (1980) and for the same but in the domain of attraction to a stable law of index $0 < \alpha < 2$ ASIP's have been proved independently by Fisher (1984) and Mijnher (1983).

We return to the case of finite second moments. Major (1979) showed that it is possible to obtain (1.6) with an error term $o(n^{\frac{1}{2}})$ (instead of $o((nLLn)^{\frac{1}{2}})$ if the approximating sequence $\{\eta_j, j \geq 1\}$ is independent but not IID.

4.1.2. Independent random variables

We drop the assumption of identical distribution.

4.1.2.1. IPID

This is due to Prohorov (1956) under the assumption of the Lindeberg condition.

4.1.2.2. ASIP

In this case and with $f(t) = o(1)$ as $t \to \infty$ Theorem 3.2 above is due to Major (1977). For Hilbert space valued random variables an ASIP was given in Morrow and Philipp (1982).

4.1.3. Martingales

4.1.3.1. IPID

There is a large literature on the subject generalizing Prohorov's (1956) result (see Section 4.1.2.1). For a survey of the literature up to about 1980, see Hall and Heyde (1980, pp. 97-100). For later results see Gaenssler and Haeusler (1985).

4.1.3.2. ASIP

The first result is due to Strassen (1967). The analogue of (1.6) for stationary ergodic martingale difference sequences with finite second moments was proved by A. K. Basu (1973) and by Heyde and Scott (1973). Many authors use V_n as norming sequence (Jain, Jogdeo and Stout (1975)). For more general norming sequences see Hall and Heyde (1980, pp. 117-118). These and several other authors state their results in the form of a FLIL. But in most of the proofs of these FLIL an ASIP is established (see e.g. Hall and Heyde (1980, p. 120, (4.45)). For the ASIP under Kolmogorov's condition see Theorem 3.2 above (Philipp and Stout (1985)). This generalizes

Major's (1977) result (see Section 4.1.2.2). All these results are established by means of the Skorohod embedding theorem (Section 3.2). The newest results for reverse martingales and martingale tails are due to Scott and Huggins (1983) and Huggins (1985a,b). To obtain these results these authors developed the method described in Section 3.3.

Under slightly more restrictive conditions than Strassen's (1967) Morrow and Philipp (1982) obtained an ASIP for $\mathbb{R}^d$-valued martingales and for Hilbert space valued martingales.

4.1.4. Mixing r.v.s.

4.1.4.1. IPID

For an extensive survey see Peligrad (1985b).

4.1.4.1.1. ϕ-mixing.

Let $\sigma_n^2 := ES_n^2 \to \infty$. Ibragimov (1975) proved the IPID for strictly stationary sequences $\{\xi_j, j \geq 1\}$ with $E\xi_1^2 < \infty$, $\Sigma\, \phi^{\frac{1}{2}}(2^n) < \infty$ and $f(0) \neq 0$. (Here f denotes the spectral density.) For sequences with $E|\xi_1|^{2+\delta} < \infty$ for some $\delta > 0$ he proved the IPID mentioned in 4.1.4.1.2 below. The best result in the present category is due to Peligrad (1985a): If $\{\xi_j, j \geq 1\}$ is strictly stationary, $E\xi_1^2 < \infty$, $\phi(n) \to 0$ (no rate required) and

$$\lim_{n \to \infty} n\sigma_n^{-2} E(\xi_1^2\, 1\{|\xi_1| > \varepsilon\sigma_n\}) = 0 \tag{4.1}$$

then IPID holds (provided that $n^{-\frac{1}{2}}$ is replaced by σ_n^{-1} in (1.1)). It is still an unsolved problem whether or not the CLT holds without the assumption (4.1). But Herrndorf (1983) showed among other things that if $\liminf n^{-1}\sigma_n^2 = 0$ then IPID cannot hold (see also Peligrad (1985a)).

An IPID for triangular arrays of Banach space valued r.v.s. has been established by Eberlein (1979).

4.1.4.1.2. ρ-mixing

Ibragimov (1975) has an IPID if $E|\xi_1|^{2+\delta} < \infty$ for some $\delta > 0$ and $\rho(n) \to 0$ (and $\sigma_n^2 \to \infty$). See also Hall and Heyde (1980), Theorem 5.6, p. 148.

4.1.4.1.3. Strong mixing

Let $\sigma_n^2 \to \infty$. Oodaira and Yoshihara (1972) showed IPID for strictly stationary sequences satisfying $E|\xi_1|^{2+\delta} < \infty$ and $\Sigma\, \alpha(n)^{\delta/(2+\delta)}$ for some $\delta > 0$.

4.1.4.1.4. Mixingales

McLeish (1975) proved several results for them. Both McLeish's as well as the Oodaira-Yoshikora results are contained in a theorem of Herrndorf (1984) as a special case. For the non-stationary case see McLeish (1977).

4.1.4.2. IPIP

4.1.4.2.1. ϕ-mixing

Theorem 4 of Philipp (1980) has been withdrawn in a recent correction note. It deals with stationary ϕ-mixing sequences in the domain of attraction to a stable law. The statement of the theorem is false for $\alpha < 2$, as was shown recently by Samur (1985). Under an additional hypothesis the Gaussian case can be saved (Dabrowski (1982a) and Samur (1985)).

Triangular arrays of B-valued ϕ-mixing random variables with stationary row sums have been treated by Samur (1984).

4.1.4.3. ASIP

Research has been carried out in two opposing directions. The first direction aims at (1.6) with an error term $O((t/LLt)^{\frac{1}{2}})$ under the assumption of logarithmic mixing rates. This error term is sharp enough to imply upper and lower class results of the form (1.8). The second direction aims at (1.6) with an error term $O(t^{\frac{1}{2}-\lambda})$ with $\lambda > 0$ as large as possible. As a rule a polynomial mixing rate suffices in the proofs of such results.

4.1.4.3.1. Logarithmic mixing rates

I thank R. Bradley for contributing this section. I only rearranged the material to preserve uniformity of exposition with the previous sections. Throughout this section $\{\xi_j, j \geq 1\}$ is a strictly stationary sequence of $\mathbb{R}$-valued r.v.s. with $E\xi_1 = 0$, $E|\xi_1|^{2+\delta} < 0$ for some $\delta > 0$ and Var $S_n \to \infty$. The conclusion is: There exists σ^2, $0 < \sigma^2 < \infty$ such that n^{-1}Var $S_n \to \sigma^2$. WLOG we assume $\sigma^2 = 1$. Then (1.6) holds with an error term $o((t/LLt)^{\frac{1}{2}})$. Berkes and Philipp (1979) proved ASIP assuming ϕ-mixing with a logarithmic mixing rate. Dabrowski (1982) refined their arguments in order to obtain essentially the same result (with a slightly larger error term) under ϕ-mixing with an even slower logarithmic rate. By modifying Dabrowski's argument, Bradley (1983) obtained a similar result under strong mixing, with a slightly faster mixing rate and a stringent moment condition. The details are as follows.

4.1.4.3.1.1. ϕ-mixing

Dabrowski (1982) with trivial modifications:

$$\phi(n) = o((Ln)^{-\tau}), \ \tau > 1 + 2/\delta .$$

4.1.4.3.1.2. Absolute regularity

Bradley (1985):

$$\beta(n) = o((Ln)^{-\tau}), \ \tau > 1 + 2/\delta,$$

$$\sup_{n \geq 1} E|S_n|^{2+\delta}/n^{1+\frac{1}{2}\delta} < \infty .$$

4.1.4.3.1.3. ρ-mixing

Bradley (1985):

$$\rho(n) = o((Ln)^{-\tau}), \ \tau > 1 + 3/\delta, \ 0 < \delta \leq 1 .$$

4.1.4.3.1.4. Strong mixing

Bradley (1983):

$$\alpha(n) = o((Ln)^{-\tau}), \ \tau > 1 + 3/\delta,$$

$$\sup_{n \geq 1} E|S_n|^{2+\delta}/n^{1+\frac{1}{2}\delta} < \infty .$$

Bradley (1985):

$$\alpha(n) = o((Ln)^{-\tau}), \ \tau > 1 + 3/\delta, \ \rho^* := \lim_{n \to \infty} \rho(n)$$

satisfies a condition of Peligrad (1985).

4.1.4.3.2. Polynomial mixing rates

We make the same assumptions as under 4.1.4.3.1. Then (1.6) holds with a specified error term.

4.1.4.3.2.1. ϕ-mixing

Heyde and Scott (1973): $\Sigma\, \phi(n)^{(1+\delta)/(2+\delta)} < \infty$ for some $\delta \geq 0$ implies (1.6). This appears to be the only ASIP for mixing sequences applying to the case $\delta = 0$.

Qiman and Chuanrong (1985): $\phi(n) \ll n^{-\alpha}$, $\alpha > 1$, $0 < \delta \leq 2$ yields (1.6) with an error term $O(n^{1/(2+\delta)}(Ln)^p)$ for some $p > 1$. Their proof is based on Philipp and Stout (1975) using Skorohod embedding.

Dehling (1983): $\phi(n) = O(n^{-\tau})$, $\tau > 2$, ξ_j B-valued and approximable by finite-dimensional r.v.s with a certain rate implies (1.6) with error term $O(n^{\frac{1}{2}-\lambda})$, $\lambda > 0$.

4.1.4.3.2.2. Absolute regularity

Dehling and Philipp (1982): $\beta(n) = O(n^{-\tau})$, $\tau > 1 + 2/\delta$, $0 < \delta \leq \frac{1}{5}$, ξ_j B-valued and approximable by finite-dimensional r.v.s with a certain rate implies (1.6).

Dehling (1983): $\beta(n) = O(n^{-\tau})$, $\tau > 1 + 2/\delta$, $0 < \delta \leq 2/3$, ξ_j B-valued and approximable by finite-dimensional r.v.s with a faster rate implies (1.6) with error term $O(n^{\frac{1}{2}-\lambda})$, $\lambda > 0$.

4.1.4.3.2.3. Strong mixing

Kuelbs and Philipp (1980): $\alpha(n) = O(n^{-\tau})$, $\tau > 1 + 2/\delta$, $0 < \delta \leq 1$, ξ_j $\mathbb{R}^d$-valued implies (1.6) with an error term $O(n^{\frac{1}{2}-\lambda})$, $\lambda > 0$. The result is best possible in the sense that there are examples (Herrndorf (1983b)) of strongly mixing sequences with rate $\alpha(n) = O(n^{-\tau})$, $\tau = 1 + 2/\delta$ which violate the CLT.

Dehling and Philipp (1982): $\alpha(n) = O(n^{-\tau})$, $\tau > 1 + 2/\delta$, $0 < \delta \leq 1$, ξ_j H-valued implies (1.6).

Dehling (1983): $\alpha(n) = O(n^{-\tau})$, $\tau > 2(1 + 2/\delta)$, ξ_j B-valued and approximable by finite-dimensional r.v.s with a certain rate implies (1.6) with an error term $o((nLLn)^{\frac{1}{2}})$.

This result was extended to non-measurable ξ_j's assuming values in B, not necessarily separable and the error term improved by Philipp (1984a). He applies this to empirical processes (see Section 4.1.10.3.2.3 below).

4.1.5. Dependence structures of mixing type

4.1.5.1. IPID

Dabrowski (1983a,b) proves IPID's for stationary sequences in the domain of attraction to a stable law and satisfying a distributional mixing condition in the sense of Leadbetter and other mixing conditions common in the theory of extremes. Dabrowski's results partially generalize early results of Skorohod (1957) for the independent case.

4.1.5.2. ASIP

Philipp and Stout (1975), Sections 7, 8, 11 proved ASIP's with error term $O(n^{\frac{1}{2}-\lambda})$, $\lambda > 0$ for functions of processes satisfying a retarded strong mixing condition, for retarded asymptotic martingales and for non-stationary strongly mixing sequences. Hofbauer and Keller (1982) prove an ASIP for sums based on piecewise monotonic functions.

Eberlein (1983) proved an ASIP with error term $O(n^{\frac{1}{2}-\lambda})$ for stationary bounded sequences which are strictly very weak Bernoulli. An ASIP with error term $o((nLLn)^{\frac{1}{2}})$ under very general conditions is given in Hall and Heyde (1980), Corollary 5.4, p. 145. An ASIP for flows under a function was proved by Denker and Philipp (1984).

4.1.6. Lacunary sequences

We consider first sums of the form $\Sigma_{k \leq n} a_k \cos 2\pi n_k \omega$, $0 \leq \omega \leq 1$. Put $A_n^2 := \frac{1}{2} \Sigma_{k \leq n} a_k^2$ and define

$$S(t) := \Sigma_{k \leq n} a_k \cos 2\pi n_k \omega \quad \text{if} \quad A_n^2 \leq t < A_{n+1}^2 .$$

4.1.6.1. IPID

Billingsley (1967): $a_n = o(A_n)$ and (2.13) imply $\frac{S(\cdot)}{\sqrt{\cdot}} \xrightarrow{\mathcal{L}} W$, in $D[0,1]$ with the Skorohod topology. This was improved by Takahashi (1980) to include sequences satisfying (2.14) with a best possible condition on $\{a_k, k \geq 1\}$.

4.1.6.2. ASIP

Berkes (1975b): $a_k = 1$, $k \geq 1$ and (2.14) with $\phi(k) \ll k^{\frac{1}{2}-\epsilon}$ for some $\epsilon > 0$ imply (1.6) with an error term $O(t^{\frac{1}{2}-\lambda})$, $\lambda > 0$.

Philipp and Stout (1975): $a_k \ll A_k^{1-\delta}$ for some $\delta > 0$ and (2.13) imply $S(t) - B(t) \ll t^{\frac{1}{2}-\lambda}$ a.s. with $\lambda > 0$.

Takahashi (1979) extends both of these results relaxing the condition on a_k but obtaining a weaker error term.

Qiman (1985) improves the first two results in a different direction.

There are easy examples showing that the CLT cannot hold for sums $\Sigma_{k \leq n} f(n_k \omega)$ for all f having period 2π. Berkes (1976) gives sufficient conditions on f such that (1.6) holds with an error term $O(n^{\frac{1}{2}-\lambda})$, $\lambda > 0$. Berkes and Philipp (1979) prove (1.6) under similar conditions. Berkes (1975a) proves an ASIP for lacunary Walsh series.

4.1.7. Gaussian sequences (Linearly growing variances)

Let $\{\xi_j, j \geq 1\}$ be a Gaussian sequence with $E\xi_j = 0$, $j \geq 1$ and such that uniformly in $m \geq 0$

$$E(\Sigma_{j=m+1}^{m+n} \xi_j)^2 = n + O(n^{1-\delta}), \quad \delta > 0$$

and

$$E\xi_m \xi_{m+n} \ll n^{-1-\epsilon}, \quad \epsilon \geq 0 .$$

4.1.7.1. ASIP

Philipp and Stout (1975): $\epsilon = 1$ (actually $\epsilon = 7/8$ did suffice) implies (1.6) with an error term $O(n^{\frac{1}{2}-\lambda})$, $\lambda > 0$.

Morrow (1980) improves this result to $\epsilon > 0$. This result is best possible in the sense that for $\epsilon = 0$ examples can be constructed violating the CLT. Moreover, Morrow (1980) obtains the same result but with n^{δ}, n^{ϵ} and n^{λ} above replaced by logarithmic factors.

Quiman (1985b): $\epsilon > 1$ implies (1.6) with an error term $O((Lt)^{\frac{1}{2}})$.

The case of variances growing non-linearly falls into the domain of long range dependence. For a recent survey of this topic, see Taqqu and Czado (1985).

4.1.8. Multipicative systems

Berkes (1973b) proves a FCLT and a FLIL for multiplicative systems using a very general theorem, due to him (Berkes (1973a)). The problem of proving an ASIP in the form (1.6) with an error term not exceeding $o((tLLt)^{\frac{1}{2}})$ is still open. At some point in time it was thought that Theorem 5 of Berkes and Philipp (1979) might be helpful in the construction of the limiting Brownian motion. The same comment applies to associated r.v.'s to be considered next.

4.1.9. Associated r.v.s.

Newman and Wright (1981) showed IPID for $\{\xi_j, j \geq 1\}$ strictly stationary with $\Sigma_{j \geq 1} E\xi_1\xi_j < \infty$. Under a more restrictive condition on the higher moments and covariances Dabrowski (1984) proved an FLIL. The problem of ASIP is still open. Recently Burton, Dabrowski and Dehling (1985) proved IPID for associated $\mathbb{R}^d$-valued and Hilbert space valued r.v.s.

4.1.10. Empirical processes

A new proof of Kiefer's (1972) theorem (see (1.9) above)) was given by M. Csörgő and Révész (1974). The error term was improved to $O(\log^2 n)$ by Komlós, Major and Tusnády (1975). Since then many papers have been written on this subject. Fortunately, there are four excellent surveys and books (Dudley (1984), Gänssler (1984), Gänssler and Stute (1979) and Pollard (1984)) on the subject of empirical processes for independent r.v.'s and their applications. The survey by Gänssler (1984) is based on the Dudley theory of weak convergence, which was developed in a series of papers culminating in his paper on the CLT (Dudley (1978)). In this approach it is essential to make several measurability assumptions (such as P-EM class) just for the formulations of the theorems to make sense. Some of these assumptions become redundant if the limit theorems are formulated as IPIP's or ASIP's (see Dudley and Philipp (1983)). In this approach, which is also the basis of Dudley's (1984) authoritative survey, empirical processes are considered as sums of random elements, not necessarily measurable with values in a Banach space not necessarily separable. The notions of independence and strong mixing have been defined in Dudley and Philipp (1983) and Philipp (1984) respectively.

4.1.10.1. IID

4.1.10.1.1. IPIP and L^p-invariance principles

Dudley and Philipp (1983), Giné and Zinn (1984).

4.1.10.1.2. ASIP

Dudley and Philipp (1983), see also the remark by Philipp (1984) concerning Giné and Zinn (1984).

4.1.10.2. Independent random variables

Alexander (1984a,b,c) proves several CLT's and LIL's.

4.1.10.3. Mixing random variables

4.1.10.3.1. IPID

4.1.10.3.1.1. ϕ-mixing

The first results are due to Billingsley (1968, Section 22) under a polynomial mixing rate. Several authors have improved upon Billingsley's results by somewhat relaxing the mixing rate (see Yokoyama (1980) and the literature cited there).

4.1.10.3.1.2. Strong mixing

The first result, due to Deo (1973), was generalized in several directions by Mehra and M. S. Rao (1975).

4.1.10.3.2. ASIP

These are all proved by using Theorem 3.3, apparently the only method available to deal with the weakly dependent case. For an outline of the proof see Berkes and Philipp (1977). In what follows let $\{x_j, j \geq 1\}$ be a strictly stationary sequence of $\mathbb{R}^q$-valued r.v.'s and common distribution function F.

4.1.10.3.2.1. ϕ-mixing

Conjecture: $q \geq 1$, F arbitrary and a logarithmic mixing rate implies (1.9) with an error term $O(n^{\frac{1}{2}}(Ln)^{-\lambda})$, $\lambda > 0$.

4.1.10.3.2.2. Absolute regularity

Dhompongsa (1984): $q \geq 1$, F continuous and $\beta(n) \ll n^{-\tau}$, $\tau > 3 + q$ implies (1.9) with and error term $O(n^{\frac{1}{2}-\lambda})$, $\lambda > 0$.

4.1.10.3.2.3. Strong mixing

The first results are due to Berkes and Philipp (1977). This was later extended to $q > 1$ and F continuous by Philipp and Pinzur (1980) which in turn was improved by Dhompongsa (1984): $q \geq 1$, F continuous and $\alpha(n) \ll n^{-\tau}$, $\tau > 3 + q$ implies (1.9) with error term $O(n^{\frac{1}{2}}(Ln)^{-\lambda})$, $\lambda > 0$. (Remark: As Dhompongsa uses the method of Philipp and Pinzur (1980) to reduce the case of general F to the case that F has uniformly distributed martingales, F has to be continuous.) The continuity of F can be dropped as was shown by Denker (1982), Bradley (1983a) and Philipp (1984) using totally different methods: $q \geq 1$ and $\alpha(n) \ll n^{-\tau}$, $\tau > 4 + 2q$ implies (1.9) with an error term $O(n^{\frac{1}{2}}(Ln)^{-\lambda})$, $\lambda > 0$. The result can be extended to empirical processes over classes of sets with metric entropy with inclusion not growing too quickly (see Philipp (1984)).

4.1.10.4 Lacunary sequences

Unless n_k grows very fast, invariance principles for the empirical process of lacunary sequences seem to be impossible to obtain. Only a few facts on empirical processes for lacunary sequences are known.

4.1.10.4.1. Bounded LIL

If $\{n_k, k \geq 1\}$ satisfies (2.13), then the sequence $\{\langle n_k\omega\rangle, k \geq 1\}$ satisfies the bounded Chung Smirnov LIL, i.e. for almost all $\omega \in [0,1]$

$$C_1 \leq \limsup_{n \to \infty} (nLLn)^{-\frac{1}{2}} \sup_{0 \leq s \leq 1} \left|\Sigma_{k \leq n} (1\{\langle n_k\omega\rangle \leq s\} - s)\right| \leq C_2/\log q \qquad (4.2)$$

where C_1 and C_2 are absolute constants (Philipp (1977)).

It is still unknown whether (2.14) with $\phi(k) \geq k^{\epsilon}$, $\epsilon > 0$ implies such a result, or whether such a result holds for $\{n_k, k \geq 1\}$ equal to the semigroup generated by a fixed number of primes and arranged according to size.

4.1.10.4.2. Uniform LIL over classes of functions

Let BV be the class of all real-valued functions of period 1, with variation not exceeding 1 and $\int_0^1 f(x)dx = 0$. Relation (4.2) implies at once

$$C_1 \leq \limsup_{n \to \infty} (nLLn)^{-\frac{1}{2}} \sup_{f \in BV} \left|\Sigma_{k \leq n} f(n_k\omega)\right| \leq C_2/\log q \qquad (4.3)$$

for almost all $\omega \in [0,1]$.

Kaufman and Philipp (1978) proved that (4.3) continues to hold if BV is replaced by Λ_α, $\alpha < \frac{1}{2}$. Here Λ_α is the class of all real-valued functions with period 1, belonging to Lip α and satisfying $\int_0^1 f(x)dx = 0$. This last result was generalized by Dhompongsa (1981) to sequences $\{n_k, k \geq 1\}$ satisfying (2.14) with $\phi(k) \geq k^\epsilon$, $\epsilon > 0$.

4.1.10.4.3. ASIP

Dhompongsa (1981): Let $n_{k+1}/n_k \geq (\log k)^{\psi(k)}$ with $\psi(k) \uparrow \infty$, $\psi(k) \ll k^\beta$, $0 < \beta < 1$, and $\psi(k)$ concave downward. Then $x_k(\omega) = \langle n_k \omega\rangle$ satisfies (1.9) with an error term $O((\psi(Ln)^{-\lambda}(nLLn)^{\frac{1}{2}})$ for any $0 < \lambda < 1/12$.

4.1.10.5. U-statistic structures

These will be considered in 4.2.2.3 below.

4.1.11. Empirical characteristic functions and random Fourier series

Let $\{\xi_j, j \geq 1\}$ be i.i.d. r.v. with characteristic function $c(u) = E\exp(iu\xi_1)$. The Fourier transform of the empirical distribution function is called the empirical characteristic function. The empirical characteristic process is defined as

$$C(u,n) = \Sigma_{j \leq n}(\exp(iu\xi_j) - c(u)), \quad |u| \leq 1 .$$

S. Csörgő (1981) showed an ASIP for C(u,n). This was later extended by him and by Marcus and Philipp (1982) to $\mathbb{R}^d$-valued sequences $\{\xi_j, j \geq 1\}$. In the latter paper this is derived from a general ASIP for B-valued r.v. which also yields ASIP for random Fourier series.

4.2. Sums of random elements indexed by lattice points

4.2.1. IID and mixing families

Goldie and Morrow (1985) have prepared a survey on limit theorems for random fields for the present volume. For the sake of completeness I will say a few things about pathwise invariance principles. Let $\{\xi_j, j \in \mathbb{N}^d\}$ be a family of independent identically formed random elements with values in a Banach space B, let μ be a measure on $[0,\infty)^d$ and let $\mathcal{A}$ be a class of measurable sets $A \subset [0,1]^d$. For $j \in \mathbb{N}^d$ denote by C_j the unit cube with "NE-vertex" in j. The problem is to approximate the sums $\Sigma_j \xi_j \mu(nA \cap C_j)$ by sums $\Sigma_j \eta_j \mu(nA \cap C_j)$ uniformly over all sets $A \in \mathcal{A}$. Here $\{\eta_j, j \in \mathbb{N}^d\}$ is a family of i.i.d. Gaussian r.v.'s with values in a subspace of B. The purpose of introducing this generality is again to include empirical processes. A typical consequence of the results obtained is the following (Morrow and Philipp (1984)). Let $\mathcal{C}$ be the collection of all convex sets contained in the unit square $[0,1]^2$. Let $\{x_j, j \in \mathbb{N}^2\}$ be i.i.d. r.v.'s uniformly distributed over $[0,1]^2$ and defined on a rich enough probability space. Then there exists a family $\{\eta_j, j \in \mathbb{N}^2\}$ of i.i.d. Gaussian processes defined on the same space, indexed by $C \in \mathcal{C}$ with $E\eta_1(C) = 0$ for all $C \in \mathcal{C}$ and $E\eta_1(C)\eta_1(D) = \lambda(C \cap D) - \lambda(C)\lambda(D)$ for all $C,D \in \mathcal{C}$ such that

$$\sup_{C,D \in \mathcal{C}} \left|\Sigma_{j \in nC}(1_D(x_j) - \lambda(D) - \eta_j(D))\right| \ll n(Ln)^{-\sigma} \text{ a.s.}$$

Here λ denotes the Lebesgue-measure and $\sigma > 0$ is a constant. There are now several papers in the literature dealing with approximation theorems for weakly dependent families. I refer the reader to the survey by Goldie and Morrow (1985) for the details.

For a very general CLT providing a link between this section and empirical processes (Section 4.1.10.2), see Alexander (1984c).

4.2.2. U-statistics

Dehling (1985) has prepared a survey on limit theorems for U-statistics for the present volume. Again for the sake of completeness I shall make a few remarks about the basic facts.

For simplicity let $\{\xi_j, j \geq 1\}$ be a sequence of i.i.d. r.v. with common distribution function F. Let $h: \mathbb{R}^2 \to \mathbb{R}$ be a measurable function. The U-statistic of the sequence with kernel h is defined as

$$U_n(h) := \Sigma_{1 \leq i < j \leq n} h(\xi_i, \xi_j) . \tag{4.4}$$

We assume that

$$\iint h^2(x,y)dF(x)dF(y) + \int h^2(x,x)dF(x) < \infty .$$

It is easy to see that if $\text{Var} \int h(\xi_1,y)dF(y) > 0$ or $\text{Var} \int h(x,\xi_1)dF(x) > 0$ then $\frac{1}{n} U_n(h)$ equals a sum of i.i.d. r.v. plus a remainder which is $O(\log n)$ a.s. This case is called the non-degenerate case and of course most invariance principles are direct applications of results of Section 4.1 above. Many tests in statistics (such as the χ^2-test of goodness of fit), however, lead to degenerate kernels which thus seems to be a misnomer since the only limit law for non-degenerate U-statistics is the Gaussian law. In the degenerate case $\{U_n(h), \sigma(\xi_1,\ldots,\xi_n), n \geq 1\}$ is a martingale and the canonical process (corresponding to the Brownian motion process in the case of sums of r.v. considered in Section 4.1) is a doubly stochastic integral with respect to Kiefer processes. I refer the reader to the original papers and to the survey by Dehling (1985) for the details.

4.2.2.1. Nondegenerate kernels

The classical result is Hoeffding's CLT (1948). Since then numerous papers have been written. I only mention a few.

4.2.2.1.1. IPID

Denker and Keller (1983) treat the case of absolutely regular, ϕ-mixing and ϕ-mixing in both direction sequences $\{\xi_j, j \geq 1\}$.

4.2.2.1.2. ASIP

Sen (1974) treats the case of independent sequences $\{\xi_j, j \geq 1\}$ and Yoshihara (1976) the case of absolutely regular sequences. Denker and Keller (1983) extend these results to sequences mentioned in 4.2.2.1.1.

4.2.2.2. Degenerate kernels

4.2.2.2.1. IPID

See Neuhaus (1977), Hall (1979) and Denker, Grillenberger and Keller (1985).

4.2.2.2.2. IPIP and ASIP

See Dehling, Denker and Philipp (1984, 1985). In the latter paper Hilbert space valued martingales play an important role.

4.2.2.3. Empirical processes

4.2.2.3.1. IPID

See Serfling (1984) and Silverman (1983).

4.2.2.3.2. ASIP

The first result is due to S. Csörgő, Horvath and Serfling (1983). Later Dehling, Denker and Philipp (1985) remove the smoothness assumptions on h by treating the empirical process as a U-statistic with a $D(-\infty,\infty)$ valued kernel.

4.3. Extremal processes

Let $\{\xi_j, j \geq 1\}$ be i.i.d. r.v. with common distribution function F and set $M_n := \max_{j \leq n} \xi_j$. Suppose there exist constants $a_n > 0$, b_n such that

$$P(M_n \leq a_n x + b_n) = F^n(a_n x + b_n) \to G(x), \quad x \in \mathbb{R} \qquad (4.5)$$

for some non-degenerate function G. By a well-known theorem of Gnedenko G must be one of three canonical types. Let
$Y_n(t) := (M_{[nt]} - b_n)/a_n$.

4.3.1. IPID

Lamperti (1964) showed that (4.5) implies $Y_n \xrightarrow{\mathcal{L}} Y$ where $Y = Y_F$ is a so-called extremal F-process which plays the same role for partial maxima as Brownian motion plays for partial sums. For its definition, for a new proof of Lamperti's theorem, for more results in this general area as well as for an illuminating discussion about the connections with point processes, see Resnick (1975). For a comprehensive survey of the results see Leadbetter, Lindgren and Rootzén (1983).

4.3.2. IPIP and ASIP

Deheuvels (1981, 1983a, 1983b) proves the almost sure approximation in the Lévy metric. Dabrowski (1982) proves IPIP and ASIP in the supremum norm and extends his results to ϕ-mixing sequences $\{\xi_j, j \geq 1\}$.

4.4. Quantile processes

Let $\{x_j, j \geq 1\}$ be a sequence of i.i.d. r.v.'s with uniform distribution over $[0,1]$. For each $n \geq 1$ let $x_{1:n} \leq \dots \leq x_{n:n}$ denote the order statistic of the sample $x_1,\dots,x_n$. The uniform quantile function is defined as

$$U_n(s) = \begin{cases} x_{k:n} & \text{if } (k-1)/n < s \leq k/n \qquad 1 \leq k \leq n \\ 0 & \text{if } s = 0 \end{cases}$$

and the uniform quantile process is defined as $n^{\frac{1}{2}}(U_n(s) - s)$, $0 \leq s \leq 1$. There are two books treating the quantile process: M. Csörgő and Révész (1981) and M. Csörgő (1983). I quote a central result, due to Csörgő and Révész (1975). If the underlying probability space is rich enough then there exists a sequence $\{B_j(\cdot), j \geq 1\}$ of independent standard Brownian bridges such that with probability 1

$$\sup_{0 \leq s \leq 1} |n(U_n(s) - s) - \Sigma_{j \leq n} B_j(s)| \ll (n(Ln)^2 LLn)^{\frac{1}{4}} \quad . \tag{4.6}$$

This result was extended to sequences $\{x_j, j \geq 1\}$ of i.i.d. r.v.'s with common distribution function F sufficiently smooth by Csörgő and Révész (1978).

(4.6) with error term $O(n^{\frac{1}{2}}(Ln)^{-\lambda})$, $\lambda > 0$ continues to hold for all sequences $\{x_j, j \geq 1\}$ such that (1.9) holds with error term $O(n^{\frac{1}{2}}(Ln)^{-\lambda})$, $\lambda > 0$. Hence for instance this result holds for strongly mixing sequences (see Section 4.1.10.3.2.3) and for lacunary sequences (see Section 4.1.10.4.3). This follows from the proof of M. Csörgő and Révész (1981), Theorem 4.5.3, pp. 146-147.

ASIP for the empirical and quantile process based on non-overlapping k-spacings are obtained by Aly, Beirlant and Horvath (1984).

4.5. Miscellaneous structures

This is a portmanteau section for results which do not readily fit into the previous sections.

4.5.1. Non-additive processes

A family $\{S_{m,n}, 0 \leq m < n, n \geq 1\}$ of r.v.'s is called stationary if $\mathcal{L}(S_{mn}) = \mathcal{L}(S_{m+1,n+1})$ for all $0 \leq m < n$.

Example. $S_{mn} := \Sigma_{j=m+1}^{n} \xi_j$ with $\{\xi_j, j \geq 1\}$ stationary defines an additive stationary process.

Wacker (1983) proves ASIP for stationary processes $\{S_{mn}\}$ satisfying a weak dependence relation and such that S_{on} can be approximated in L^1 by sums $\Sigma_{0 \leq j \leq n} \xi_j$ with $\{\xi_j, j \geq 1\}$ stationary. He then applies these results to products of random matrices and to the range of a random walk.

4.5.2. Stochastic approximation processes

4.5.2.1. IPID

There are several papers by Walk (1977, 1978, 1980) treating the Robbins-Monroe process.

4.5.2.2. ASIP

Mark (1982) proves a very general ASIP for stochastic approximation processes and applies it to the Robbin-Monroe process on a separable Banach space and for the Kiefer-Wolfowitz process on $\mathbb{R}^d$.

4.5.3. Renewal processes

Applications of the ASIP for stationary sequences to renewal processes are given by Horvath (1984).

4.5.4. Markov chains and processes

As is well-known, Markov chains with stationary transition probabilities satisfying Doeblin's condition are ϕ-mixing with exponentially fast decreasing rate. Sums of functionals of Markov chains that can be approximated by sums of i.i.d. r.v.s have been treated in Philipp and Stout (1975). Recently Bhattacharya (1982) obtained a FCLT and a FLIL for $n^{-\frac{1}{2}} \int_0^{nt} f(X_s)ds$ where $\{X_s, s \geq 0\}$

is a stationary ergodic Markov process and f is in the range of its infinitesimal generator.

4.5.5. The "other LIL"

For FLIL concerning the so-called other LIL and for more results on FLIL, see Bingham (1985).

References

A.P. = Annals of Probability

Z.W. = Zeitschrift für Wahrscheinlichkeitstheorie verw. Geb.

Alexander, K. S. (1984a). Probability inequalities for empirical processes and a law of the iterated logarithm, A.P. 12, 1041-1067.

Alexander, K. S. (1984b). The central limit theorem for empirical processes on Vapnik-Červonenkis classes, preprint.

Alexander, K. S. (1984c). Central limit theorems for stochastic processes under random entropy conditions, preprint.

Aly, E.-E.A.A., Beirlant, J. and Horváth, L. (1984). Strong and weak approximations of k-spacings processes, Z.W. 66, 461-484.

Basu, A. K. (1973). A note on Strassen's law of the law of the iterated logarithm, Proc. AMS 41, 596-601.

Berbee, H. C. P. (1979). Random walks with stationary increments and renewal theory. Amsterdam: Mathematical Centre tract 112.

Berbee, H. C. P. (1984). Convergence rates in the strong law for bounded mixing sequences, preprint.

Berkes, I. (1973a). A functional law of the iterated logarithm for dependent random variables, Z.W. 26, 245-258.

Berkes, I. (1973b). On Strassen's version of the log log law for multiplicative systems, Studia Sci. Math. Hungar. 8, 425-431.

Berkes, I. (1975a). Approximation of lacunary Walsh series with Brownian motion, Studia Sci. Math. Hungar. 9, 111-124.

Berkes, I. (1975b). An almost sure invariance principle for lacunary trigonometric series, Acta Math. Acad. Sci. Hungar. 26, 209-220.

Berkes, I. (1976). On the asymptotic behaviour of $\Sigma f(n_k x)$, I and II, Z.W. 34, 319-345 and Z.W. 34, 347-365.

Berkes, I., Dabrowski, A., Dehling, H. and Philipp, W. (1985). A strong approximation theorem for sums of random vectors in the domain of attraction to a stable law, Acta Math. Acad. Sci. Hungar. (to appear).

Berkes, I. and Philipp, W. (1977). An almost sure invariance principle for the empirical distribution function of mixing random variables, Z.W. 41, 115-137.

Berkes, I. and Philipp, W. (1979a). Approximation theorems for independent and weakly dependent random vectors, A.P. 7, 29-54.

Berkes, I. and Philipp, W. (1979b). An a.s. invariance principle for lacunary series $f(n_k x)$, Acta Math. Acad. Sci. Hungar. 34, 141-155.

Berger, E. (1982a). Fastsichere Approximation von Partialsummen unabhängiger und stationärer ergodischer Folgen von Zufallsvektoren, Diss. Univ. Göttingen.

Berger, E. (1982b) (personal communication).

Bhattacharya, R. N. (1982). On the functional central limit theorem for Markov processes, Z.W. 60, 185-201.

Billingsley, P. (1967) (personal communication).

Billingsley, P. (1968). Convergence of probability measures, Wiley, N.Y.

Bingham, N. H. (1985). Variants on the law of the iterated logarithm, Bull. London Math. Soc.

Blum, J. R., Hanson, D. L. and Koopmans, L. (1963). On the strong law of large numbers for a class of stochastic processes, Z.W. 2, 1-11.

Bradley, R. C. (1983a) (personal communication).

Bradley, R. C. (1983b). Approximation theorems for strongly mixing random variables, Michigan Math. J. 30, 69-81.

Bradley, R. C. (1984). On a very weak Bernoulli condition, Stochastics 13, 61-81.

Bradley, R. C. (1985a) (personal communication).

Bradley, R. C. (1985b). Basic properties of strong mixing conditions, this volume.

Bryc, W. (1982). On the approximation theorem of I. Berkes and W. Philipp, Demonstratio Math. 15, 807-816.

Brown, B. M. (1971). Martingale central limit theorems, Ann. Math. Stat. 42, 59-66.

Burton, R., Dabrowski, A. and Dehling, H. (1985) (personal communication).

Chiang, I. S. and Hsiung, C. A. (1983). Strassen's invariance principle for random subsequences, Z.W. 64, 401-409.

Csaki, E. and Révész, P. (1983). Strong invariance for local time, Z.W. 62, 263-278.

Csörgő, M. (1983). Quantile processes with statistical applications, SIAM, Philadelphia, PA.

Csörgő, M. and Révész, P. (1975a,b). A new method to prove Strassen type laws of invariance principle, I and II. Z.W. 31, 255-260 and Z.W. 31, 261-269.

Csörgő, M. and Révész, P. (1975c). Some notes on the empirical distribution function and the quantile process, Coll. Math. Soc. J. Bolyai 11, Limit theorems in probability theory (P. Révész, Ed.).

Csörgő, M. and Révész, P. (1978). Strong approximations of the quantile process, Ann. Statist. 6, 882-894.

Csörgő. M. and Révész, P. (1981). Strong approximations in probability and statistics, Academic Press, New York, San Francisco, London.

Csörgő, S. (1981). Limit behaviour of the empirical characteristic function, A.P. 9, 130-144.

Csörgő, S. and Hall, P. (1982). Upper and lower classes for triangular arrays, Z.W. 61, 207-222.

Csörgő, S., Horváth, L. and Serfling, R. (1983). An approximation for the empirical process of U-statistic structure, Technical Report, Johns Hopkins Univ.

Dabrowski, A. R. (1982a). Invariance principles for random processes generated by extrema and partial sums of random variables, Ph.D. thesis, Univ. of Illinois, Urbana.

Dabrowski, A. R. (1982b). A note on a theorem of Berkes and Philipp for dependent sequences, Statist. Prob. Letters 1, 53-55.

Dabrowski, A. R. (1984). A functional law of the iterated logarithm for associated sequences, preprint.

Dabrowski, A., Dehling, H. and Philipp, W. (1984). An almost sure invariance principle for triangular arrays of Banach space valued random variables, Z.W. 65, 483-491.

Davydov, Yu. A. (1970). The invariance principle for stationary processes. Theor. Prob. Appl. 15, 487-498.

deAcosta, A. (1982). An invariance principle in probability for triangular arrays of B-valued random vectors, A.P. 10, 346-373.

Deheuvels, P. (1981). The strong approximation of extremal processes, Z.W. 58, 1-6.

Deheuvels, P. (1983a). The strong approximation of extremal processes II, Z.W. 62, 7-15.

Deheuvels, P. (1983b). The complete characterization of the upper and lower class of the record and inter-record times of an i.i.d. sequence, Z.W. 62, 1-6.

Dehling, H. (1983a). A note on a theorem of Berkes and Philipp, Z.W. 62, 39-43.

Dehling, H. (1983b). Limit theorems for sums of weakly dependent Banach space valued random variables, Z.W. 63, 393-432.

Dehling, H. (1985). Almost sure approximations for U-statistics, this volume.

Dehling, H., Denker, M. and Philipp, W. (1984a). Versik processes and very weak Bernoulli processes with summable rates are independent, Proc. AMS, 61, 618-624.

Dehling, H., Denker, M. and Philipp, W. (1984b). Invariance principles for von Mises and U-statistics, Z.W. 67, 139-167.

Dehling, H., Denker, M. and Philipp, W. (1985a). A bounded law of the iterated logarithm for Hilbert space valued martingales and its application to U-statistics, Z.W.

Dehling, H., Denker, M. and Philipp, W. (1985b). The almost sure invariance principle for the empirical process of U-statistic structure, preprint.

Dehling, H. and Philipp, W. (1982). Almost sure invariance principles for weakly dependent vector-valued random variables, A.P. 10, 689-701.

Denker, M. (1982) (personal communication) (An ASIP for emp. d.f. of strongly mixing vectors).

Denker, M. and Keller, G. (1983). U-statistics and v. Mises' statistics for weakly dependent processes, Z.W. 64, 505-522.

Denker, M. and Philipp, W. (1984). Approximation by Brownian motion for Gibbs measures and flows under a function, Ergod. Th. & Dynam. Syst. 4, 541-552.

Deo, Ch. M. (1973). A note on empirical processes of strong mixing sequences, A.P. 1, 870-875.

Dhompongsa, S. (1981). Limit theorems for weakly dependent random vectors, Ph.D. thesis, Univ. of Illinois, Urbana.

Dhompongsa, S. (1984). A note on the almost sure approximation of empirical processes of weakly dependent random vectors, Yokohama Math. J. 32, 113-121.

Donsker, M. (1951). An invariance principle for certain probability limit theorems, Memoirs AMS, 6.

Donsker, M. (1952). Justification and extension of Doob's heuristic approach to the Kolmogorov-Smirnov theorems, Ann. Math. Stat. 23, 277-283.

Dudley, R. M. (1968). Distances of probability measures and random variables, Ann. Math. Stat. 39, 1563-1572.

Dudley, R. M. (1978). Central limit theorems for empirical measures, A.P. 6, 899-929, Correction A.P. 7, 909-911 (1979).

Dudley, R. M. (1984). A course on empirical processes, Lecture Notes in Math.

Dudley, R. M. and Philipp, W. (1983). Invariance principles for sums of Banach space valued random elements and empirical processes, Z.W. 62, 509-552.

Eberlein, E. (1979). An invariance principle for lattices of dependent random variables, Z.W. 50, 119-133.

Eberlein, E. (1983). Strong approximation of very weak Bernoulli processes, Z.W. 62, 17-37.

Eberlein, E. (1985). On strong invariance principles under dependence assumptions, A.P. to appear.

Einmahl, U. (1984). Starke Approximationen von Partialsummen unabhängiger Zufallsvektoren, Diss. Univ. Köln.

Erdös, P. and Kac, M. (1946). On certain limit theorems of the theory of probability, Bull. AMS, 52, 292-302.

Erdös, P. and Kac, M. (1947). On the number of positive sums of independent random variables, Bull. AMS, 53, 1011-1020.

Fisher, E. (1984). An almost sure invariance principle for random variables in the domain of attraction to a stable law, Z.W. 67, 461-471.

Freedman, D. (1971). Brownian motion and diffusion, Holden-Day, San Francisco.

Gaenssler, P. (1984). Empirical processes, IMS Lecture Notes, Monograph Series 3.

Gaenssler, P. and Haeusler, E. (1985). On martingale central limit theory, this volume.

Gaenssler, P. and Stute, W. (1979). Empirical processes: A survey of results for independent and identically distributed random variables, A.P. 7, 193-243.

Giné, E. and Zinn, J. (1984). Some limit theorems for empirical processes, A.P. 12, 929-989.

Goldie, Ch. M. and Morrow, G. J. (1985). Central limit questions for random fields, this volume.

Hall, P. (1979). On the invariance principle for U-statistics, Stochastic Proc. Appl. 9, 163-174.

Hall, P. and Heyde, C. C. (1980). Martingale limit theory and its applications, Academic Press, New York.

Heath, D. (1977). Interpolation of martingales, A.P. 5, 804-806.

Herrndorf, N. (1983a). The invariance principle for ϕ-mixing sequences, Z.W. 63, 97-109.

Herrndorf, N. (1983b). Strongly mixing sequences not satisfying the central limit theorem, A.P. 11, 809-813.

Herrndorf, N. (1984). A functional central limit theorem for weakly dependent sequences of random variables, A.P. 12, 141-153.

Herrndorf, N. (1985). A functional central limit theorem for strongly mixing sequences of random variables, Z.W.

Heyde, C. C. and Scott, D. J. (1973). Invariance principles for the law of the iterated logarithm for martingales and processes with stationary increments, A.P. 1, 428-436.

Hoeffding, W. (1948). A class of statistics with asymptotically normal distribution, Ann. Math. Stat. 19, 293-325.

Hofbauer, F. and Keller, G. (1982). Ergodic properties of invariant measures for piecewise monotonic transformations, Math. Zeitschrift, 180, 119-140.

Horvath, L. (1984). Strong approximation of extended renewal processes, A.P. 12, 1149-1166.

Huggins, R. M. (1985a). On functional laws of the iterated logarithm, Z.W.

Huggins, R. M. (1985b). A strong invariance principle for reverse martingales, Acta Math. Acad. Sci. Hungar. 46

Ibragimov, I. A. (1962). Some limit theorems for stationary processes, Theor. Prob. Appl. 7, 349-382.

Ibragimov, I. A. (1975). A note on the central limit theorem for dependent random variables, Theor. Prob. Appl. 20, 135-141.

Jain, N. C. (1982). A Donsker-Varadhan type of invariance principle, Z.W. 59, 117-138.

Jain, N. C., Jogdeo, K. and Stout, W. (1975). Upper and lower functions for martingales and mixing processes, A.P. 3, 119-145.

Jain, N. C. and Pruitt, W. E. (1984). An invariance principle for the local time of a recurrent random walk, Z.W. 66, 141-156.

Kaufman, R. and Philipp, W. (1978). A uniform law of the iterated logarithm for classes of functions, A.P. 6, 930-952.

Kiefer, J. (1972). Skorohod embedding of multivariate r.v.'s and the sample df, Z.W. 24, 1-35.

Knight, F. B. (1971). A reduction of continuous square-integrable martingales to Brownian motion, Lecture Notes in Math. 190, 19-31.

Komlós, J., Major, P. and Tusnády, G. (1975, 1976). An approximation of partial sums of independent RV's and the sample DF. I and II, Z.W. 32, 111-131 and Z.W. 34, 33-58.

Kuelbs, J. and Philipp, W. (1980). Almost sure invariance principles for partial sums of mixing B-valued random variables, A.P. 8, 1003-1036.

Kunita, H. and Watanabe, S. (1967). On square integrable martingales, Nagoya Math. J. 30, 209-245.

Lamperti, J. (1964). On extreme order statistics, Ann. Math. Stat. 35, 1726-1737.

Leadbetter, M. R., Lindgren, G. and Rootzén, H. (1983). Extremes and related properties of random sequences and processes, Springer Verlag, New York, Heidelberg, Berlin.

Major, P. (1976a). Approximation of partial sums of i.i.d. r.v.s. when the summands have only two moments, Z.W. 35, 221-229.

Major, P. (1976b). The approximation of partial sums of independent r.v.s. Z.W. 35, 213-220.

Major, P. (1977). A note on Kolmogorov's law of the iterated logarithm, Studia Scient. Math. Hungar. 12, 161-167.

Major, P. (1979). An improvement on Strassen's invariance principle, A.P. 7, 55-61.

Marcus, M. B. and Philipp, W. (1982). Almost sure invariance principles for sums of B-valued random variables with applications to random Fourier series and the empirical characteristic process, Trans. AMS, 269, 67-90.

Mark, G. (1982). Loglog-Invarianzprinzipien für Prozesse der stochastischen Approximation, Mitteilungen Math. Seminar Giessen, 153.

McLeish, D. L. (1975). Invariance principles for dependent random variables, Z.W. 32, 165-178.

McLeish, D. L. (1977). On the invariance principle for non-stationary mixingales, A.P. 5, 616-621.

Mehra, K. L. and Rao, M. S. (1975). Weak convergence of generalized empirical processes relative to d_q under strong mixing, A.P. 3, 979-991.

Mijnher, J. (1980). A strong approximation of partial sums of i.i.d. random variables with infinite variance, Z.W. 52, 1-7.

Mijnher, J. (1983). Strong approximations of partial sums of i.i.d. random variables in the domain of attraction of a symmetric stable distribution, Trans. Ninth Prague Conf. Information Theor., Statist. Decision Fcts, Random Proc. Prague 1982, 83-89.

Millar, P. W. (1968). Martingale integrals, Trans. AMS, 133, 145-166.

Morrow, G. J. (1980). Invariance principles for Gaussian sequences, Z.W. 52, 115-126.

Morrow, G. J. and Philipp. W. (1982). An almost sure invariance principle for Hilbert space valued martingales, Trans. AMS, 273, 231-251.

Morrow, G. J. and Philipp, W. (1984). Invariance principles for partial sum processes and empirical processes indexed by sets, preprint.

Müller, C. (1983). Strassen's law for local time, Z.W. 63, 29-41.

Neuhaus, G. (1977). Functional limit theorems for U-statistics in the degenerate case, J. Multivariate Anal. 7, 424-439.

Newman, C. M. and Wright, A. L. (1981). An invariance principle for certain dependent sequences, A.P. 9, 671-675.

Oodaira, H. and Yoshihara, K. I. (1972). Functional central limit theorems for strictly stationary processes satisfying the strong mixing condition, Kodai Math. Sem. Rep. 24, 259-269.

Peligrad, M. (1985a). An invariance principle for ϕ-mixing sequences, A.P.

Peligrad, M. (1985b). Recent advances in the central limit theorem and its weak invariance principle for mixing sequences of random variables (A survey), this volume.

Philipp, W. (1977). A functional law of the iterated logarithm for empirical distribution functions of weakly dependent random variables, A.P. 5, 319-350.

Philipp, W. (1979). Almost sure invariance principles for sums of B-valued random variables, Probability in Banach Spaces (Proc. Conf. Oberwolfach 1978), Lecture Notes in Math. 709, 171-193, Springer, New York, Heidelberg, Berlin.

Philipp, W. (1980). Weak and L^p-invariance principles for sums of B-valued random variables, A.P. 8, 68-82, Correction, ibid. to appear.

Philipp, W. (1984a). Invariance principles for sums of mixing random elements and the multivariate empirical process, Coll. Math. Soc. J. Bolyai 36. Limit Theorems in Prob. and Statist. Verzprem (Hungary), 1982.

Philipp, W. (1984b). Discussion of a paper by Giné and Zinn, A.P. 12, 993-995.

Philipp, W. and Stout, W. (1975). Almost sure invariance principles for partial sums of weakly dependent random vectors, Memoirs AMS, 161.

Philipp, W. and Stout, W. (1985). Invariance principles for martingales and sums of independent random variables, Math. Zeitschrift.

Philipp, W. and Pinzur, L. (1980). Almost sure approximation theorems for the multivariate empirical process, Z.W. 54, 1-13.

Pollard, D. (1984). Convergence of stochastic processes, Springer Verlag, New York.

Prohorov, Yu. V. (1956). The convergence of random processes and limit theorems in probability theory, Theor. Prob. Appl. 1, 157-214.

Qiman, S. (1985a). Strong approximations of lacunary trigonometric series with weights, preprint.

Qiman, S. (1985b). An almost sure invariance principle for partial sums of Gaussian sequences, preprint.

Qiman, S. and Chuanrong, L. (1985). Strong approximations for partial sums of weakly dependent random sequences, preprint.

Resnick, S. I. (1975). Weak convergence of extremal processes, A.P. 3, 951-960.

Rosenblatt, M. (1956). A central limit theorem and a strong mixing condition, Proc. Natl. Acad. Sci. USA, 42, 43-47.

Samur, J. D. (1984). Convergence of sums of mixing triangular arrays of random vectors with stationary rows, A.P. 12, 390-426.

Samur, J. D. (1985). On the invariance principle for stationary ϕ-mixing triangular arrays with infinitely divisible limits, preprint.

Schwarz, G. (1980). Finitely determined processes--an indiscrete approach, J. Math. Anal. Appl. 76, 146-158.

Scott, D. J. and Huggins, R. M. (1983). On the embedding of processes in Brownian motion and the law of the iterated logarithm for reverse martingales, Bull. Austral. Math. Soc. 27, 443-459.

Sen, P. K. (1974). Almost sure behaviour of U-statistics and v. Mises differentiable statistical functions, Ann. Statist. 2, 387-395.

Serfling, R. (1984). Generalized L-, M- and R-statistics, Ann. Statist. 12, 76-86.

Silverman, B. W. (1983). Convergence of a class of empirical distribution functions of dependent random variables, A.P. 11, 745-751.

Simon, G. and Stout, W. (1978). A weak invariance principle with applications to domains of attraction, A.P. 6, 294-315.

Skorohod, A. V. (1957). Limit theorems for stochastic processes with independent increments, Theor. Prob. Appl. 2, 138-171.

Skorohod, A. V. (1965). Studies in the theory of random processes, Addison-Wesley, Reading, MA.

Stout, W. F. (1979). Almost sure invariance principles when $EX_1^2 = \infty$, Z.W. 44, 23-32.

Strassen, V. (1964). An almost sure invariance principle for the law of the iterated logarithm, Z.W. 3, 211-226.

Strassen, V. (1965). The existence of probability measures with given marginals, Ann. Math. Statist. 36, 423-439.

Strassen, V. (1967). Almost sure behaviour of sums of independent random variables and martingales, Proc. Fifth Berkeley Symp. Math. Stat. Prob. Vol. II, Part 1, 315-343.

Strittmatter, W. (1982). Metriken für stochastische Prozesse und schwache Abhängigkeit, Diplomarbeit, Univ. Freiburg im Breisgau.

Takahashi, S. (1979). Almost sure invariance principles for lacunary trigonometric series, Tohoku Math. J. (2), 31, 439-451.

Takahashi, S. (1980). The functional central limit theorem for lacunary trigonometric series, Acta Math. Acad. Sci. Hungar. 35, 185-192.

Taqqu, M. and Czado, C. (1985). A survey of functional laws of the iterated logarithm for self-similar processes, Stochastic Models 1, 77-115.

Wacker, U. (1983). Grenzwertsätze für nichtadditive, schwach-abhängige Prozesse, Diss. Univ. Göttingen.

Walk, H. (1977). An invariance principle for the Robbins-Monroe process in a Hilbert space, Z.W. 39, 135-150.

Walk, H. (1978). Martingales and the Robbins-Monroe procedure in D[0,1], J. Multivariate Anal. 8, 430-452.

Walk, H. (1980). A functional central limit theorem for martingales in C(K) and its application to sequential estimates, J. Reine Angew. Math. 314, 117-135.

Yokoyama, R. (1980). A note on weak convergence of empirical processes for ϕ-mixing random variables, Yokohama Math. J. 28, 37-40.

Yoshihara, K. (1976). Limiting behaviour of U-statistics for stationary absolutely regular processes, Z.W. 35, 237-352.

Department of Mathematics
University of Illinois
Urbana, IL 61801

UNIFORM INTEGRABILITY AND THE CENTRAL LIMIT THEOREM FOR STRONGLY MIXING PROCESSES

by

MANFRED DENKER

ABSTRACT: It is shown that for a strongly mixing sequence the central limit theorem holds if and only if the squares of the normalised partial sums $\{\sigma_n^{-2}S_n^2 : n \geq 1\}$ are uniformly integrable.

Let $X_1, X_2, \ldots$ be a (strictly) stationary sequence of square integrable real valued random variables satisfying Rosenblatt's strong mixing condition with mixing coefficients

$$\alpha(n) := \sup\{|P(A\cap B)-P(A)P(B)| : A\in F_o^m, B\in F_{m+n}^{\infty}, m\in \mathbb{N}\} \to o$$

where F_s^t denotes the σ-algebra generated by $X_s, \ldots, X_{t-1}$. Denote by $S_n = \sum_{i\leq n} X_i$ the partial sums and by $\sigma_n^2 = \mathrm{Var}\, S_n$ the variance of S_n $(n\geq 1)$. We shall always assume the $EX_1 = o$ and $\sigma_n^2 \to \infty$.

The first result relating the uniform integrability of $\{\sigma_n^{-2}S_n^2 : n\geq 1\}$ with the central limit theorem was obtained by Volkonskii and Roszanov (1959) (cf. also [5, Theorem 18.4.2.]).

<u>Theorem 1:</u> ([8]) Assume that $\sigma_n^2 = \sigma^2 n(1+o(1))$ $(n\geq 1)$ for some $\sigma^2 > o$.

Then the sequence $\{\sigma_n^{-1}S_n : n\geq 1\}$ satisfies the central limit theorem if and only if $\{\sigma_n^{-2}S_n^2 : n\geq 1\}$ is uniformly integrable.

Ibragimov (1962) proved the following characterization for the central limit theorem:

Theorem 2: ([4]) The following are equivalent:

(1) $\{\sigma_n^{-1}S_n : n\geq 1\}$ satisfies the central limit theorem.

(2) $\sigma_n^2 = nh(n)$, where $h(x)$ is a slowly varying function of the continuous variable $x>o$, and

there exist sequences $p=p(n)$ and $q=q(n)$ $(n\geq 1)$ of integers such that

(a) $p\to\infty$, $q\to\infty$, $q=o(p)$, $p=o(n)$ (as $n\to\infty$),

(b) $\lim_{n\to\infty} n^{1-b}q^{1+b}p^{-2}=o$ for all $b>o$,

(c) $\lim_{n\to\infty} np^{-1}\alpha(q)=o$,

(d) $\lim_{n\to\infty} np^{-1}\sigma_n^{-2}\int_{\{|S_p|>\varepsilon\sigma_n\}} S_p^2 dP=o$ for every $\varepsilon>o$.

Note that the proof of this result has a small gap, which, however, easily can be fixed, (cf. [5, Theorem 18.4.1.]). Using this theorem one can extend Theorem 1; the proof is given below.

Theorem 3: The following are equivalent:

(1) $\{\sigma_n^{-1}S_n : n\geq 1\}$ satisfies the central limit theorem.

(2) $\{\sigma_n^{-2}S_n^2 : n\geq 1\}$ is uniformly integrable.

It follows immediately from this theorem that $\sigma_n^{-1}S_n$ is asymptotically $N(0,1)$ if $E|S_n|^{2+\delta}=O(\sigma_n^{2+\delta})$ for some $\delta>o$. Another immediate consequence of the theorem concerns symmetrizations. Recall that the symmetrized process $\hat{X}_1, \hat{X}_2, \ldots$ is defined by $\hat{X}_n = X_n^1 - X_n^2$ $(n\geq 1)$ where $\{X_n^i : n\geq 1\}$ $(i=1,2)$ are two independent copies of the process $\{X_n : n\geq 1\}$. Let $\hat{S}_n$ denote the partial sums of the symmetric process.

Corrollary: $\{\sigma_n^{-1}S_n : n\geq 1\}$ satisfies the central limit theorem if and only if $\{2^{-1/2}\sigma_n^{-1}\hat{S}_n : n\geq 1\}$ does.

In general, it is a non-trivial problem to show uniform integrability. From [2] we obtain the following equivalent statement.

Theorem 4: [2] Let $\sigma_n^2=nh(n)$ $(n\geq 1)$ where h is a slowly varying function on the positive reals. then the following are equivalent:

(1) $\{\sigma_n^{-1}S_n:n\geq 1\}$ is uniformly integrable.

(2) $\limsup_{n\to\infty} \sigma_n^2(E|S_n|)^{-2}\leq \pi/2.$

Examples of strongly mixing sequences for which $\{\sigma_n^{-1}S_n:n\geq 1\}$ is not uniformly integrable are given e.g. by Bradley (1980) and Herrndorff (1983). In the uniformly mixing case, i.e. if

$\varphi(n)=\sup\{|P(A|B)-P(A)|:B\in F_o^m, A\in F_{m+n}^{\infty}, m\geq 1\}\to 0,$

M. Peligrad ([7]) recently obtained the following remarkable result:

Theorem 5: [7] Assume that $\varphi(1)<1$. Then the following are equivalent

(1) $\{\sigma_n^{-2}S_n^2:n\geq 1\}$ is uniformly integrable.

(2) $\{X_n:n\geq 1\}$ satisfies the Lindeberg condition

$$\lim_{n\to\infty} \frac{n}{\sigma_n^2} EX_1^2 1\{|X_1|>\varepsilon\sigma_n^2\}=o \quad \text{for every } \varepsilon>o \quad .$$

(3) $\{\max_{k\leq n} \sigma_k^{-2}S_k^2:n\geq 1\}$ is uniformly integrable.

(4) $\{\max_{k\leq n} \sigma_k^{-2}X_k^2:n\geq 1\}$ is uniformly integrable.

This theorem may be an important step in proving Ibragimov's conjecture that the central limit theorem always holds in the uniformly mixing case.

We now prove Theorem 3 and its corollary.

Proof of Theorem 3[+]: There are two completely different proofs of the theorem. The first is based on Kersting's method ([6,p.1015]) to prove the central limit theorem in the i.i.d. case. The proof is el-

---

[+] Theorem 3 was obtained in 1978. In the meantime a similar argument appeared in Withers (1981, Theorem 2.3(b)).

ementary and does not use Theorem 2. Here we shall give the proof using this theorem.

"(1) ⇒ (2)" is obvious (cf. [5,p. 339]).

"(2) ⇒ (1)" First note that (2) implies by Theorem 17.2.1 of [5] and a truncation argument that for every $\eta>o$ there exists a $q\in\mathbb{N}$ such that

(*) $ES_n(S_{n+m+q}-S_{n+q})\leq\eta\sigma_n\sigma_m \qquad (n,m\geq 1)$.

Therefore, by the remark on page 330 in [5], $\sigma_n^2=nh(n)$ where h is slowly varying on $[1,\infty)$.

In the notation of Theorem 2, let $p=p(n)$ and $q=q(n)$ satisfy conditons (a),(b) and (c). In addition we may choose p and q in such a way that $\lim np^{-1}\eta(q)=o$ where $\eta(q)$ is defined in ($_*$). Fix $\varepsilon>o$, and it is left to prove (d).

Given $\delta>o$, by (2) there exist $R>o$ and n_o such that

(**) $\int_{\{|S_n|>R\sigma_n\}}S_n^2dP<\delta\sigma_n^2 \qquad (n\geq n_o)$.

As noted before, Theorem 18.2.3 and Lemma 18.2.4 of [5] apply in the present situation. Therefore

$$\sigma_p^2\sigma_n^{-2}=\frac{ph(p)}{nh(n)}\leq(pn^{-1})^{1/2}=o(1) \qquad (\text{as } n\to\infty) .$$

If n is sufficiently large $\sigma_pR<\varepsilon\sigma_n$ and thus by (**)

$$\int_{\{|S_p|>\varepsilon\sigma_n\}}S_p^2dP\leq\int_{\{|S_p|>R\sigma_p\}}S_p^2dP<\delta\sigma_p^2 .$$

If in addition p and q are chosen in such a way that

$$n^2p^{-2}\sigma_q^2\sigma_p^{-2}=n^2qp^{-3}h(q)h(p)^{-1}\to o$$

it follows that

$$np^{-1}\int_{\{|S_p|>\varepsilon\sigma_n\}}S_p^2dP<\delta np^{-1}\sigma_p^2<2\delta\sigma_n^2$$

since by (*), (c) and the decompostition of S_n as on pages 358/359 in [4]

$$\sigma_n^2 = \sum_{i,j=o}^{k-1} E\xi_i\xi_j + \sum_{i=o}^{k-1}\sum_{j=o}^{k} E\xi_i\eta_j + \sum_{i,j=o}^{k} E\eta_i\eta_j =$$

$$= np^{-1}\sigma_p^2(1+o(1)).$$

<u>Proof of the corollary:</u> Suppose that $\{(2\sigma_n^2)^{-1}\hat{S}_n^2 : n\geq 1\}$ is uniformly integrable and let $R>0$. Let $S'_n = \sum_{i\leq n} X_i^1$ and $S''_n = \sum_{i\leq n} X_i^2$.

Since the median of the distribution of S''_n is bounded by $\sigma_n\sqrt{2}$ we have

$$\int_{\{|\hat{S}_n|\geq\sigma_n R\sqrt{2}\}} \hat{S}_n^2 dP \geq \int_{\{S'_n\geq\sigma_n\sqrt{2}\,(R+1),\ S''_n\leq\sigma_n\sqrt{2}\}} \hat{S}_n^2 dP +$$

$$+ \int_{\{S'_n\leq -\sigma_n\sqrt{2}(R+1),\ S''_n\geq -\sigma_n\sqrt{2}\}} \hat{S}_n^2 dP \geq$$

$$\geq \frac{1}{2}\int_{\{|S'_n|\geq\sigma_n\sqrt{2}(R+1)\}} S_n'^2 dP - 4\sigma_n \int_{\{|S'_n|\geq\sigma_n\sqrt{2}(R+1)\}} |S'_n|\, dP \geq$$

$$\geq \left(\frac{1}{2} - \frac{4}{\sqrt{2}(R+1)}\right) \int_{\{|S'_n|\geq\sigma_n\sqrt{2}(R+1)\}} S_n'^2 dP .$$

It follows that $\{\sigma_n^{-2} S_n'^2 : n\geq 1\}$ is uniformly integrable. The converse is shown similarly.

<u>REFERENCES:</u>

[1] BRADLEY, R.C.: A remark on the central limit question for dependent random variables. J.Appl.Prob. <u>17</u>, 94-101, (1980)

[2] DEHLING,H. and M. DENKER and W. PHILIPP: Central limit theorems for mixing sequences of random variables under minimal conditions. Annals Prob. to appear

[3] HERRNDORFF, N.: Stationary strongly mixing sequences not satisfying the central limit theorem. Annals Prob. <u>11</u>, 809-813, (1983)

[4] IBRAGIMOV, I.A.: Some limit theorems for stationary processes. Theory Prob. Appl. <u>7</u>, 349-382, (1962)

[5] IBRAGIMOV, I.A. and Yu.V. LINNIK: Independent and stationary sequences of random variables. Walters-Noordhoff Publ., Groningen 1971

[6] KERSTING, G.: A weak convergence theorem with application to the

Robbins-Monro process. Annals Prob. 6, 1015-1025, (1978)

[7] PELIGRAD, M.: An invariance principle for φ-mixing sequences. Annals Prob. to appear

[8] VOLKONSKII, V.A. and Yu.A. ROZANOV: Some limit theorems for random functions. Teor. Veroyatnost. i. Prem. 4, 186-207, (1959)

[9] WITHERS, C.S.: Central limit theorems for dependent variables I. Z. Wahrscheinlichkeitstheorie verw. Geb. 57, 509-534, (1981).

M. DENKER
Institut für Mathematische Stochastik
Georg-August-Universität
Lotzestr. 13
D-3400 Göttingen, West Germany

CENTRAL LIMIT QUESTIONS FOR RANDOM FIELDS

by

CHARLES M. GOLDIE, GREGORY J. MORROW

1. Introduction

The purpose of this paper is to survey some recent progress in extending the classical central limit theory for weakly stationary random fields $\{x_n; n \in \mathbb{Z}^d\}$.

There are two basic fronts of development. One area is motivated by the study of equilibrium (Gibbs) states of lattice systems in statistical physics. There it is assumed that the random field is governed by a Gibbs state (measure). In that way the dependence of the field is defined only implicitly. The field is renormalized in the form $\{z_N\}$ with $z_N = \sum_{n \in Q(N)} x_N/A_N$ for some index sets $Q(N)$ which increase to $\mathbb{Z}^d$ in the limit $N \to \infty$. The limiting finite dimensional distributions of the renormalized field are studied. The limits depend on a temperature parameter, and may be non-Gaussian with non-classical norming A_N at the so-called critical temperature, but the limits are independent Gaussian at high temperatures. See the survey [10] for an account of this model and of this critical phenomenon in particular.

Throughout the sequel however we shall consider only weak dependence structures and, thus, only classical normalizations. In fact much of our attention will be devoted to another area of results which are only partially motivated by Gibbs fields, namely, limit theorems that are uniform over a large class of sets. Besides their interest above, these theorems are applicable to the problem of detecting an index set, say, that corresponds to a simple alternative in the sample distribution of spatially indexed observations.

We discuss first the central limit problem for finite dimensional distributions of a renormalized field of random variables $\{x_n\}$ that are in some sense close to independent. The dependence structures are motivated primarily by models of Gibbs fields with short range interactions [30]. We then pass to a discussion of the problem of uniform limits in the independent case as well as recent invariance principles for certain mixing cases.

2. Finite Dimensional Limits

In dimension $d=1$ we may consider a stationary irreducible aperiodic Markov chain on a finite state space. Then it is well known ([13]) that the following uniform strong mixing condition holds.

$$\sup_{S,S' \,:\, d(S,S') \ge t} \quad \sup_{\substack{A \in \sigma(x_n; n \in S) \\ B \in \sigma(x_n; n \in S')}} |P(AB) - P(A)P(B)| \to 0, \quad \text{as } t \to \infty. \tag{2.1}$$

Here S, S' are subsets of the universe of indices, $d(\cdot,\cdot)$ is the minimal Euclidean distance between them, and, for the present situation, the convergence is in fact exponentially fast in t. This condition has played an important role for the central limit theory in case $d=1$. However, as pointed out by Dobrushin [12] condition (2.1) is too strong in general for applications to Gibbs fields when $d \geq 2$. For this reason much of the finite-dimensional limit theory centers on the kinds of dependence conditions to impose and, in turn, on estimates of the remainder in the central limit theorem. Several variations on the mixing condition (2.1) relevant to the physical context have appeared in [11], [16], [21], [22], [29], [30].

The following result, formulated in terms of a non-uniform strong mixing condition, serves to illustrate these points. The boundary of a set S is denoted ∂S.

THEOREM 2.1 [11]. Suppose $\{x_n;\ n \in \mathbb{Z}^d\}$ is strictly stationary with zero means and $E|x_0|^{2+\delta} < \infty$ for some $\delta > 0$. Let $a_{2,\infty}(t)$ denote the supremum in (2.1) under the added condition card $(S) \leq 2$. Suppose that

$$\sum_{t\geq 1} t^{d-1} a_{2,\infty}^{\delta/(2+\delta)}(t) < \infty \quad \text{and} \quad \sigma^2 := \sum_{n\in\mathbb{Z}^d} Ex_0 x_n \neq 0 .$$

Let $Q_N \uparrow \mathbb{Z}^d$ such that card $(\partial Q_N)/\text{card}\ (Q_N) \to 0$. Then

$$z_N := \sum_{n\in Q_N} x_n/(\sigma\sqrt{\text{card}(Q_N)}) \to N(0,1).$$

The proof is based on the following identity involving the characteristic function $f_N(t) := E\exp(itz_N)$.

$$f_N'(t) + tf_N(t) = E((t+iz_N)\exp(itz_N)). \tag{2.2}$$

Denote the right side of this identity by $h_N(t)$ and note that $|h_N(t)| \leq t + \sup_N E|z_N| = t + \text{constant}$. Hence, since the solution of (2.2) with $f_N(0) = 1$ is

$$f_N(t) = \exp(-t^2/2)(1 + \int_0^t \exp(u^2/2)h_N(u)\,du),$$

it suffices to show that $h_N(t) \to 0$ for each fixed t. This is done by expanding $(t + iz_N)\exp(itz_N)$ so as to exploit the weak dependence structure ([11], [19], [31]). The idea of writing a differential equation for $f_N(t)$ is due apparently to Tikhomirov although the idea of exploiting weak dependence in a similar spirit appears in work of Stein [26]. By the same method, rates may be obtained in the central limit theorem ([19], [29], [31]); cf. also [22].

All the above results assume polynomial decay of the relevant mixing coefficient ($a_{2,\infty}(t)$ in the theorem above). However when $d=1$ a logarithmic rate often suffices, or no rate at all if suitable other mixing or moment conditions are imposed. (See M. Peligrad's paper in the present volume.) Goldie and Greenwood [17] obtain a CLT in case $d > 1$, assuming logarithmic rate of decay of the maximal correlation coefficient

$$\rho(t) := \sup_{\substack{S,S' \\ d(S,S') \geq t}} \quad \sup_{\substack{X \in L_2(\sigma(x_n;\, n \in S)) \\ Y \in L_2(\sigma(x_n;\, n \in S'))}} |\mathrm{corr}(X,Y)|.$$

For the stationary case their result is

THEOREM 2.2. Suppose $\{x_n : n \in \mathbb{Z}^d\}$ is strictly stationary with zero means and finite variances. For all Borel sets B in $(0,1]^d$ set

$$S_N(B) := N^{-d/2} \sum_{n \in \mathbb{Z}^d} x_n \cdot \lambda(NB \cap (n+(0,1]^d)) \qquad (N \geq 1),$$

where λ is Lebesgue measure. Assume

(a) $\Sigma_{j=1}^{\infty} \rho(2^j) < \infty$,

(b) for every $t \in (0,1]^d$, $ES_N^2((0,t]) \to t \quad (N \to \infty)$.

Then each finite collection $\{S_N(B_j);\ 1 \le j \le k\}$ converges weakly to the Gaussian law with means zero and covariances $\lambda(B_i \cap B_j)$.

3. Uniform Limits

(i) independent case. Let $\{x_n;\ n \in \mathbb{Z}_+^d\}$ be i.i.d. with zero means and unit variances. Let $\mathcal{A}$ be a collection of Lebesgue measurable subsets of $(0,1]^d$ and define $d_\lambda(A,B) := \lambda(A\Delta B)$. Thus $(\mathcal{A},d_\lambda)$ is a pseudo-metric space and we define $C(\mathcal{A})$ to be the space of continuous functions on $(\mathcal{A},d_\lambda)$ equipped with the supremum norm. Consider $S_N(A)$ as above, with A in $\mathcal{A}$. Thus in case $d=1$ and $\mathcal{A}$ is the collection of intervals $\{(0,t] : t \in (0,1]\}$, $S_N(\cdot)$ becomes the familiar linearly interpolated partial-sum process, and Donsker's theorem asserts:

(3.1) $S_N(\cdot)$ converges weakly in $C(\mathcal{A})$ to a standard Brownian process on $\mathcal{A}$.

The limit process is of course that whose finite-dimensional laws are the limit laws of Theorem 2.2 above.

How large can we take the class $\mathcal{A}$ and still preserve (3.1)? It turns out that this question is closely connected with the limit theory of empirical processes (see [2], [25]). Recently Alexander and Pyke [3] established the following generalization of Donsker's theorem, and Bass [4] subsequently found another proof leading also to the law of the iterated logarithm.

THEOREM 3.1. Let $N_I(\mathcal{A};u)$ be the least number k so that there exist pairs of sets A_-^i, $A_+^i \in \mathcal{A}$ with $A_-^i \subset A_+^i$ and $\lambda(A_+^i \backslash A_-^i) < u$, $i=1,\ldots,k$, so that for every $A \in \mathcal{A}$, $A_-^i \subset A \subset A_+^i$ for some i. Suppose

$$(3.2) \qquad \int_{0+} \sqrt{u^{-1}\log N_I(\mathcal{A};u)}\, du < \infty.$$

Then (3.1) holds.

The condition (3.2) is known as the condition on metric entropy with bracketing. Although the order of integrability of $N_I(\mathcal{A};u)$ in (3.2) is sharp for continuity of the limiting Brownian process, one can consider different classes of sets $\mathcal{A}$.

Let $N(\mathcal{A};u)$ denote the least number k so that there are measurable sets A^i, $i = 1,\ldots,k$ such that for every $A \in \mathcal{A}$, $d_\lambda(A,A^i) < u$ for some i. Note that Brownian processes exist in $C(\mathcal{A})$ for the larger classes of sets which satisfy (3.2) with $N(\mathcal{A};u)$ in place of $N_I(\mathcal{A};u)$. Denote the so-called exponent of metric entropy:

$$r := \limsup_{u \to 0} \ \log\log N(\mathcal{A};u)/\log(1/u).$$

The convex subsets of $(0,1]^d$ have exponent $r = (d-1)/2$ ([14]), so this class is too big for continuity of the Brownian process if $d \geq 3$ ([15]). Related classes of sets with uniformly smooth boundaries up to order q have exponent $r = (d-1)/q$. The polyhedra with a bounded number of vertices, or all ellipsoidal regions, are examples of classes with $r=0$. There is a combinatorial property shared by these last two classes, namely, there is a finite index v so that the power set of no v-element subset E of $(0,1]^d$ is realized as intersections of E with sets from this class. This property defines the so-called Vapnik-Červonenkis (V-C) classes, which turn out to have exponent $r=0$, yet for which even $N_I(\text{V-C};u) = \infty$, all small u, is possible ([15]). Nevertheless the analogue of Theorem 3.1 holds with (3.2) replaced by the condition that $\mathcal{A}$ be a V-C class ([2]).

The next result gives a uniform central limit theorem based on r for variables with higher than second moments. It is due to Pyke and co-workers, in that one can use the tightness proof in [24], modified by later improvements, as in the outline following.

THEOREM 3.2. Let $0 < r < 1$ be the exponent of metric entropy and suppose that $E|x_1|^s < \infty$ for some $s > 2/(1-r)$. Then (3.1) holds.

PROOF OUTLINE. Theorem 2.2 of [3] gives finite-dimensional weak convergence. For tightness we need first the following truncation (aural communication of R. Pyke). Fix $a > 0$ and let

$$U_N(a) := N^{-d/2} \sum_{n\in N(0,1]^d} |x_n| \, 1\{|x_n| > aN^{\frac{1}{2}d/(s-1)}\} ,$$

then

$$\begin{aligned} EU_N(a) &= N^{d/2} E(|x_1| \, 1\{|x_1| > aN^{\frac{1}{2}d/(s-1)}\}) \\ &< a^{-(s-1)} E(|x_1|^s \, 1\{|x_1 > aN^{\frac{1}{2}d/(s-1)}\}) \\ &\to 0 \quad (N \to \infty) ; \end{aligned}$$

thus $U_N(a)$ tends to zero as $N \to \infty$ in L_1, hence in probability. Let $S_N^{(a)}(A)$ be defined like $S_N(A)$ but with x_n replaced by

$$x_n 1\{|x_n| \le aN^{\frac{1}{2}d/(s-1)}\} - E(x_n 1\{|x_n| \le aN^{\frac{1}{2}d/(s-1)}\}) .$$

The uniform distance between $S_N(\cdot)$ and $S_N^{(a)}(\cdot)$ is at most $U_N(a) + EU_N(a)$. So to show $\{S_N(\cdot)\}$ tight it suffices to show $\{S_N^{(a)}(\cdot)\}_{N\ge1}$ tight. In fact it is enough to prove that

$$\lim_{a,\delta\downarrow 0} \limsup_{N\to\infty} P(\sup_{\substack{A,B\in\mathcal{A} \\ d_\lambda(A,B)<\delta}} |S_N^{(a)}(A) - S_N^{(a)}(B)| > \eta) = 0$$

for each $\eta > 0$. This is accomplished by the proof in [24] §2, with the truncation modified as above. □

Under stronger conditions, Morrow and Philipp proved (3.1) by a (blocking) method that is applicable also to vector-valued variables:

THEOREM 3.3. [20]. Let $0 < r < 1$ be the exponent of metric entropy and suppose that $E|x_1|^s < \infty$ for some $s > 2(1+r)/(1-r)$. Suppose also that for all $A \in \mathcal{A}$,

$$\text{(3.3)} \qquad \lambda(A^\varepsilon \cap \widetilde{A^\varepsilon}) \le \text{const. } \varepsilon^\beta$$

for some $0 < \beta < 1$, where $A^\varepsilon := \{b : d(b,A) < \varepsilon\}$. Then (3.1) holds.

Remark that the proof in [20] actually yields more, namely a uniform central limit theorem with remainder and corresponding almost-sure invariance principle as well as the same Donsker-type theorem for the case of Banach-space-valued random vectors. Note that whereas the smoothing factor λ in the definition of $S_N(\cdot)$ is needed for the proofs of Theorems 3.1 and 3.2 it is not needed for the blocking method.

CENTRAL LIMIT QUESTIONS FOR RANDOM FIELDS

We have neglected comments about the functional law of the iterated logarithm until now. The analogue of Strassen's original result [27] can be stated as follows.

$$(3.4)\quad \{A \to S_N(A)/\sqrt{2\ \text{loglogN}}\} \twoheadrightarrow \{f \in C(\mathcal{A}) : f(A) = \int gd\lambda \ \text{ with } \int_{(0,1]^d} g^2 d\lambda \le 1\}\ .$$

This is due to [4] under the hypotheses of Theorem 3.1, and to [5] under those of Theorem 3.2 together with the assumption that (3.3) holds with $\beta=1$. An appropriate Banach-space version of (3.4) is obtained in [20] for the context of Thecrem 3.3.

Finally we mention two forms of central limit theorem with limits other than those above. The first is where the underlying variables are in the domain of attraction of a stable law, so that the limit process is a set-indexed stable process, suitably defined. Here the space D in which the sample paths live has first to be defined and developed; see [1], [6], [7]. Secondly there is the so-called 'Lévy's (multidimensional) Brownian motion' and weak convergence to it ([23]).

(ii) mixing case. We describe next what is known for a weakly stationary random field satisfying the strong mixing condition (2.1) with a polynomial rate of decay and having uniformly bounded s^{th} moments for some $s > 2$. Let $\mathcal{A}_0$ be the collection of intervals $\{(0,t] : t \in (0,1]^d\}$. If the field is independent then Donsker's theorem can be strengthened to a central limit theorem with remainder depending on the size of $A \in \mathcal{A}_0$. For convenience we formulate this result as an almost-sure invariance principle: for some $\gamma > 0$ and

Brownian sheet $W(\cdot)$, with probability 1 we have

$$(3.5) \qquad \left| \sum_{n \in NA} x_n - W(NA) \right| = O(\text{card}(NA)^{\frac{1}{2}-\gamma}) ,$$

uniformly in $A \in \mathcal{A}_0$ as $N \to \infty$, $\text{card}(NA) \to \infty$. However in the mixing case the variance of $\sum_{n \in NA} x_n$ need not be asymptotic to $\sigma \cdot \text{card}(NA)$ for a fixed σ, so (3.5) can fail. But (3.5) does hold if, say, one restricts $\mathcal{A}_0$ to contain only rectangles with side lengths that are comparable on a logarithmic scale ([9]), while an approximation with error term $O(\text{card}(NA)^{\frac{1}{2}}/(\text{loglogcard}(NA)^{\gamma})$ valid over all $\mathcal{A}_0$ is possible by taking the approximating Gaussian process to have the same covariance structure as the partial-sum process ([8]).

Nevertheless a Donsker-type relation with Brownian limit can be obtained with larger classes of sets, but so far not (under strong mixing with polynomial rate) for any classes with exponent of metric entropy $r > 0$. In his thesis, Strittmatter [28] obtains a version of Theorem 3.3 for the strong mixing case, assuming $\limsup_{u \to 0} (\log N(\mathcal{A};u))/(-\log u) < \infty$. Also under a strong-mixing condition with exponential rate the methods of [28] yield a strong approximation result with some exponent $r > 0$.

As the reader may already know, good exponential bounds are the key to obtaining uniform central-limit theorems for large classes of sets. A basic trick in the weakly dependent case is to approximate sums of dependent random variables by sums of, say, independent variables for which these bounds are known. This method has worked even in the Banach-space setting for the following so-called absolutely regular mixing condition which is stronger than (2.1).

$$(3.6) \qquad \sup_{d(S,S')\geq t} \| L(x_n; n \in S \cup S') - L(x_n; n \in S) \times L(x_n; n \in S')\|_{var} \to 0.$$

Goldie and Greenwood [18] employ (3.6) by taking S and S' to be blocks of indices; on the replacement process, which has between-blocks independence, they use Bass's [4] method of obtaining an exponential inequality to derive (3.1) (using also Theorem 2.2) assuming polynomial decay in (3.6) in terms of the exponent of metric entropy with bracketing, r'. Here r' is defined like r but with $N_I(A;u)$ in place of $N(A;u)$. The mixing rate $o(t^{-b})$ needed depends on the number $s > 2$ of moments and on r', as follows: $b \geq ds/(s-2)$ and $b > d(1+r')/(1-r')$. Alternatively the usual exponent of metric entropy r can be used if the conditions are strengthened to $s > 2/(1-r)$ and $b > ds(1+r)/(s(1-r)-2)$.

An extension of the a.s. invariance principle, Theorem 3.3, for values in Banach space is derived in [28], also with polynomial decay in (3.6) but with a more restrictive relation between r, s and b. Without describing this relation, suffice it to say that one can just take $r < 1$ if the variables are uniformly bounded and there is an exponential rate of decay in (3.6).

4. Suggested Problems

It will be apparent that the broad front of results presented above has many possibilities for advance. Taking first §2, a challenging problem is to extend either of its theorems to the Banach-space-valued case. Equally, it would be good to have a random-field central limit theorem employing a physically-relevant non-uniform mixing coefficient, but with no more than logarithmic rate.

Turning to uniform central limit theorems, in the case of i.i.d. fields an extension of Theorem 3.1 to the Banach-space-valued case is hoped for. Similarly it is suspected that the analogue of Theorem 3.1 for V-C classes can be extended to this case. An open problem for the real-valued case is to determine an optimal moment condition for a uniform central limit theorem for unsmoothed partial sums indexed by convex subsets in the plane. The result of [20, Cor. 1] requires higher than tenth moments. Also in the real-valued case, one would like to show that (3.1) implies (3.4). It is open whether $s \geq 2/(1-r)$ is necessary in Theorem 3.2.

For uniform central limit theorems, nothing weaker than polynomial mixing rate has yet produced an adequate exponential bound for tightness or approximation purposes, even over a V-C class. One might nevertheless consider whether mixing conditi ; are always the best formulation of weak dependence for these purposes. Conditions in terms of some form of asymptotic martingale, mixingale or the Doléans measure (cf. the article by E. Eberlein in this volume) are attractive but the key problem is to identify the asymptotic variance structure (hence the normalisation) in the absence of a time-indexed filtration.

References

1. R.J. ADLER and P.D. FEIGIN. On the cadlaguity of random measures. *Ann. Probab. 12* (1984), 615-630.

2. K.S. ALEXANDER. Central limit theorems for stochastic processes under random entropy conditions. Preprint, Univ. of Washington, 1984.

3. K.S. ALEXANDER and R. PYKE. A uniform central limit theorem for set-indexed partial-sum processes with finite variance. *Ann. Probab.* (to appear).

4. R.F. BASS. A law of the iterated logarithm for set-indexed partial-sum processes with finite variance. *Z. Wahrscheinl. verw. Geb.* (to appear).

5. R.F. BASS and R. PYKE. Functional law of the iterated logarithm and uniform central limit theorem for partial-sum processes indexed by sets. *Ann. Probab. 12* (1984), 13-34.

6. R.F. BASS and R. PYKE. The existence of set-indexed Lévy processes. *Z. Wahrscheinl. verw. Geb. 66* (1984), 157-172.

7. R.F. BASS and R. PYKE. The space D(A) and weak convergence for set-indexed processes. *Ann. Probab. 13* (1985), 860-884.

8. I. BERKES. Gaussian approximation of mixing random fields. *Acta Math. Hung. 43* (1984), 153-185.

9. I. BERKES and G.J. MORROW. Strong invariance principles for mixing random fields. *Z. Wahrscheinl. verw. Geb. 57* (1981), 13-37.

10. P.M. BLEHER and P. MAJOR. Critical phenomena and universal exponents in statistical physics: on Dyson's hierarchical model. *Comm. Math. Physics* (to appear).

11. E. BOLTHAUSEN. On the central limit theorem for stationary random fields. *Ann. Probab. 10* (1982), 1047-1050.

12. R.L. DOBRUSHIN. The description of a random field by its conditional distribution and its regularity condition. *Theory Probab. Appl. 13* (1968), 197-227.

13. J.L. DOOB. *Stochastic Processes.* Wiley, New York, 1953.

14. R.M. DUDLEY. Metric entropy of some classes of sets with differentiable boundaries. *J. Approx. Th.* *10* (1974), 227-236.

15. R.M. DUDLEY. A course in empirical processes. In: R.M. Dudley, H. Kunita, F. Ledrappier, *Ecole d'Eté de Probabilités de Saint Flour XII*. Springer Verlag, Lecture Notes in Math. *1097*, 1984.

16. E. EBERLEIN. An invariance principle for lattices of dependent random variables. *Z. Wahrscheinl. verw. Geb.* *50* (1979), 119-133.

17. C.M. GOLDIE and P.E. GREENWOOD. Characterisations of set-indexed Brownian motion and associated conditions for finite-dimensional convergence. *Ann. Probab.* (to appear).

18. C.M. GOLDIE and P.E. GREENWOOD. Variance of set-indexed sums of mixing random variables and weak convergence of set-indexed processes. *Ann. Probab.* (to appear).

19. X. GUYON and S. RICHARDSON. Vitesse de convergence du théorème de la limite centrale pour des champs faiblement dépendants. *Z. Wahrscheinl. verw. Geb.* *66* (1984), 297-314.

20. G.J. MORROW and W. PHILIPP. Invariance principle for partial-sum processes and empirical processes indexed by sets. Preprint, Washington Univ., St. Louis, 1984.

21. B.S. NAHAPETIAN. The central limit theorem for random fields with mixing property. In: *Multicomponent Random Systems* (ed. R.L. Dobrushin, Ya-G. Sinai). *Advances in Probability*, vol. 6, Marcel Dekker, New York, 1980, pp. 531-547.

22. C.C. NEADERHOUSER. Some limit theorems for random fields. *Commun. Math. Phys.* *61* (1981), 293-305.

23. M. OSSIANDER and R. PYKE. Lévy's Brownian motion as a set-indexed process and a related central limit theorem. Preprint, Univ. of Washington, 1984.

24. R. PYKE. A uniform central limit theorem for partial-sum processes indexed by sets. *Probability, Statistics and Analysis*, ed. J.F.C. Kingman, G.E.H. Reuter, 219-240. Cambridge University Press, Cambridge, 1983.

25. R. PYKE. Asymptotic results for empirical and partial-sum processes: a review. *Canad. J. Statist. 12* (1984), 241-264.

26. C. STEIN. A bound for the error in the normal approximation of a sum of dependent random variables. *Proc. Berkeley Symp. Math. Stat. and Probab. 2* (1972), 583-603.

27. V. STRASSEN. An invariance principle for the law of the iterated logarithm. *Z. Wahrscheinl. verw. Geb. 3* (1964), 211-226.

28. W. STRITTMATTER. *Invariance principles for partial-sum processes indexed by sets in the case of mixing random fields.* Thesis, Universität Freiburg, 1985.

29. H. TAKAHATA. On the rates in the central limit theorem for weakly dependent random fields. *Z. Wahrscheinl. verw. Geb. 64* (1983), 445-456.

30. H. TAKAHATA. The central limit problems for energy in the Gibbs random fields. A short survey. *Bull. Tokyo Gakugei Univ., Sect.* IV, *36* (1984), 1-15.

31. A.N. TIKHOMIROV. On the rate of convergence in the central limit theorem for weakly dependent random variables. *Theory Probab. Appl. 25* (1980), 790-809.

C.M. GOLDIE
Mathematics Division
University of Sussex
Brighton BN1 9QH
England.

G.J. MORROW
Department of Mathematics
Washington University
Campus Box 1146
St. Louis, Missouri 63130
U.S.A.

SUMMABILITY METHODS AND DEPENDENT STRONG LAWS

N. H. BINGHAM

1. Introduction

We recall the classical Kolmogorov strong law of large numbers (see e.g. [16],[17]):

THEOREM 1 (Kolmogorov's LLN). If $X, X_0, X_1, \ldots$ are independent and identically distributed random variables, the following are equivalent:

(i) $E|X| < \infty$ & $EX = \mu$,

(ii) $\frac{1}{n+1} \Sigma_0^n X_k \to \mu$ a.s. $(n \to \infty)$.

We wish to generalise this in two ways - by weakening independence and generalising the averaging process - while keeping the two-way link between moment conditions and almost-sure convergence.

To set the stage, we recall (see e.g. Hardy [23]) the classical Cesaro (C), Abel (A), Borel (B) and Euler ($E_p, p \in (0,1)$) summability methods, defined as follows:

$s_n \to s$ (C) means $\frac{1}{n+1} \Sigma_0^n s_k \to s \ (n \to \infty)$,

$s_n \to s$ (A) means $(1-r) \Sigma_0^\infty r^k s_k \to s \ (r \uparrow 1)$,

$s_n \to s$ (B) means $\Sigma_0^\infty e^{-x} \frac{x^k}{k!} s_k \to s \ (x \to \infty)$,

$s_n \to s$ (E_p) means $\Sigma_0^n \binom{n}{k} p^k (1-p)^{n-k} s_k \to s \ (n \to \infty)$.

We will again suppose the Xs independent and identically distributed (iid). The two basic results are the following:

THEOREM 2 (Lai's LLN, [30]). For $X, X_0, X_1, \ldots$ iid, equivalent are:

(i) $E|X| < \infty$ & $EX = \mu$,

(ii) $X_n \to \mu$ a.s. (C),

(iii) $X_n \to \mu$ a.s. (A).

THEOREM 3 (Chow's LLN, [13]). For $X, X_0, X_1, \ldots$ iid, equivalent are:

(i) $E(X^2) < \infty$ & $EX = \mu$,

(ii) $X_n \to \mu$ a.s. (E_p) for some (all) $p \in (0,1)$,

(iii) $X_n \to \mu$ a.s. (B).

2. Riesz and Valiron means. Chow went further, and linked (i) - (iii) in Theorem 3 with a.s. convergence of certain 'delayed averages', to which we now turn. For a real sequence $\lambda_n \uparrow \infty$, recall ([12],[24],[23, §§4.16,5.16]) the Riesz (typical) mean $R(\lambda_n, 1)$: with $s_n = \Sigma_0^n a_k$, we write

$$s_n \to s \quad R(\lambda_n, 1)$$

to mean

$$\frac{1}{x} \int_0^x \{\sum_{n:\lambda_n \leq y} a_n\}\, dy \to s \quad (x \to \infty).$$

For the special case $\lambda_n = \exp(n^{1-\beta})$, $0 < \beta < 1$, one has the

PROPOSITION ([10],[8]). For $0 < \beta < 1$, equivalent are:

(i) $s_n \to s \quad R(\exp(n^{1-\beta}), 1)$,

(ii) $\frac{1}{un^\beta} \sum_{n \leq k < n+un^\beta} s_k \to s \quad (n \to \infty)$ for all $u > 0$,

(iii) There exists $\varepsilon_n \to 0$ with $\frac{1}{n+1} \Sigma_0^n (s_k + \varepsilon_k) = s + o(1/n^{1-\beta})$.

Here (ii) is a 'delayed average' statement, (iii) is 'perturbed Cesàro convergence with rate'. We recall also ([23], Th. 58) that $R(n,1)$ is the Cesàro method, C. It will be convenient to handle the range $0 < \beta \leq 1$ by using $R(\exp(\int_1^n x^{-\beta} dx), 1)$, which is $R(n,1) = C$ for $\beta = 1$ and equivalent to $R(\exp(n^{1-\beta}), 1)$ for $0 < \beta < 1$.

The Valiron means V_β, $0 < \beta < 1$, are defined by writing

$$s_n \to s \quad V_\beta$$

for

$$\frac{1}{x^\beta (2\pi)^{\frac{1}{2}}} \; \Sigma_0^\infty \, s_k \, \exp\{-\tfrac{1}{2}(x-k)^2/x^{2\beta}\} \to s \;\; (x \to \infty).$$

The 'delayed-average' part of the next result is in Chow [13]; for the rest see [9].

THEOREM 3'. In Theorem 3, equivalent to (i)-(iii) are

(iv) $\quad X_n \to \mu$ a.s. $\; R(e^{\sqrt{n}},1)$,

(v) $\quad X_n \to \mu$ a.s. $\; V_{\frac{1}{2}}$.

The next result provides a continuum of LLNs including - and linking - those of Lai and Chow above. The delayed-average part is from Chow [13]; for the full result see [10].

THEOREM 4. For $X, X_1, \ldots$ iid, $p \geq 1$, the following are equivalent:

(i) $\quad E(|X|^p) < \infty$ & $EX = \mu$,

(ii) $\quad X_n \to \mu$ a.s. $\; R(\exp(\int_1^n x^{-1/p} dx),1)$.

For $p > 1$, (i) and (ii) are equivalent to

(iii) $\quad X_n \to \mu$ a.s. $\; V_{1/p}$.

One may interpret Theorem 4 as saying that in Theorems 3,3' it is the Riesz and Valiron summability methods which are crucially relevant, rather than those of Borel and Euler. For a detailed study of the links between the methods B, E_p and $V_{\frac{1}{2}}$, we refer to [4], [5], [6].

The Riesz means are more basic (from Theorem 4) - and perhaps more intuitive - than the Valiron, and for simplicity we shall focus on Riesz methods in what follows. We regard the Riesz method $R(\exp(\int_1^n x^{-1/p} dx),1)$ as the natural summability method to use in strong laws with pth moments finite, and will denote it by R_p for brevity.

It is always true that convergence to μ under R_p implies convergence to μ under R_q for $q < p$, by the second consistency theorem for Riesz means ([22], [12]) - just as, in Theorem 4, $E|X|^p < \infty$ implies $E|X|^q < \infty$.

3. Marcinkiewicz-Zygmund laws and their extensions. Theorem 4 extends Theorem 1 by using extra moment conditions to strengthen the a.s. summability method. A different extension to Kolmogorov's LLN is provided by the Marcinkiewicz-Zygmund LLN (see e.g. [16], §5.2). Write $S_n = \Sigma_1^n X_k$: for $\{X_n\}$ iid and $p \in (0,2)$,

there exists c with $(S_n - nc)/n^{1/p} \to 0$ a.s. iff $E|X|^p < \infty$ (and then c can be taken as EX if $1 \leq p < 2$, 0 if $0 < p < 1$). This result has a different character to those above in that no summability method is involved. Also we cannot isolate $\mu = EX$ in advance, so we may as well centre at means and take $\mu = 0$.

While Theorem 4 successfully handles arbitrarily large p, the Marcinkiewicz-Zygmund law is restricted to $p < 2$: no norming function can be as small as $\sqrt{n}$, in view of the central limit theorem. To combine arbitrarily large moments with Marcinkiewicz-Zygmund character, one needs to introduce a second parameter α, in addition to p. This was done by Baum & Katz [2], following earlier work of Hsu & Robbins, Erdős, and Spitzer; see also Chow & Lai [14].

THEOREM 5. If $\alpha > \frac{1}{2}$, $\alpha p > 1$, $EX = 0$ if $\alpha \leq 1$, $X, X_1, \ldots$ are iid, equivalent are:

(i) $E|X|^p < \infty$,

(ii) $\Sigma\, n^{\alpha p-2}\, P(|S_n| > \varepsilon n^{\alpha}) < \infty$ for all $\varepsilon > 0$,

(iii) $\Sigma\, n^{\alpha p-2}\, P(\max_{j \leq n} |S_j| > \varepsilon n^{\alpha}) < \infty$ for all $\varepsilon > 0$.

We shall re-formulate this in the language of r-quick convergence due to Strassen [37]; cf. Lai [31]. If $T_n \to 0$ a.s., $U := \sup\{n: |T_n| > \varepsilon\} < \infty$ a.s. for each $\varepsilon > 0$. For $r > 0$, we say that $T_n \to 0$ r-quickly if this random variable has finite rth moment, that is, if $\Sigma\, n^{r-1} P(U > n) < \infty$ for some (all) $\varepsilon > 0$. Thus (iii) above says that

$$S_n/n^{\alpha} \to 0 \quad (p\alpha-1)\text{-quickly}.$$

This is an 'r-quick version of a Marcinkiewicz-Zygmund strong law', or more briefly a 'Lai strong law' in the terminology of [7]. Interpreting r-quick convergence with $r = 0$ as a.s. convergence, the limiting case $\alpha p = 1$ gives the classical strong law. Note the trade-off involved in choosing α: the smaller α, the stronger the statement '$S_n/n^{\alpha} \to 0$', but the smaller $p\alpha-1$, so the weaker the qualifier '($p\alpha$-1)-quickly'.

A related idea is that of complete convergence, due to Hsu & Robbins; see Heathcote [25] for a review, Asmussen & Kurtz [1] for results in which complete convergence is tied to 'moment' conditions $Ef(X) < \infty$, with functions f more general than the powers considered here.

4. Dependence. We consider now how one may weaken the 'iid' assumption in the results above. One could, for instance, keep independence but weaken identical distribution (as is done, for example, in Petrov's book [35]). But the clear-cut link with moment conditions is now lost, as there is now not one but a whole sequence of distributions to consider.

Instead, we shall keep identical distribution; assume for the moment, for simplicity, that $\{X_n\}$ is stationary ergodic. If we simply drop independence, all the above results fail badly. For instance, in Theorem 1 the implication from (i) to (ii) still holds, by the ergodic theorem (see e.g. Stout [36], §3.5). But the converse fails: $S_n/n \to 0$ a.s. may hold with $E|X| = \infty$ (see e.g. Halmos [21],32). Again, with $p > 1$, $S_n/n \to 0$ (p-1)-quickly implies $E|X|^p < \infty$ in the iid case by Theorem 5, but nothing more than $E|X|^{p-1} < \infty$ in the stationary ergodic case (Baum & Katz [2], 118-119). In the opposite direction, one may have the distribution of X_n of compact support (so with all moments finite) but fail to have $\Sigma n^{-1} P(|S_n| \geq n) < \infty$, so in particular, fail to have $S_n/n \to 0$ r-quickly for any $r > 0$ (Baum & Katz [2],117-118). In the same context, one may lose the Marcinkiewicz-Zygmund character by failing to have $S_n/n^{\alpha} \to 0$ a.s. for any $\alpha < 1$ (Jain, Jogdeo & Stout [28], 135-137).

Thus to have any hope of obtaining dependent versions of the results above, one needs some type of weak-dependence or mixing condition to eliminate such examples. We may take $\{X_n\}$ ergodic as above, or specialise successively to $\{X_n\}$ mixing (in the sense of ergodic theory: [36], 173, [26], 302), strong (or Rosenblatt) mixing ([26],305-6), or ϕ- (Ibragimov, uniform) mixing ([26], 308): writing F_i^j for $\sigma(\{X_k\}: i \leq k \leq j)$, we require

$$\phi(n) := \sup_{m} \sup_{A\in F_1^m,\ B\in F_{m+n}^{\infty},\ P(A)>0} |P(B|A)-P(B)| \to 0 \ (n\to\infty)$$

(Bradley [11] shows that for $\{X_n\}$ stationary and mixing, $\phi(n) \downarrow 0$ or $\phi(n) \equiv 1$).

Kolmogorov's strong law extends to the ϕ-mixing case, as follows:

THEOREM 6. Let $\{X_n\}$ be
(i) identically distributed, and ϕ-mixing with $\Sigma\ \phi(n)^{\frac{1}{2}} < \infty$, or
(ii) stationary, and ϕ-mixing with any rate.
Then $S_n/n \to c$ (finite) a.s. iff $E|X| < \infty$, and then $EX = c$.

Proof. That $E|X| < \infty$ and $EX = \mu$ implies $S_n/n \to \mu$ a.s. follows from the ergodic theorem in case (ii) ([36], Th. 3.5.7) and a result of Cohn [18] in case (i) (Iosifescu & Theodorescu [27], 19). Conversely, if $S_n/n \to c$ a.s., then

$$X_n/n = S_n/n - (S_{n-1}/(n-1))(n-1)/n \to 0 \text{ a.s.};$$

in particular,

$$P(|X_n|/n \geq 1 \text{ i.o.}) = 0.$$

By the Borel-Cantelli lemma for ϕ-mixing sequences ([27], 4, or O'Brien [33], Lemma 2),

$$\Sigma\ P(|X_1|/n \geq 1) = \Sigma\ P(|X_n|/n \geq 1) < \infty.$$

Thus $E|X_1| < \infty$, and then $c = ES_1$ as above.

The mixing-rate $\Sigma\ \phi(n)^{\frac{1}{2}} < \infty$ holds typically if $\phi(n)$ decreases like a power. That the Baum-Katz theorem may be extended to the ϕ-mixing case under a logarithmic mixing-rate is shown by the following important recent result of Peligrad [34]:

THEOREM 7. If $\alpha > \frac{1}{2}$, $\alpha p > 1$, $EX = 0$ if $\alpha \leq 1$, $\{X_n\}$ is stationary and ϕ-mixing with

$$\Sigma\ \phi^{1/k}(2^i) < \infty$$

(where k is 0 if $0 < p < 1$, 2 if $1 \leq p < 2$, $[(\alpha p-1)/(\alpha-\frac{1}{2})] + 1$ if $p \geq 2$), then statements (i)-(iii) of Theorem 5 are still equivalent.

We note that no rate of ϕ-mixing need be assumed to show that (iii) implies (i), by a result of Lai ([32], Lemma 5 (ii)).

Thus under the conditions of Theorem 7, one has

$$n^{-\alpha} S_n \to 0 \quad (p\alpha-1)\text{-quickly},$$

so also

$$n^{-\alpha} S_{[n+un^{1/p}]} \to 0 \ (p\alpha-1)\text{-quickly},$$

whence easily

$$n^{-\alpha}(S_{[n+un^{1/p}]} - S_n) \to 0 \quad (p\alpha-1)\text{-quickly}.$$

This is a delayed-average type result, which may be compared with Theorem 4.

One may also compare the results above with those of Berbee ([3],Th.1.1,6.1). There, the $\{X_n\}$ are uniformly bounded, so the link with moment conditions is lost. But, statements like Theorem 5 (iii) are obtained for absolutely regular $\{X_n\}$ ('β-mixing', which is weaker than ϕ-mixing), under sharp mixing rates.

5. <u>Remarks</u>.

1. It would be of interest to know whether Theorem 6(i) can be obtained without any rate-condition on $\phi(n)$, and if not, what type of rate on $\phi(n)$ is needed.

2. Similarly, one may seek a ϕ-mixing version of Theorem 4, and the classical Marcinkiewicz-Zygmund strong law, under suitable mixing rates.

3. Peligrad [33] obtains part of Theorem 6 with ϕ-mixing replaced by the weaker ρ-mixing (or 'maximal-correlation mixing': [26], 313).

4. Etemadi's proof [20] of Kolmogorov's strong law dispenses with Kolmogorov's inequality and uses a Borel-Cantelli argument; it is valid even in the pairwise-independent case. It would be interesting to know whether or not the results above hold under pairwise-independence, and whether the ϕ-mixing version of the Borel-Cantelli lemma can be combined with Etemadi's method.

5. In Theorems 2,3,3' and 4, different summability methods appear. In each case the first-named implies the second, while the second implies the first under suitable (Tauberian) conditions.

Best-possible Tauberian conditions are known; see [5],[10].

6. Theorem 5 is linked to certain 'Paley-type inequalities', for which see Chow & Lai [15].

7. The laws of large numbers considered here have analogues of iterated-logarithm type, for which see e.g. [29], [30], [9].

8. Intermediate between these laws of large numbers and laws of the iterated logarithm are certain 'laws of the single logarithm', which are in a sense limiting cases of Theorem 5 as $\alpha \to \frac{1}{2}$; see Lai [29], Chow & Lai [14].

9. Some results for the weak law of large numbers analogous to those considered here for the strong law have recently been given by Esseen & Janson [19].

Acknowledgement

I am indebted to Professor George O'Brien for the argument in Theorem 6(ii).

References

1. S. ASMUSSEN & T.G. KURTZ: Necessary and sufficient conditions for complete convergence in the law of large numbers. Ann. Probab. 8 (1980), 176-182.

2. L.E. BAUM & M. KATZ: Convergence rates in the law of large numbers. Trans.Amer.Math.Soc. 120 (1965),108-123.

3. H.C.P. BERBEE: Convergence rates in the strong law for bounded mixing sequences. Report MS-R8412, Amsterdam, 1984.

4. N.H. BINGHAM: On Valiron and circle convergence. Math. Z. 186 (1984), 273-286.

5. ———————: On Euler and Borel summability. J.London Math.Soc. (2) 29 (1984), 141-146.

6. ———————: Tauberian theorems for summability methods of random-walk type. J.London Math.Soc. (2) 30 (1984), 281-287.

7. ——————— & C.M. GOLDIE: Probabilistic and deterministic averaging. Trans.Amer.Math.Soc. 269 (1982), 453-480.

8. ———————: On one-sided Tauberian conditions. Analysis 3 (1983), 159-188.

9. ——————— & M. MAEJIMA: Summability methods and almost sure convergence. Z. Wahrschein. 68 (1985), 383-392.

10. ——————— & G. TENENBAUM: Riesz and Valiron means and fractional moments. Math.Proc.Camb.Phil.Soc., to appear.

11. R.C. BRADLEY: On the ϕ-mixing condition for stationary random sequences. Duke Math.J. 47 (1980), 421-433.

12. K. CHANDRASEKHARAN & S. MINAKSHISUNDARAM: Typical means. Oxford Univ.Press, 1952.

13. Y.S. CHOW: Delayed sums and Borel summability of independent identically distributed random variables. Bull.Inst.Math. Acad. Sinica 1 (1973), 207-220.

14. ——————& T.L.LAI: Some one-sided theorems on the tail-distribution of sample sums with applications to the last time and largest excess of boundary crossings. Trans. Amer.Math.Soc. 208 (1975), 51-72.

15. —————— : Paley-type inequalities and convergence rates related to the law of large numbers and extended renewal theory. Z. Wahrschein. 45 (1978), 1-19.

16. Y.S. CHOW & H. TEICHER: Probability theory: independence, interchangeability, martingales. Springer, New York, 1978.

17. K.L. CHUNG: A course in probability theory. Academic Press, New York, 1974.

18. H. COHN: On a class of dependent random variables. Rev.Roum. Math.Pures Appl. 10 (1965), 1593-1606.

19. C.G. ESSEEN & S. JANSON: On moment conditions for normed sums of independent random variables and martingale differences. Stoch.Proc.Appl. 19 (1985), 173-182.

20. N. ETEMADI: An elementary proof of the strong law of large numbers. Z. Wahrschein. 55 (1981), 119-122.

21. P.R. HALMOS: Lectures on ergodic theory. Chelsea, New York,1960.

22. G.H. HARDY: The second consistency theorem for summable series. Proc.London Math. Soc. (2) 15 (1916), 72-88 (Collected Works of G.H. Hardy VI, 588-605, Oxford Univ. Press, 1974).

23. ——————: Divergent series. Oxford Univ. Press, 1949.

24. —————— & M. RIESZ: The general theory of Dirichlet series. Cambridge Univ. Press, 1952.

25. C.R. HEATHCOTE: Complete exponential convergence and some related topics. J.Appl. Probab. 4 (1967), 217-256 (Methuen Review Series in Applied Probability Vol. 7).

26. I.A. IBRAGIMOV & Yu. V. LINNIK: Independent and stationary sequences of random variables. Walters-Noordhoff, Groningen, 1971.

27. M. IOSIFESCU & R. THEODORESCU: Random processes and learning. Springer, New York, 1969.

28. N. C. JAIN, K. JOGDEO & W.F. STOUT: Upper and lower functions for martingales and mixing processes. Ann. Probab. 3 (1975), 119-145.

29. T.L. LAI: Limit theorems for delayed sums. Ann. Probab. 2 (1974), 432-440.

30. ——————: Summability methods for independent, identically distributed random variables. Proc.Amer. Math.Soc. 45 (1974), 253-261.

31. ——————————: On r-quick convergence and a conjecture of Strassen. Ann.Probab. 4 (1976), 612-627.

32. ——————————: Convergence-rates and r-quick versions of the strong law for stationary mixing sequences. Ann. Probab. 5 (1977), 693-706.

33. G.L. O'BRIEN: Path properties of successive sample maxima from stationary processes. Z. Wahrschein. 38 (1977), 313-327.

34. M. PELIGRAD: Convergence rates of the strong law for stationary mixing sequences. Z.Wahrschein. 70 (1985), 307-314.

35. V. V. PETROV: Sums of independent random variables. Springer, New York, 1975.

36. W.F. STOUT: Almost sure convergence. Academic Press, New York, 1974.

37. V. STRASSEN: Almost sure behaviour of sums of independent random variables and martingales. Proc. Fifth Berkeley Symp. Math.Statist. Probab. 2 (1967), 315-343.

Mathematics Department, Royal Holloway & Bedford New College, (University of London), Egham Hill, Egham, Surrey, TW20 OEX, UK.

SECTION 3

MARTINGALE DEPENDENCE STRUCTURES

ON MARTINGALE CENTRAL LIMIT THEORY

by

Peter Gaenssler and Erich Haeusler, Munich

Summary. The present paper is expository in nature focussing on an effective way of proving ordinary and functional central limit theorems (CLT's and FCLT's) for martingales starting with a martingale version of Lindeberg's proof of the classical CLT and going up to FCLT's for continuous time local martingales known through the work of Rebolledo, Liptser and Shiryayev, and Helland.

1. Introduction

Martingale central limit theory is surely one of the most successful attempts of extending the classical limit theory for sums of independent random variables (rv's) to cover also dependent variables. As is well known, the classical CLT's and FCLT's have analogues for martingale difference arrays (mda's) which constitute a versatile tool in applied probability and statistics; cf. the monograph of Hall and Heyde [17] for some of the many applications. Furthermore, the theory has been extended to sequences of continuous time local martingales and semimartingales through the important work of Rebolledo [35,36], Liptser and Shiryayev [28], Helland [18], and many others, and in this form it has proved its value in extensive applications, e.g. to the statistical analysis of counting processes; cf. Helland [19] and references therein.

The present paper provides an exposition of what the authors consider as key results in ordinary and functional martingale central limit theory, trying to indicate a coherent and effective treatment of both the discrete and continuous time case. The discrete time theory for mda's will be discussed in Section 2. Starting from a martingale version of Lindeberg's [26] proof for the classical CLT and an appropriately simplified version of Brown's [4] and McLeish's [32] treatment of the martingale FCLT we want to demonstrate that the basic martingale results may be obtained by almost the same proofs as the classical results for independent rv's. It will be seen that exactly the same estimates which are fundamental for the classical reasoning hold true in the martingale case, too. One has to give only slightly different arguments to justify them. The basic sufficient conditions arising naturally out of this approach are the conditional form of Lindeberg's condition and a norming condition based on conditional variances. In the literature there occur a large number of different sufficient conditions for the martingale CLT and FCLT to hold. A second norming condition, known as Raikov's condition, is of particular interest, namely, norming by the sums of the squares of the variables. Many papers, e.g. [18,32,39,40], reveal a close connection between all these different sets of sufficient conditions. In the second part of Section 2 we are going to make this connection as plain as possible by formulating a set of sufficient conditions which is implied by most of the other sets occurring in the literature, and which in turn implies that the underlying array of rv's is asymptotically equivalent to a mda that satisfies the formally strongest set of sufficient conditions. Thus, most of the known martingale CLT's and FCLT's may easily be transformed into each other so that they may be viewed as being tantamount to each other, not in a logical sense, but as far as their mathematical content is concerned.

The understanding of the discrete time theory mentioned so far requires only some basic knowledge about discrete time martingales and, for the FCLT, of course, aquaintance with the elements of convergence of distributions on function spaces. Regarding the continuous time theory the situation is different. The approach of Rebolledo and Liptser and Shiryayev has the conceptual advantage of covering the discrete time theory as a special case, but at the same time presupposes familiarity with the general theory of stochastic processes, of continuous time local martingales and of semimartingales, and with stochastic integration. Thus, the non-expert in these fields might have difficulties to follow. It is therefore of interest to have a treatment that clearly reveals the analogies between the discrete and continuous time case, and a derivation of the continuous time theory from the discrete time one which reduces the necessary prerequisites as far as possible. Such an approach has been developed by Helland [18] and will be further analyzed in Section 3 of the present paper. Helland's method consists in discretizing continuous time martingales and then applying the discrete time CLT or FCLT to the 'discrete skeletons'. He created special arguments for three different sets of sufficient conditions. Reformulating parts of Section 2 in the language of continuous time processes we will see that as in the discrete case all the different sufficient conditions occurring in Rebolledo [36] are tantamount to each other up to asymptotic equivalence of the underlying processes. Consequently, to deduce the whole continuous time theory from the discrete time one, it is enough to employ a single discretization argument, and, what of course is the point, one then can choose the most tractable one. This approach leads to the same unified view of the different sets of sufficient conditions in continuous time martingale central limit theory as in discrete time and, in our opinion, it also leads to a rather effective proof of Rebolledo's [36] results.

From the description of the aim of the present paper it is clear that it partly coincides with the aim of the two recent expositions of martingale central limit theory by Helland [18] and Rootzén [39]. In fact, we were greatly inspired by these two papers and build heavily upon them as well as on [36]. Due to our goal to concentrate on effective methods of proof of basic results we will not go into the details of further topics of interest but will only mention some of them briefly. Necessity of the sufficient conditions in martingale central limit theory has been studied by several authors. For the FCLT in discrete time, rather satisfactory results are obtained in Rootzén [37] and [38] and Gaenssler and Haeusler [12]. Corresponding results for the continuous time case are given by Jacod [21] and Liptser and Shiryayey [29] and [30]. For the CLT the problem is much harder. Jeganathan [22] and [23] deals with this problem, even for more general limiting distributions than the standard normal, but his results do not seem to be the final answer. As far as rates of weak convergence are concerned, the situation is also a bit unsatisfactory. The number of authors dealing with this problem is too large to quote all of them here. In most cases bounds on the rate of convergence are given without investigating their optimality. Bolthausen [3] is an outstanding exception. His results show that the rate in the martingale CLT is rather poor in general and that one has to impose rather restrictive conditions to obtain optimal rates comparable to the rate in the independent case. Bolthausen [3] is not concerned with the most general situation as considered, for example, in Heyde and Brown [20] and Liptser and Shiryayev [31], and it would be worthwhile to know whether the bounds obtained in these papers are optimal, or not. As far as the FCLT is concerned, the situation is better as long as one restricts oneself to the case of martingales having finite moments of low order only. The rate obtained by Hall and Heyde [17] is optimal up to a logarithmic factor, and even this additional logarithmic factor can be removed under stronger assumptions leading to the same optimal rate in the martingale FCLT as in the FCLT for independent rv's; see Haeusler [14].

Since our intention is to present what can now be regarded as a rather complete body of knowledge we will not discuss here the far-reaching generalizations of martingale central limit theory to more general processes, e.g. semimartingales, and to more general limiting processes than Brownian motion. These more general theories are still in rapid development.

2. Ordinary and functional CLT's for martingale difference arrays

In this section we will always consider real-valued rv's ξ_{nk} for $n,k \in \mathbb{N} = \{0,1,2,\ldots\}$, all defined on some basic p-space $(\Omega,\mathcal{F},P)$. Let $\mathcal{F}_{n,k}$, $n,k \in \mathbb{N}$, be sub-σ-fields of $\mathcal{F}$ with $\mathcal{F}_{n,k-1} \subset \mathcal{F}_{nk}$ for all n and k. The array $(\xi_{nk},\mathcal{F}_{nk})$ is called adapted iff ξ_{nk} is measurable wrt $\mathcal{F}_{nk}$ for all $n,k \in \mathbb{N}$, and an adapted array of integrable rv's is called a martingale difference array (mda) iff $E(\xi_{nk}|\mathcal{F}_{n,k-1}) = 0$ a.s. for all $n \in \mathbb{N}$ and $k = 1,2,\ldots$ Centered and row-wise independent rv's, i.e. when $\xi_{n0},\xi_{n1},\ldots$ have zero means and are independent for each n, are a special case of this notion if we set $\mathcal{F}_{nk} := \sigma(\xi_{n0},\ldots,\xi_{nk})$.

We always set $\mathcal{F}_{n,-1} := \mathcal{F}_{n0}$, and the conditional expectation operator $E(\cdot|\mathcal{F}_{n,k-1})$ will be abbreviated by $E_{k-1}(\cdot)$. For convenience we write $\sigma^2_{nk} = E_{k-1}(\xi^2_{nk})$ provided that ξ_{nk} is square integrable. $\overset{P}{\to}$ and $\overset{\mathcal{L}}{\to}$ denote convergence in probability and in distribution (in law), respectively, $\overset{\mathcal{L}}{=}$ denotes equality in distribution (in law), and limits are always to be understood as $n \to \infty$ if not stated otherwise. In general, all equations and inequalities between random quantities are meant to hold with probability one only also when not stated explicitely. By an abuse of notation $N(\mu,\sigma^2)$ denotes the normal law with mean μ and variance σ^2 or a rv having this distribution, as the case may be.

Through the work of numerous authors it is well known that the martingale analogues of the classical CLT's for partial sums of row-wise independent arrays (with finite rows) are obtained if in the classical conditions the probabilities and expectations, respectively, are replaced by the corresponding conditional quantities, where conditioning is wrt the past. In the martingale case it is appropriate to consider infinite rows and row sums of the form $\Sigma_{k=0}^{\tau(n)} \xi_{nk}$ where $\tau(n)$ is an a.s. finite stopping time wrt $(F_{nk})_{k\in\mathbb{N}}$ for each n. Then the martingale analogue of Lindeberg's fundamental CLT reads as follows:

<u>Theorem 2.1.</u> Let (ξ_{nk},F_{nk}) be a mda with square integrable rv's, and let $\tau(n)$ for each n be an a.s. finite stopping time wrt $(F_{nk})_{k\in\mathbb{N}}$. If

$$\Sigma_{k=0}^{\tau(n)} \sigma_{nk}^2 \xrightarrow{P} 1 \tag{N}$$

and if the conditional Lindeberg condition

$$\Sigma_{k=0}^{\tau(n)} E_{k-1}(\xi_{nk}^2 I(|\xi_{nk}| > \varepsilon)) \xrightarrow{P} 0 \quad \text{for each } \varepsilon > 0 \tag{CL}$$

is satisfied, then

$$S_n := \Sigma_{k=0}^{\tau(n)} \xi_{nk} \xrightarrow{L} N(0,1) \; . \tag{CLT}$$

Theorem 2.1 as formulated here is essentially due to Brown [4] and Dvoretzky [10] whereas a martingale CLT under Ljapunov's condition already appears in Billingsley [1] who mentions his debt to Lévy, e.g. [25]. The proofs in [1,4,10] are based on characteristic function techniques. On the other hand, Lindeberg's [26] original proof for independent rv's involved a direct estimate of the distance between $L\{S_n\}$ and $N(0,1)$ in terms of an upper bound of

$$\Delta_n := |E(f(S_n)) - E(f(N(0,1)))|$$

for every function $f:\mathbb{R} \to \mathbb{R}$ having bounded and continuous derivatives of any order: If one successively replaces the ξ_{nk}'s in their sum S_n

by independent normal rv's N_{nk} with mean zero and appropriate variances, then (N) and (CL) guarantee that $\Delta_n \to 0$ which is tantamount to (CLT) by the definition of convergence in distribution for real-valued rv's. This intuitively appealing idea yields an elementary proof of asymptotic normality which avoids the use of characteristic functions completely; cf. the monographs of Billingsley [2] and Pollard [34]. In the martingale context Lindeberg's approach has been used by Lévy [25] under rather restrictive assumptions, but as remarked in Haeusler [15], it may also be used to prove Theorem 2.1. In a first step it is convenient to consider $\tau(n)$'s which are non-random, i.e. $\tau(n) = k(n)$ for non-negative integers $k(n)$, and to suppose that

$$V_n^2 := \Sigma_{k=0}^{k(n)} \sigma_{nk}^2 = 1 \quad \text{a.s.} \quad \text{for all} \quad n \in \mathbb{N} . \tag{2.1}$$

Furthermore, we assume wlog that on $(\Omega, \mathcal{F}, P)$ there exist standard normal rv's N_{nk}, $0 \leq k \leq k(n)$, $n \in \mathbb{N}$, such that

$$N_{n0}, \ldots, N_{n,k(n)} \quad \text{and} \quad \mathcal{F}_{n,k(n)} \quad \text{are independent for all} \quad n . \tag{2.2}$$

Then we have

$$\Sigma_{k=0}^{k(n)} \sigma_{nk} N_{nk} \stackrel{\mathcal{L}}{=} N(0,1) . \tag{2.3}$$

If $\xi_{n0}, \ldots, \xi_{n,k(n)}$ are independent, then the σ_{nk}^2's are constants summing up to 1, and (2.3) is immediate. In the martingale case the σ_{nk}^2's may be random, but by (2.1) their sum still equals 1 a.s., and since by (2.2) the conditional distribution $\mathcal{L}\{\Sigma_{k=0}^{k(n)} \sigma_{nk} N_{nk} | \mathcal{F}_{n,k(n)}\}$ equals $N(0,1)$ a.s. on $\{V_n^2 = 1\}$, this is enough to entail (2.3) in this case, too. Setting

$$\zeta_{nk} := \Sigma_{j=0}^{k-1} \xi_{nj} + \Sigma_{j=k+1}^{k(n)} \sigma_{nj} N_{nj}$$

for $0 \leq k \leq k(n)$ and $n \in \mathbb{N}$ we have $\zeta_{n,k+1} + \sigma_{n,k+1} N_{n,k+1} = \zeta_{nk} + \xi_{nk}$ and thus obtain for all n and f

$$f(S_n) - f(\Sigma_{k=0}^{k(n)} \sigma_{nk} N_{nk}) = \Sigma_{k=0}^{k(n)} (f(\zeta_{nk} + \xi_{nk}) - f(\zeta_{nk} + \sigma_{nk} N_{nk})) .$$

Taking expectations and using (2.3) we get

$$\Delta_n \leq \Sigma_{k=0}^{k(n)} |E(f(\zeta_{nk} + \xi_{nk}) - f(\zeta_{nk} + \sigma_{nk} N_{nk}))| . \tag{2.4}$$

If $\xi_{n0}, \ldots, \xi_{n,k(n)}$ are independent, then the σ_{nk}^2's are constants, and ζ_{nk} is independent of ξ_{nk} so that

$$E(f'(\zeta_{nk})\xi_{nk}) = 0 \ , \ E(f'(\zeta_{nk})\sigma_{nk}N_{nk}) = 0 \ \text{ and}$$
$$E(f''(\zeta_{nk})(\xi_{nk}^2 - \sigma_{nk}^2 N_{nk}^2)) = 0 \ . \tag{2.5}$$

Therefore, the usual method is to employ Taylor's expansion of f around ζ_{nk} up to order three on the rhs of (2.4): The zero order terms cancel, the first order terms vanish separately by (2.5), and the second order terms also cancel by (2.5). The remainder terms are forced to converge to zero by the classical Lindeberg condition

$$\Sigma_{k=0}^{k(n)} E(\xi_{nk}^2 I(|\xi_{nk}| > \varepsilon)) \to 0 \quad \text{for each} \quad \varepsilon > 0 \tag{L}$$

and elementary arguments. In the martingale case (2.1) and (CL) entail (L) via dominated convergence, and the remainder terms can be dealt with exactly as in the independent case. Therefore, it only remains to establish (2.5) in the martingale case, too. The equality in the middle is still obvious from independence of $(\zeta_{nk}, \sigma_{nk})$ and N_{nk} following from (2.2). To verify the other two parts of (2.5), one again argues by means of conditional laws as in the proof of (2.3) observing that (2.2) implies

$$L\{\zeta_{nk} | F_{n,k(n)}\} = N(\Sigma_{j=0}^{k-1}\xi_{nj} \ , \ \Sigma_{j=k+1}^{k(n)}\sigma_{nj}^2) \quad \text{a.s.} \tag{2.6}$$

If we assume that each F_{n0} contains all P-null sets from F , which causes no loss of generality because events of probability zero do not affect distributional properties, then it follows from (2.1) and (2.6) that there exist $F_{n,k-1}$ - measurable versions of $E_{k(n)}(f'(\zeta_{nk}))$ and of $E_{k(n)}(f''(\zeta_{nk}))$. But this yields

$$E(f'(\zeta_{nk})\xi_{nk}) = E\{E_{k(n)}(f'(\zeta_{nk}))E_{k-1}(\xi_{nk})\} = 0$$

by the mda property of (ξ_{nk}, F_{nk}) and

$$E(f''(\zeta_{nk})\xi_{nk}^2) = E\{E_{k(n)}(f''(\zeta_{nk}))E_{k-1}(\xi_{nk}^2)\} = E(f''(\zeta_{nk})\sigma_{nk}^2 N_{nk}^2)$$

proving (2.5) and Theorem 2.1 under (2.1) for $\tau(n) = k(n)$.

Retaining the assumption $\tau(n) = k(n)$, condition (2.1) can be easily removed by a stopping time technique which at least traces back to Lévy [25]. The version that is appropriate in the present situation has been given by Dvoretzky [10] as part of the proof of his Theorem 2.2 so that we may skip the details here. In a last step it remains to remove the assumption that the $\tau(n)$'s are non-random. But since they are assumed to be a.s. finite, there exists a sequence $(k(n))_{n\in\mathbb{N}}$ of non-negative integers such that $P(\tau(n) > k(n)) \to 0$. The rv's $\eta_{nk} := \xi_{nk} I(k \leq \tau(n))$ define a mda (η_{nk}, F_{nk}) which satisfies (CL) and (N) with $k(n)$ instead of $\tau(n)$ so that (CLT) follows from the previous part of the proof and $\Sigma_{k=0}^{\tau(n)} \xi_{nk} - \Sigma_{k=0}^{k(n)} \eta_{nk} \overset{P}{\to} 0$. This concludes the proof of Theorem 2.1.

For row-wise independent ξ_{nk}'s with zero means and finite variances Donsker and Prokhorov developed the celebrated functional version of Theorem 2.1. In our notation this result may be formulated as follows: Assume $\limsup_{k\to\infty} \Sigma_{j=0}^{k} E(\xi_{nj}^2) > 1$ for each $n \in \mathbb{N}$ and set

$$\tau(n,t) := \inf\{k : \Sigma_{j=0}^{k} E(\xi_{nj}^2) > t\} \quad \text{for each} \quad 0 \leq t \leq 1 . \tag{2.7}$$

Then the partial sum processes

$$S_{(n)}(t) := \Sigma_{k=0}^{\tau(n,t)} \xi_{nk} \ , \quad 0 \leq t \leq 1 , \tag{2.8}$$

are random functions belonging to the Skorokhod space $D[0,1]$ of all real-valued functions on $[0,1]$ which are right-continuous and have left-hand limits. Under (L) the sequence $(S_{(n)})_{n\in\mathbb{N}}$ converges in distribution to a standard Brownian motion B_1 with time interval $[0,1]$, written as

$$S_{(n)} \overset{L}{\to} B_1 . \tag{FCLT}$$

Here convergence in distribution of random functions in $D[0,1]$ is meant in the sense of Billingsley [2].

The natural martingale analogue of the non-random time scales of (2.7) are the random time scales

$$\tau(n,t) := \inf\{k : \Sigma_{j=0}^{k} \sigma_{nj}^2 > t\} \quad \text{for} \quad 0 \leq t \leq 1 \tag{2.9}$$

where we assume $\limsup_{k\to\infty} \Sigma_{j=0}^{k} \sigma_{nj}^2 > 1$ a.s. for each n . The martingale version of the Donsker-Prokhorov theorem then states that (FCLT) remains to be true with the time scales (2.9) used in (2.8) for any mda which satisfies (CL) (with $\tau(n)$ replaced by $\tau(n,1)$; obvious adaptions like this will not be mentioned explicitely in the sequel). Actually, the special structure of the time scales in (2.9) is not even needed, only the fact that $\tau(n,t)$ is a stopping time wrt $(F_{nk})_{k\in\mathbb{N}}$ for each n and t , and that

$$\Sigma_{k=0}^{\tau(n,t)} \sigma_{nk}^2 \overset{P}{\to} t \quad \text{for each} \quad 0 \leq t \leq 1 \qquad (N_t)$$

holds which for the time scales in (2.9) is obviously implied by (CL). Therefore, from now on we shall always consider partial sum processes $S_{(n)}$ as in (2.8) that are constructed with a general sequence $(\tau(n,\cdot))_{n\in\mathbb{N}}$ of time scales which by definition is a sequence of random processes $(\tau(n,t))_{0\leq t\leq 1}$ such that $\tau(n,\cdot)$ is a non-decreasing integer-valued element of $D[0,1]$ with $\tau(n,0) = 0$ a.s. and $\tau(n,t)$ is an a.s. finite stopping time wrt $(F_{nk})_{k\in\mathbb{N}}$ for each t and n. For this setting we have the following basic martingale FCLT:

Theorem 2.2. Let (ξ_{nk}, F_{nk}) be a mda with square integrable rv's, and let $(\tau(n,\cdot))_{n\in\mathbb{N}}$ be a sequence of time scales. If (CL) and (N_t) are satisfied, then (FCLT) holds true.

Several methods of proof for this and related results have been developed in the literature. Drogin's [8] approach is based entirely upon results about fair coin tossing. For proofs involving martingale embeddings into Brownian motion see Freedman [11], Section 1.9, Hall [16], Hall and Heyde [17] and Scott [40]. Durrett and Resnick [9] and Rootzén [37,38,39] start from FCLT's under quite restrictive conditions and obtain the most general results via random change of time arguments. The proofs given by Brown [4] and McLeish [32] follow the classical two step pattern: In a first step convergence in law of the finite dimensional distributions (fidi convergence for short) is established and then uniform tightness is proven by an application of fidi convergence and an appropriate maximal inequality. In the present situation this sort of reasoning requires a minimum number of prerequisites and, up to notational changes due to the fact that the $\tau(n,t)$'s are random, it is the same as in the classical case of row-wise independent rv's: Since fidi's and joint distributions of increments can be continuously transformed into each other, it is enough to show

$$(S_{(n)}(0), S_{(n)}(t_1) - S_{(n)}(t_0), \ldots, S_{(n)}(t_\ell) - S_{(n)}(t_{\ell-1})) \overset{L}{\to}$$

$$(B_1(0), B_1(t_1) - B_1(t_0), \ldots, B_1(t_\ell) - B_1(t_{\ell-1}))$$

for each $\ell \geq 1$ and $0 = t_0 < t_1 < \ldots < t_\ell = 1$. According to the Cramér-Wold device this is equivalent to verify for all real $\lambda_0, \lambda_1, \ldots, \lambda_\ell$ (observe that $B_1(0) = 0$ a.s.)

$$\lambda_0 S_{(n)}(0) + \Sigma_{j=1}^{\ell}\lambda_j(S_{(n)}(t_j) - S_{(n)}(t_{j-1})) \xrightarrow{L} \Sigma_{j=1}^{\ell}\lambda_j(B_1(t_j) - B_1(t_{j-1})) \ . \tag{2.10}$$

Setting

$$\eta_{nk} := \lambda_0 I(k \leq \tau(n,0))\xi_{nk} + \Sigma_{j=1}^{\ell}\lambda_j I(\tau(n,t_{j-1}) < k \leq \tau(n,t_j))\xi_{nk}$$

we obtain a mda (η_{nk}, F_{nk}) for which

$$\lambda_0 S_{(n)}(0) + \Sigma_{j=1}^{\ell}\lambda_j(S_{(n)}(t_j) - S_{(n)}(t_{j-1})) = \Sigma_{k=0}^{\tau(n,1)}\eta_{nk} \ .$$

(η_{nk}, F_{nk}) satisfies (CL) because of $|\eta_{nk}| \leq \Sigma_{j=0}^{\ell}|\lambda_j||\xi_{nk}|$, and

$$\Sigma_{k=0}^{\tau(n,1)} E_{k-1}(\eta_{nk}^2) = \lambda_0^2\Sigma_{k=0}^{\tau(n,0)}\sigma_{nk}^2 + \Sigma_{j=1}^{\ell}\lambda_j^2\Sigma_{k=\tau(n,t_{j-1})+1}^{\tau(n,t_j)}\sigma_{nk}^2 \xrightarrow{P} \Sigma_{j=1}^{\ell}\lambda_j^2(t_j - t_{j-1})$$

by (N_t) for $0 \leq t \leq 1$. Theorem 2.1 therefore implies

$$\Sigma_{k=0}^{\tau(n,1)}\eta_{nk} \xrightarrow{L} N(0, \Sigma_{j=1}^{\ell}\lambda_j^2(t_j - t_{j-1})) \overset{L}{=} \Sigma_{j=1}^{\ell}\lambda_j(B_1(t_j) - B_1(t_{j-1}))$$

proving (2.10).

To establish uniform tightness of the distributions of $S_{(n)}$ via Theorem 15.5 in [2] we have to show that for each $\varepsilon > 0$

$$\lim_{m\to\infty} \limsup_{n\to\infty} P(n,m,\varepsilon) = 0 \tag{2.11}$$

with $P(n,m,\varepsilon) := P(\sup\{|S_{(n)}(s) - S_{(n)}(t)| : 0 \leq s,t \leq 1 \ ; \ |s-t| \leq 1/m\} \geq \varepsilon)$. The appropriate maximal inequality for doing this is

$$P(\max_{0\leq k\leq\tau}|S_k| \geq 2\varepsilon) \leq \varepsilon^{-1}E(|S_\tau| I(|S_\tau| \geq \varepsilon)) \tag{2.12}$$

being valid for each $\varepsilon > 0$, any uniformly integrable martingale $(S_k, F_k; k \in \mathbb{N})$ and any a.s. finite stopping time τ wrt $(F_k)_{k\in\mathbb{N}}$. Brown [4] derived a sharper result from Doob's upcrossing inequality, but (2.12) is all what is needed here. It follows easily from Doob's maximal inequality for non-negative submartingales; cf. Dellacherie and Meyer [6], p. 18: Setting $M_m := \max_{0\leq k\leq m}|S_{k\wedge\tau}|$ and observing that $(|S_{k\wedge\tau}|, F_k; k \in \mathbb{N})$ is a non-negative submartingale, we obtain for all $m \in \mathbb{N}$ and $\varepsilon > 0$

$$2\varepsilon P(M_m \geq 2\varepsilon) \leq E(|S_{m\wedge\tau}| I(M_m \geq 2\varepsilon))$$

$$= E(|S_{m\wedge\tau}| I(M_m \geq 2\varepsilon, |S_{m\wedge\tau}| \geq \varepsilon)) + E(|S_{m\wedge\tau}| I(M_m \geq 2\varepsilon, |S_{m\wedge\tau}| < \varepsilon))$$

$$\leq E(|S_{m\wedge\tau}| I(|S_{m\wedge\tau}| \geq \varepsilon)) + \varepsilon P(M_m \geq 2\varepsilon)$$

whence $P(M_m \geq 2\varepsilon) \leq \varepsilon^{-1} E(|S_{m\wedge\tau}| I(|S_{m\wedge\tau}| \geq \varepsilon))$. Letting m tend to infinity yields (2.12); for this, observe that the rv's $S_{m\wedge\tau}$ are uniformly integrable.

The classical proof for (2.11) now works under the additional assumption

$$\Sigma_{k=0}^{\tau(n,1)} \sigma_{nk}^2 \leq c < \infty \text{ a.s. for all } n \in \mathbb{N} \text{ and some } c < \infty, \quad (2.13)$$

for then one can use the familiar discretization technique and inequality (2.12), applied to the uniformly integrable martingales $(\Sigma_{i=0}^{k} \xi_{ni} I(\tau(n,\frac{j-1}{m}) < i \leq \tau(n,\frac{j}{m})), F_{nk}; k \in \mathbb{N})$, to obtain

$$P(n,m,\varepsilon) \leq \Sigma_{j=1}^{m} P(\max_{0\leq k\leq\tau(n,j/m)} |\Sigma_{i=0}^{k} \xi_{ni} I(\tau(n,\tfrac{j-1}{m}) < i \leq \tau(n,\tfrac{j}{m}))| \geq \tfrac{\varepsilon}{3})$$

$$\leq \frac{6}{\varepsilon} \Sigma_{j=1}^{m} E(|\Sigma_{i=\tau(n,(j-1)/m)+1}^{\tau(n,j/m)} \xi_{ni}| I(|\Sigma_{i=\tau(n,(j-1)/m)+1}^{\tau(n,j/m)} \xi_{ni}| \geq \tfrac{\varepsilon}{6})) .$$

Applying the Cauchy-Schwarz inequality and (2.13) we get

$$P(n,m,\varepsilon) \leq \frac{6}{\varepsilon} c^{1/2} [\Sigma_{j=1}^{m} P(|S_{(n)}(\tfrac{j}{m}) - S_{(n)}(\tfrac{j-1}{m})| \geq \tfrac{\varepsilon}{6})]^{1/2}$$

whence by fidi convergence

$$\limsup_{n\to\infty} P(n,m,\varepsilon) \leq \frac{6}{\varepsilon} c^{1/2} [\Sigma_{j=1}^{m} P(|B_1(\tfrac{j}{m}) - B_1(\tfrac{j-1}{m})| \geq \tfrac{\varepsilon}{6})]^{1/2}$$

$$\leq (\frac{6}{\varepsilon})^3 c^{1/2} [\Sigma_{j=1}^{m} E(|B_1(\tfrac{j}{m}) - B_1(\tfrac{j-1}{m})|^4)]^{1/2}$$

$$= (\frac{6}{\varepsilon})^3 c^{1/2} E(|N(0,1)|^4) m^{-1/2}$$

which proves (2.11) under (2.13). To remove this additional assumption one can employ a simple truncation procedure: Consider the mda (η_{nk}, F_{nk}) with $\eta_{nk} := \xi_{nk} I(\Sigma_{j=0}^{k} \sigma_{nj}^2 \leq 2)$ which fulfills (2.13) with $c = 2$ by construction. Furthermore, it is easy to see that (η_{nk}, F_{nk}) also satisfies (CL) and (N_t) so that (FCLT) holds for the array (η_{nk}, F_{nk}). (FCLT) for the array (ξ_{nk}, F_{nk}) now follows from the fact that the partial sum processes based on (ξ_{nk}, F_{nk}) and (η_{nk}, F_{nk}) are asymptotically equivalent in the sense that

$$\sup_{0\le t\le 1}\left|\Sigma_{k=0}^{\tau(n,t)}\xi_{nk} - \Sigma_{k=0}^{\tau(n,t)}\eta_{nk}\right| \overset{P}{\to} 0 \tag{2.14}$$

so that (FCLT) holds for (ξ_{nk},F_{nk}) iff it holds for (η_{nk},F_{nk}). But (2.14) is immediate from the definition of η_{nk} and (N_t) for the array (ξ_{nk},F_{nk}). This concludes the proof of Theorem 2.2.

In the remaining part of this section we want to compare the basic conditions (CL) and (N_t) with other sets of conditions occurring in the literature and implying (FCLT) for mda's or even adapted arrays (ξ_{nk},F_{nk}) with arbitrary dependence structure. Our aim is to show that most of these conditions imply existence of a mda (η_{nk},F_{nk}) which satisfies (FCLT) by virtue of Theorem 2.2 and which is asymptotically equivalent to (ξ_{nk},F_{nk}) in the sense of (2.14) so that (FCLT) is satisfied for (ξ_{nk},F_{nk}), too. As an important tool we need two results which link together convergence to zero of the partial sums and the partial sums of the conditional expectations for an arbitrary adapted array (ξ_{nk},F_{nk}) of non-negative rv's. If $\tau(n)$ is an a.s. finite stopping time wrt $(F_{nk})_{k\in\mathbb{N}}$ for each n, then

$$\Sigma_{k=0}^{\tau(n)}E_{k-1}(\xi_{nk}) \overset{P}{\to} 0 \quad \text{implies} \quad \Sigma_{k=0}^{\tau(n)}\xi_{nk} \overset{P}{\to} 0\,, \tag{2.15}$$

and if $\{\max_{0\le k\le\tau(n)}\xi_{nk} : n\in\mathbb{N}\}$ is uniformly integrable, then

$$\Sigma_{k=0}^{\tau(n)}\xi_{nk} \overset{P}{\to} 0 \quad \text{implies} \quad \Sigma_{k=0}^{\tau(n)}E_{k-1}(\xi_{nk}) \overset{P}{\to} 0\,. \tag{2.16}$$

Special cases of these results have been used by many authors, often in an implicite form via truncation arguments like the one employed at the end of the proof of Theorem 2.2. The present explicit formulation turns out to be extremely useful for comparing different sets of conditions in CLT's for adapted arrays. Together with its simple proof this is a quite recent contribution due to Rootzén [39]. To verify (2.15) consider the stopping times $\nu(n) := \max\{k : \Sigma_{j=0}^{k}E_{j-1}(\xi_{nj}) \le 1\} \wedge \tau(n)$ wrt $(F_{nk})_{k\in\mathbb{N}}$. Then we have $\Sigma_{k=0}^{\nu(n)}E_{k-1}(\xi_{nk}) \le 1$ a.s. for all n and

$$\Sigma_{k=0}^{\nu(n)}E_{k-1}(\xi_{nk}) \le \Sigma_{k=0}^{\tau(n)}E_{k-1}(\xi_{nk}) \overset{P}{\to} 0$$

whence by dominated convergence

$$E(\Sigma_{k=0}^{\nu(n)}\xi_{nk}) = E(\Sigma_{k=0}^{\nu(n)}E_{k-1}(\xi_{nk})) \to 0\,.$$

In view of

$$P(\nu(n) \neq \tau(n)) = P(\Sigma_{j=0}^{\tau(n)} E_{j-1}(\xi_{nj}) > 1) \rightarrow 0$$

this implies (2.15). Using the stopping times $\nu(n) := \min\{k : \Sigma_{j=0}^{k}\xi_{nj} > 1\} \wedge \tau(n)$ the argument may be copied to prove (2.16) except that one has to guarantee uniform integrability of $\Sigma_{k=0}^{\nu(n)}\xi_{nk}$ by the additional assumption on $\max_{0\leq k\leq \tau(n)}\xi_{nk}$.

Implication (2.16) can be used, for example, to show that

$$E(\max_{0\leq k\leq \tau(n)}\xi_{nk}^2) \rightarrow 0 \qquad (M_2)$$

always implies (CL): Observe that for $\varepsilon > 0$, $0 < \delta < \varepsilon^2$ and $n \in \mathbb{N}$

$$P(\Sigma_{k=0}^{\tau(n)}\xi_{nk}^2 I(|\xi_{nk}| > \varepsilon) > \delta) = P(\max_{0\leq k\leq \tau(n)}|\xi_{nk}| > \varepsilon) \qquad (2.17)$$

so that (M_2) implies $\Sigma_{k=0}^{\tau(n)}\xi_{nk}^2 I(|\xi_{nk}| > \varepsilon) \overset{P}{\rightarrow} 0$. (CL) is therefore immediate from (M_2) and (2.16).

Now we are prepared to derive from Theorem 2.2 the following FCLT for arbitrary adapted arrays.

<u>Theorem 2.3.</u> Let (ξ_{nk}, F_{nk}) be an adapted array and let $(\tau(n,\cdot))_{n\in\mathbb{N}}$ be a sequence of time scales. Suppose that there exists a $0 < a < \infty$ such that for $\eta_{nk} := \xi_{nk}I(|\xi_{nk}| \leq a) - E_{k-1}(\xi_{nk}I(|\xi_{nk}| \leq a))$

$$\Sigma_{k=0}^{\tau(n,t)}\eta_{nk}^2 \overset{P}{\rightarrow} t \text{ for each } 0 \leq t \leq 1 \text{ and} \qquad (R_{a,t})$$

$$\sup_{0\leq t\leq 1}|\Sigma_{k=0}^{\tau(n,t)}[\xi_{nk}I(|\xi_{nk}|>a) + E_{k-1}(\xi_{nk}I(|\xi_{nk}|\leq a))]| \overset{P}{\rightarrow} 0 . \qquad (A_a)$$

Then the mda (η_{nk}, F_{nk}) satisfies (M_2), (N_t) and (2.14). Theorem 2.2 and '$(M_2) \Rightarrow$ (CL)' therefore imply (FCLT) for the given array (ξ_{nk}, F_{nk}).

For the proof of Theorem 2.3 observe that $\Sigma_{k=0}^{\tau(n,t)}\eta_{nk}^2$ is nondecreasing in t so that for $m = 1,2,\ldots$

$$\sup_{0\leq t\leq 1}|\Sigma_{k=0}^{\tau(n,t)}\eta_{nk}^2 - t| \leq \max_{0\leq i\leq m}|\Sigma_{k=0}^{\tau(n,i/m)}\eta_{nk}^2 - \frac{i}{m}| + \frac{1}{m} .$$

Consequently, $(R_{a,t})$ holds uniformly in $0 \leq t \leq 1$. Therefore

$$\max_{0\leq k\leq \tau(n,1)}\eta_{nk}^2 \leq \sup_{0\leq t\leq 1}|\Sigma_{k=0}^{\tau(n,t)}\eta_{nk}^2 - \Sigma_{k=0}^{\tau(n,t-)}\eta_{nk}^2| \leq 2\sup_{0\leq t\leq 1}|\Sigma_{k=0}^{\tau(n,t)}\eta_{nk}^2 - t|$$

with $\tau(n,t-) := \lim_{s\uparrow t}\tau(n,s)$ implies that (M_2) holds for the array (η_{nk}, F_{nk}) (observe that $|\eta_{nk}| \leq 2a$ a.s. by construction). Setting

$\nu(n,t) := \min\{k : \Sigma_{j=0}^{k}\eta_{nj}^2 > t+1\} \wedge \tau(n,t)$ we obtain stopping times wrt $(F_{nk})_{k\in\mathbb{N}}$ for which $P(\nu(n,t) \neq \tau(n,t)) \to 0$ by $(R_{a,t})$. Thus

$$\Sigma_{k=0}^{\tau(n,t)}\eta_{nk}^2 - \Sigma_{k=0}^{\nu(n,t)}\eta_{nk}^2 \overset{P}{\to} 0$$

and

$$\Sigma_{k=0}^{\tau(n,t)}E_{k-1}(\eta_{nk}^2) - \Sigma_{k=0}^{\nu(n,t)}E_{k-1}(\eta_{nk}^2) \overset{P}{\to} 0 .$$

Furthermore,

$$E\{[\Sigma_{k=0}^{\nu(n,t)}(\eta_{nk}^2 - E_{k-1}(\eta_{nk}^2))]^2\} = E\{\Sigma_{k=0}^{\nu(n,t)}(\eta_{nk}^2 - E_{k-1}(\eta_{nk}^2))^2\}$$

$$\leq 4E(\Sigma_{k=0}^{\nu(n,t)}\eta_{nk}^4) \leq 4E((\max_{0\leq k\leq \nu(n,t)}\eta_{nk}^2)\Sigma_{k=0}^{\nu(n,t)}\eta_{nk}^2)$$

$$\leq 4(t+1+4a^2)E(\max_{0\leq k\leq \tau(n,1)}\eta_{nk}^2) \to 0$$

by (M_2). Combining with $(R_{a,t})$ we get (N_t) for the array (η_{nk},F_{nk}). (2.14) is tantamount to (A_a) so that the proof is complete.

The conditions $(R_{a,t})$ and (A_a) are somewhat artificial because they are just what is needed to make the proof of Theorem 2.3 work. On the other hand, they constitute a very basic set of sufficient conditions for (FCLT). To explain this fact, let us compare them with some other conditions occurring in the literature. For this, consider

$$E(\max_{0\leq k\leq \tau(n,1)}|\xi_{nk}|) \to 0 \qquad (M_1)$$

and

$$\Sigma_{k=0}^{\tau(n,1)}E_{k-1}(|\xi_{nk}|I(|\xi_{nk}|>\varepsilon)) \overset{P}{\to} 0 \text{ for each } \varepsilon > 0 \qquad (CL_1)$$

which correspond to (M_2) and (CL), respectively, for integrable rv's that are not necessarily square integrable. In connection with (M_1) and (CL_1) norming by the Raikov condition

$$\Sigma_{k=0}^{\tau(n,t)}\xi_{nk}^2 \overset{P}{\to} t \text{ for each } 0 \leq t \leq 1 \qquad (R_t)$$

is appropriate instead of norming by (N_t). The following relations hold between all the conditions introduced so far:

$$\begin{array}{ccccc} (M_2) \text{ and } (N_t) & \Rightarrow & (M_1) \text{ and } (R_t) & & \\ \Downarrow & & \Downarrow & & \\ (CL) \text{ and } (N_t) & \Rightarrow & (CL_1) \text{ and } (R_t) & \underset{(mda)}{\Rightarrow} & (A_a) \text{ and } (R_{a,t}). \end{array} \qquad (2.18)$$

For these results the mda property is needed only where indicated, but there it is essential. The conditions in the left column require square integrable rv's, in the middle integrability is sufficient, and on the rhs no assumptions on moments are needed at all. As seen in the proof of Theorem 2.3, (R_t) implies

$$\max_{0\leq k\leq \tau(n,1)} |\xi_{nk}| \overset{P}{\to} 0 \tag{M}$$

so that in (2.18) condition (M_1) may be weakened to uniform integrability of $\{\max_{0\leq k\leq \tau(n,1)} |\xi_{nk}| : n \in \mathbb{N}\}$. '$(M_2) \Rightarrow (CL)$' in (2.18) has been shown already, and the proof of '$(M_1) \Rightarrow (CL_1)$' runs analogously. '$(CL) \Rightarrow (CL_1)$' is obvious. Under (CL) the norming conditions (N_t) and (R_t) are equivalent. This well known fact follows by arguments similar to those applied in the second half of the proof of Theorem 2.3, using (2.15) and (2.16); cf. Rootzén [39] and Scott [40] for details. For the proof of '(CL_1) and $(R_t) \Rightarrow (A_a)$ and $(R_{a,t})$' if (ξ_{nk}, F_{nk}) is a mda observe that (R_t) implies (M) whence

$$\sup_{0\leq t\leq 1} |\Sigma_{k=0}^{\tau(n,t)} \xi_{nk} I(|\xi_{nk}| > a)| \leq \Sigma_{k=0}^{\tau(n,1)} |\xi_{nk}| I(|\xi_{nk}| > a) \overset{P}{\to} 0$$

for all $0 < a < \infty$; cf. (2.17). Using $|E_{k-1}(\xi_{nk} I(|\xi_{nk}| \leq a))| = |E_{k-1}(\xi_{nk} I(|\xi_{nk}| > a))|$ by the mda property, (A_a) follows for each $0 < a < \infty$ from (CL_1), and

$$\Sigma_{k=0}^{\tau(n,t)} \eta_{nk}^2 - \Sigma_{k=0}^{\tau(n,t)} \xi_{nk}^2 \overset{P}{\to} 0$$

is easily established for $0 \leq t \leq 1$, proving $(R_{a,t})$ for each $0 < a < \infty$.

The self-evident consequence of (2.18) and of Theorem 2.3 is the fact that for a mda any set of conditions occurring in (2.18) implies (FCLT). Furthermore, by arguments of the same type as above it can be shown that all the different conditions implying (FCLT) for discrete time arrays which are formulated in [18,28,39] either imply or are equivalent to (A_a) and $(R_{a,t})$. For example, it is easy to see that (A_a) and $(R_{a,t})$ together with (M) (which allows one to omit the first half of the bracket in (A_a)) are equivalent to the "three series condition" given by Liptser and Shiryayev [28] and Helland [18], namely,

$$\Sigma_{k=0}^{\tau(n,1)} P_{k-1}(|\xi_{nk}| > \varepsilon) \overset{P}{\to} 0 \text{ for each } \varepsilon > 0 \text{ ,} \tag{S1}$$

$$\Sigma_{k=0}^{\tau(n,1)} E_{k-1}(\xi_{nk} I(|\xi_{nk}| \leq 1)) \overset{P}{\to} 0 \tag{S2}$$

and

$$\Sigma_{k=0}^{\tau(n,t)} \mathrm{Var}_{k-1}(\xi_{nk} I(|\xi_{nk}| \le 1)) \overset{P}{\to} t \quad \text{for each} \quad 0 \le t \le 1 , \tag{S3}$$

where $P_{k-1}(\cdot)$ and $\mathrm{Var}_{k-1}(\cdot)$ denote the conditional probability and variance, respectively, wrt $F_{n,k-1}$. From an aesthetic point of view, one might argue that these conditions are the fundamental ones, mainly because of the fact that they reduce to the well known sufficient conditions of classical central limit theory; cf. Gnedenko and Kolmogorov [13], Chapter 5, Theorem 2. On the other hand, conditions (A_a) and $(R_{a,t})$ are more convenient for us when passing to the continuous time case since the formulation of the continuous time versions of (S1) - (S3) requires the concept of stochastic integration, cf. Liptser and Shiryayev [28], whereas the one for (A_a) and $(R_{a,t})$ does not. In any case, Theorem 2.3 shows that (A_a) and $(R_{a,t})$ as well as (S1) - (S3) are in a certain sense tantamount to the strongest set of conditions considered above, i.e. to (M_2) and (N_t). This demonstrates unification of a large part of martingale central limit theory.

In (2.18) and Theorem 2.3 we discussed only the functional form of the CLT. It is quite obvious that a similar treatment can be carried out for the ordinary CLT. One only has to observe that (R_t) implies (M) whereas the Raikov condition for a single time point does not.

3. Functional CLT's for continuous time local martingales

For a discrete time mda (ξ_{nk}, F_{nk}) the partial sum processes $S_{(n)}(t) := \Sigma_{k=0}^{[nt]} \xi_{nk}$, $0 \le t < \infty$, are continuous time martingales wrt the filtrations $F_{n,[nt]}$, $0 \le t < \infty$; here $[x]$ denotes the integer part of x . According to the results in Section 2, under certain conditions on the 'discrete skeletons' $(\Sigma_{j=0}^{k} \xi_{nj}, F_{nk} ; k \in \mathbb{N})$ of $S_{(n)}$ these martingales converge in distribution to a Brownian motion. Therefore, it is only natural to start from a sequence $(M_n)_{n \in \mathbb{N}}$ of arbitrary continuous time martingales and look for intrinsic properties of M_n which entail convergence in distribution to a Brownian motion again. As mentioned in the introduction, such a generalization is not merely interesting in itself, it also has important impacts towards applications, and, of course, this is its real significance. To meet the needs of these applications in full it is furthermore necessary to include local martingales because processes of interest with time interval $[0,\infty)$ often are local martingales, but not martingales.

As explained in the introduction, we aim to reduce the necessary technicalities for an understanding of continuous time martingale central limit theory as far as possible, but a knowledge of some basic concepts and results is indispensable. Due to limitations of space we will not discuss them here in detail but use Dellacherie and Meyer [5,6] as a reference text. To fix the framework, let $M_n = (M_n(t))_{0\leq t<\infty}$ for each $n \in \mathbb{N}$ be a cadlag local martingale wrt the filtration $\mathbb{F}_n = (F_n(t))_{0\leq t<\infty}$, all defined on the basic p-space (Ω,F,P) ; cf. [5], IV.11, 12; [6], VI.27. The filtrations are assumed to satisfy the 'usual conditions' in the sense of Dellacherie, i.e. $F_n(t)$ contains all P-null sets from F , and $F_n(t) = \bigcap_{s>t} F_n(s)$ for all $t \geq 0$; cf. [5], IV.47, 48. These conditions are needed to avoid certain technical complications e.g. in connection with the definition of stopping times, but they are not a severe restriction because they can always be achieved by a slight change of the σ-fields; cf. [5], IV.48. The French abbreviation 'cadlag' (continue à droite, limites à gauche) is used to describe processes having right continuous paths with left hand limits. In the following we always suppose $M_n(0) = 0$ a.s. and $F_n(0-) = F_n(0)$; cf. [5], IV.47.

By definition of local martingales, for each n there exists a sequence $(T_{nk})_{k\in\mathbb{N}}$ of $\mathbb{F}_n$-stopping times with $T_{nk} \uparrow \infty$ as $k \to \infty$ such that each stopped process $(M_n(T_{nk} \wedge t))_{0\leq t<\infty}$ is a uniformly integrable martingale wrt $\mathbb{F}_n$. According to this definition it is often possible to reduce the investigation of local martingales to that of martingales by appropriate stopping time arguments. Consequently, we will have to deal basically with continuous time martingales in the sequel. We will discuss the functional form of the CLT only, a parallel and somewhat easier treatment of one-dimensional results is obviously possible. As far as the limiting processes are concerned, we restrict ourselves to standard Brownian motion. In [28,35,36] occur more general limiting processes, but they can be obtained from a standard Brownian motion by a non-random continuous change of time so that it presents no difficulty to cover them by the present approach, too: one only has to replace the identity map $t \to t$ on $[0,\infty)$ by a continuous, non-decreasing function $t \to A(t)$ in all what follows.

Continuous time versions of the sufficient conditions for (FCLT) considered in Section 2 are readily formulated. Let ΔM_n denote the jump process of M_n, i.e. $\Delta M_n(t) := M_n(t) - M_n(t-)$ with $M_n(t-) := \lim_{s\uparrow t} M_n(s)$. For $n \in \mathbb{N}$ and $\varepsilon > 0$ set

$$\sigma^{\varepsilon}[M_n](t) := \Sigma_{s\leq t}(\Delta M_n(s))^2 I(|\Delta M_n(s)| > \varepsilon) , \quad 0 \leq t < \infty .$$

Compared with the conditional Lindeberg condition in discrete time this process corresponds to the process occurring there if the conditional expectations are cancelled. Taking conditional expectations in discrete time corresponds to forming the predictable compensator of $\sigma^{\varepsilon}[M_n]$ in continuous time which by definition is the uniquely determined predictable increasing process that has to be subtracted from $\sigma^{\varepsilon}[M_n]$ to obtain a local martingale; cf. [6], VI.80. Let $\tilde{\sigma}^{\varepsilon}[M_n]$ denote the predictable compensator of $\sigma^{\varepsilon}[M_n]$. For its existence M_n has to be supposed to be locally square integrable in the sense of [6], VII.40, since then $\sigma^{\varepsilon}[M_n]$ is a locally integrable increasing process (adopt the proof of Lemma 1 in Meyer [33]) which is sufficient for the existence of $\tilde{\sigma}^{\varepsilon}[M_n]$; cf. [6], VI.80. The conditional Lindeberg condition for locally square integrable local martingales now simply requires

$$\tilde{\sigma}^{\varepsilon}[M_n](t) \xrightarrow{P} 0 \quad \text{for each } \varepsilon > 0 \text{ and } 0 \leq t < \infty . \qquad (CL_t^c)$$

Here we have to consider each $t \geq 0$ because we want to cover the whole right half axis $[0,\infty)$.

The same sort of reasoning applies to the other conditions discussed in Section 2: (M_1) and (M_2) become

$$E(\max_{0\leq s\leq t}|\Delta M_n(s)|) \to 0 \quad \text{for each } 0 \leq t < \infty , \qquad (M_{1,t}^c)$$

and

$$E(\max_{0\leq s\leq t}(\Delta M_n(s))^2) \to 0 \quad \text{for each } 0 \leq t < \infty , \qquad (M_{2,t}^c)$$

respectively, and (CL_1) becomes

$$\tilde{\alpha}^{\varepsilon}[M_n](t) \xrightarrow{P} 0 \quad \text{for each } \varepsilon > 0 \text{ and } 0 \leq t < \infty , \qquad (CL_{1,t}^c)$$

where $\tilde{\alpha}^{\varepsilon}[M_n]$ is the predictable compensator of the process

$$\alpha^{\varepsilon}[M_n](t) := \Sigma_{s\leq t}|\Delta M_n(s)| I(|\Delta M_n(s)| > \varepsilon) , \quad 0 \leq t < \infty ,$$

which is increasing and locally integrable for local martingales M_n which need not be locally square integrable, cf. [33], so that $\tilde{\alpha}^{\varepsilon}[M_n]$

does exist by [6], VI.80. As to the norming conditions, we get the following: If M_n is locally square integrable, then its predictable quadratic variation $\langle M_n \rangle$ exists as the unique predictable increasing process such that $M_n^2 - \langle M_n \rangle$ is a local martingale which is zero at $t = 0$; cf. [6], VII.39,40. Therefore, $\langle M_n \rangle$ corresponds to the partial sums of conditional variances in discrete time, and the continuous time version of (N_t) is

$$\langle M_n \rangle(t) \overset{P}{\to} t \quad \text{for each} \quad 0 \leq t < \infty . \tag{N_t^c}$$

If M_n is merely a local martingale, then there exists a unique increasing process $[M_n]$ such that $M_n^2 - [M_n]$ is a local martingale and $\Delta[M_n] = (\Delta M_n)^2$; cf. [6], VII.42. $[M_n]$ is called the optional quadratic variation of M_n and corresponds to the partial sums of the squares of the variables in discrete time so that the continuous time version of the Raikov condition is

$$[M_n](t) \overset{P}{\to} t \quad \text{for each} \quad 0 \leq t < \infty . \tag{R_t^c}$$

Table (2.18) suggests that for appropriate combinations of all the preceding conditions, e.g. (CL_t^c) and (N_t^c) or $(CL_{1,t}^c)$ and (R_t^c), it is reasonable to suppose that they imply the natural continuous time analogue of (FCLT), namely,

$$M_n \overset{L}{\to} B . \tag{$FCLT^c$}$$

Here B is a standard Brownian motion with time interval $[0,\infty)$, and convergence in distribution is to be understood as weak convergence of p-measures on the space $D[0,\infty)$ of all real-valued functions on $[0,\infty)$ which are right-continuous and have left-hand limits, endowed with its Skorokhod topology; cf. Lindvall [27] and Whitt [41]. As mentioned in the introduction, Rebolledo [35,36] and Liptser and Shiryayev [28] proved that such results actually hold true, and Helland [18] showed that for each of the sets of conditions '(CL_t^c) and (N_t^c)', '$(M_{1,t}^c)$ and (R_t^c)' and '$(CL_{1,t}^c)$ and (R_t^c)' there is a discretization method that can be used to obtain $(FCLT^c)$ from the corresponding discrete time statement. This approach which avoids any discussion of fidi-convergence and tightness in continuous time will be further discussed in the sequel. First, observe that by the properties of convergence of distributions on $D[0,\infty)$ the desired result $(FCLT^c)$ follows from $(M_n(t))_{0 \leq t \leq T} \overset{L}{\to} (B(t))_{0 \leq t \leq T}$ in $D[0,T]$ for each $0 < T < \infty$; cf. [27], Theorem 3 and

[41], Theorem 2.8. For notational convenience, assume $T = 1$. Suppose that each M_n is a martingale and consider the rv's

$$\xi_{mnk} := M_n((k+1)/2^m) - M_n(k/2^m) \ , \quad m,n,k \in \mathbb{N} \ , \tag{3.1}$$

and the σ-fields $F_{mnk} := F_n(k/2^m)$. Obviously, $(\xi_{mnk}, F_{mnk} \ ; n,k \in \mathbb{N})$ constitutes a mda for each fixed m. Form the partial sum processes

$$S_{mn}(t) := \Sigma_{k=0}^{[2^m t]} \xi_{mnk} \ , \quad 0 \leq t \leq 1 \ ,$$

in $D[0,1]$. Helland's method consists in constructing a sequence of indices $m(n)$ with $m(n) \to \infty$ such that

$$\sup_{0 \leq t \leq 1} |M_n(t) - S_{m(n),n}(t)| \overset{P}{\to} 0 \quad \text{as} \quad n \to \infty \tag{3.2}$$

and such that the continuous time sufficient conditions under consideration imply the corresponding discrete time conditions for the mda $(\xi_{m(n),n,k}, F_{m(n),n,k} \ ; n,k \in \mathbb{N})$. The discrete time FCLT then implies $S_{m(n),n} \overset{L}{\to} B_1$ in $D[0,1]$, and (3.2) yields $(M_n(t))_{0 \leq t \leq 1} \overset{L}{\to} B_1$ in $D[0,1]$. For the sets '(CL_t^c) and (N_t^c)' and '$(CL_{1,t}^c)$ and (R_t^c)' of sufficient conditions this argument involves some technicalities because these conditions contain predictable compensators. Their discretization, however, presents the following difficulty: Consider, for example, for a square integrable martingale M_n the discrete skeleton

$$\Sigma_{k=0}^{[2^m t]} E(\xi_{mnk}^2 | F_{mnk}) = \Sigma_{k=0}^{[2^m t]} E\{[M_n((k+1)/2^m) - M_n(k/2^m)]^2 | F_n(k/2^m)\}$$

of $\langle M_n \rangle(t)$. In general, it converges to $\langle M_n \rangle(t)$ as $m \to \infty$ only weakly in L_1, but not strongly, hence not in probability; cf. [6], VII.43. Weak convergence in L_1, however, does not match well with the convergence in probability occurring in the condition (N_t) for the discretized process. For the discrete skeleton

$$\Sigma_{k=0}^{[2^m t]} \xi_{mnk}^2 = \Sigma_{k=0}^{[2^m t]} [M_n((k+1)/2^m) - M_n(k/2^m)]^2$$

of $[M_n](t)$ the situation is much neater: Theorem 2 of Doléans [7], cf. also [6], VII.43 and VIII.20, says that for fixed n and M_n being a martingale we have

$$\Sigma_{k=0}^{[2^m t]} \xi_{mnk}^2 \overset{P}{\to} [M_n](t) \quad \text{as} \quad m \to \infty \quad \text{for each} \quad 0 \leq t < \infty \tag{3.3}$$

which matches well with (R_t) and therefore leads to the by far simplest of Helland's discretization methods. This fact together with the fact that Rebolledo and Liptser and Shiryayev have established a number of relations between their different sufficient conditions for ($FCLT^c$) suggests the following procedure for obtaining the continuous time theory from the discrete time one with the least possible effort: First establish the continuous time versions of (2.18) and of Theorem 2.3, and then apply Helland's discretization argument to the conditions ($M^c_{1,t}$) and (R^c_t) which do not contain predictable compensators. That procedure will be sketched in the sequel.

The key notion for comparing different sets of sufficient conditions in the continuous time theory is the following concept of domination between non-negative cadlag processes X and Y, both being adapted to the same filtration $\mathbb{F}$. It was introduced by Lenglart [24]. X is said to be $\mathbb{F}$-dominated by Y, written $X \prec Y$, iff $E(X(T)) \leq E(Y(T))$ for all bounded stopping times T wrt $\mathbb{F}$. The main result that will be needed later on is

<u>Proposition 3.1.</u> Let $(X_n)_{n\in\mathbb{N}}$ and $(Y_n)_{n\in\mathbb{N}}$ be two sequences of non-negative, $\mathbb{F}_n$-adapted and increasing processes with $X_n(0) = 0 = Y_n(0)$ and $X_n \prec Y_n$ for all n. Suppose that one of the following two hypotheses holds:

(i) Y_n is $\mathbb{F}_n$-predictable for each n;

(ii) $\{\sup_{0\leq s\leq t} \Delta Y_n(s) : n \in \mathbb{N}\}$ is uniformly integrable for some $t > 0$.

Then $Y_n(t) \overset{P}{\to} 0$ implies $X_n(t) \overset{P}{\to} 0$.

This result is essentially due to Lenglart [24] who derived it from an appropriate maximal inequality. The formulation given above appears in Liptser and Shiryayev [29]. Adapting the argument for establishing (2.15) and (2.16) to continuous time yields a very simple proof: Consider case (i) and fix $t > 0$. Setting $R_{nm} := \inf\{s : Y_n(s) \geq 1\} \wedge (t + m^{-1})$ we obtain $\mathbb{F}_n$-predictable stopping times, cf. [5], IV.87(d), which are a.s. positive because of $Y_n(0) = 0$ and right continuity of Y_n. Consequently there exist $\mathbb{F}_n$-stopping times R_{nmk}, $k \in \mathbb{N}$, announcing R_{nm}, i.e. $R_{nmk} \uparrow R_{nm}$ as $k \to \infty$ and $R_{nmk} < R_{nm}$ a.s., cf. [5], IV.77. For each $\varepsilon > 0$ we obtain

$$P(X_n(t) \geq \varepsilon)$$

$$\leq P(X_n(t) \geq \varepsilon, R_{nmk} > t) + P(R_{nmk} \leq t, R_{nm} = t + m^{-1}) + P(R_{nm} < t + m^{-1})$$

$$\leq P(X_n(R_{nmk} \wedge t) \geq \varepsilon) + P(|R_{nmk} - R_{nm}| \geq m^{-1}) + P(Y_n(t+m^{-1}) \geq 1) \quad .$$

We have $Y_n(t+m^{-1}) \overset{P}{\to} Y_n(t)$ as $m \to \infty$ by right continuity of Y_n and $Y_n(t) \overset{P}{\to} 0$ as $n \to \infty$ by assumption as well as $R_{nmk} \uparrow R_{nm}$ a.s. as $k \to \infty$. Thus we can construct $m = m(n)$ and $k = k(n)$ such that the second and third summand on the rhs converge to zero as $n \to \infty$. The first summand is less than $\varepsilon^{-1}E(X_n(R_{nmk} \wedge t)) \leq \varepsilon^{-1}E(Y_n(R_{nmk} \wedge t))$ by $X_n \prec Y_n$ where the rhs converges to zero because of $Y_n(R_{nmk} \wedge t) \leq Y_n(t) \overset{P}{\to} 0$, $Y_n(R_{nmk} \wedge t) \leq 1$ by definition of R_{nmk}, and dominated convergence. This proves part (i) of the proposition, and part (ii) is proven similarly using the stopping times $R_n := \inf\{s : Y_n(s) \geq 1\}$, cf. [5], IV.50, which are not predictable in general so that one has to take care of a possible jump at time R_n by the additional assumption made for that part.

Proposition 3.1 will mostly be applied in the situation of

<u>Proposition 3.2.</u> Let A be an $\mathbb{F}$-adapted and locally integrable increasing process with $A(0) = 0$ and with predictable compensator $\tilde{A}$. Then $E(A(T)) = E(\tilde{A}(T))$ for each bounded $\mathbb{F}$-stopping time T so that $A \prec \tilde{A}$ and $\tilde{A} \prec A$.

Proposition 3.2 is an immediate consequence of the fact that $M := A - \tilde{A}$ is a local martingale with $M(0) = 0$ a.s. and the optional sampling theorem; cf. [6], VI.10.

The truncation of a mda used in Theorem 2.3 carries over to continuous time as follows; cf. [35,36]: For $a > 0$ set

$$A^a[M_n](t) := \Sigma_{s \leq t} \Delta M_n(s) I(|\Delta M_n(s)| > a) \quad , \quad 0 \leq t < \infty \; .$$

According to the proof of Theorem VI.85 in [6] the process $A^a[M_n]$ is locally of integrable variation so that its predictable compensator $\tilde{A}^a[M_n]$ exists. The processes

$$\overline{M}_n^a := A^a[M_n] - \tilde{A}^a[M_n] \quad \text{and} \quad \underline{M}_n^a := M_n - \overline{M}_n^a$$

are local martingales, and $|\Delta \underline{M}_n^a| \leq 2a$ a.s. The continuous time versions of $(R_{a,t})$ and (A_a) are

$$[\underline{M}_n^a](t) \overset{P}{\to} t \quad \text{for each} \quad 0 \leq t < \infty \tag{$R_{a,t}^c$}$$

and

$$(\overline{M}_n^a)^*(t) \overset{P}{\to} 0 \quad \text{for each} \quad 0 \leq t < \infty , \tag{$A_{a,t}^c$}$$

where for a cadlag process X the process X^* is defined by $X^*(t) := \sup_{s\leq t}|X(s)|$. Now we can formulate the following table corresponding to (2.18):

$$\begin{array}{ccccc} (M_{2,t}^c) \text{ and } (N_t^c) & \underset{(3)}{\Rightarrow} & (M_{1,t}^c) \text{ and } (R_t^c) & & \\ \Downarrow(1) & & \Downarrow(2) & & \\ (CL_t^c) \text{ and } (N_t^c) & \underset{(4)}{\Rightarrow} & (CL_{1,t}^c) \text{ and } (R_t^c) & \underset{(5)}{\Rightarrow} & (A_{a,t}^c) \text{ and } (R_{a,t}^c) . \end{array} \tag{3.4}$$

Recall that the conditions in the left-hand column require locally square integrable local martingales whereas the conditions in the other two columns make sense already for local martingales. The same argument as in the discrete time case shows that (R_t^c) holds uniformly in t on compact intervals, and because of

$$\max_{0\leq s\leq t}(\Delta\underline{M}_n^a(s))^2 = \max_{0\leq s\leq t}\Delta[\underline{M}_n^a](s) \leq 2 \sup_{0\leq s\leq t}|[\underline{M}_n^a](t) - t|$$

(R_t^c) implies $\max_{0\leq s\leq t}|\Delta\underline{M}_n^a(s)| \overset{P}{\to} 0$ for each $0 \leq t < \infty$. Consequently, $(M_{1,t}^c)$ may be weakened in (3.4) to uniform integrability of the rv's $\max_{0\leq s\leq t}|\Delta M_n(s)|$, $n \in \mathbb{N}$, for each $0 \leq t < \infty$.

The proofs of some of the implications in (3.4) are a bit more involved than the corresponding proofs for the discrete time case. Some relations of domination between certain processes are required which have been formulated and proved by Rebolledo in Lemma I.1.3 of [35] and Lemma 5 and Lemma 7, respectively, of [36]. We shall summarize them in the following proposition and shall also repeat some of the details of proof here since these proofs reveal how to deal with jumps in continuous time.

<u>Proposition 3.3.</u> For each $\varepsilon > 0$ and $n \in \mathbb{N}$ we have

$$[\overline{M}_n^\varepsilon] \prec 4 \,\tilde{\sigma}^\varepsilon[M_n] , \tag{3.5}$$

$$\langle \overline{M}_n^\varepsilon \rangle \prec 4 \,\tilde{\sigma}^\varepsilon[M_n] , \tag{3.6}$$

$$[\underline{M}_n^\varepsilon,\overline{M}_n^\varepsilon]^* \prec 4\varepsilon\, \tilde{\alpha}^\varepsilon[M_n] \;, \tag{3.7}$$

$$\langle \underline{M}_n^\varepsilon,\overline{M}_n^\varepsilon\rangle^* \prec 4\varepsilon\, \tilde{\alpha}^\varepsilon[M_n] \;, \tag{3.8}$$

$$([\underline{M}_n^\varepsilon]-\langle \underline{M}_n^\varepsilon\rangle)^* \prec 64\varepsilon^2\, [\underline{M}_n^\varepsilon] \;, \tag{3.9}$$

$$([\underline{M}_n^\varepsilon]-\langle \underline{M}_n^\varepsilon\rangle)^* \prec 64\varepsilon^2\, \langle \underline{M}_n^\varepsilon\rangle \tag{3.10}$$

and

$$(\overline{M}_n^\varepsilon)^* \prec 2\tilde{\alpha}^\varepsilon[M_n] \;. \tag{3.11}$$

(For the definition of $[\underline{M}_n^\varepsilon,\overline{M}_n^\varepsilon]$ and $\langle \underline{M}_n^\varepsilon,\overline{M}_n^\varepsilon\rangle$ see [6], VII.39,42.)

The following facts will be needed throughout the proof of Proposition 3.3: If X is an adapted cadlag process, then there exists a sequence $(T_m)_{m\in\mathbb{N}}$ of stopping times such that

$$\{(t,\omega)\in[0,\infty)\times\Omega : X_t(\omega) \neq X_{t-}(\omega)\} = \bigcup_{m=1}^\infty [\![T_m]\!] \;, \tag{3.12}$$

where $[\![T_m]\!] := \{(t,\omega)\in[0,\infty)\times\Omega : T_m(\omega)=t\}$ denotes the graph of the stopping time T_m. In (3.12) the stopping times can be chosen such that their graphs are disjoint, and if X is predictable, then they can also be chosen to be predictable; cf. [5], IV.88B. Furthermore, if X is arbitrary, then it follows from [5], IV.81, that the T_m can be chosen with disjoint graphs such that T_m is either predictable or totally inaccessible (defined in [5], IV.80).

For the proof of (3.5) - (3.11), fix n and ε. Let $(T_m)_{m\in\mathbb{N}}$ be a sequence of stopping times with disjoint graphs which 'exhaust' the jumps of $\overline{M}_n^\varepsilon$ in the sense of (3.12) and which are either predictable or totally inaccessible. $\overline{M}_n^\varepsilon$ is a martingale of finite variation, cf. [6], VI.85, whence for $0\le t<\infty$ by [6], VII.37,

$$\begin{aligned}[\overline{M}_n^\varepsilon](t) &= \Sigma_{s\le t}(\Delta\overline{M}_n^\varepsilon(t))^2 \\ &= \Sigma_{m\in N_1}(\Delta\overline{M}_n^\varepsilon(T_m))^2 I(T_m\le t) + \Sigma_{m\in N_2}(\Delta\overline{M}_n^\varepsilon(T_m))^2 I(T_m\le t)\end{aligned} \tag{3.13}$$

with $N_1 := \{m\in\mathbb{N} : T_m \text{ totally inaccessible}\}$ and $N_2 := \{m\in\mathbb{N} : T_m \text{ predictable}\}$. Consider $m\in N_1$ first. Observe that in $\overline{M}_n^\varepsilon = A^\varepsilon[M_n] - \tilde{A}^\varepsilon[M_n]$ the process $\tilde{A}^\varepsilon[M_n]$ is predictable so that its jumps are exhausted in the sense of (3.12) by a sequence of predictable stopping times. By [5], IV.81(b), this implies $\Delta\tilde{A}^\varepsilon[M_n](T_m)=0$ a.s. so that with probability one

$$\Delta\overline{M}_n^\varepsilon(T_m)^2 = \Delta A^\varepsilon[M_n](T_m)^2 = (\Delta M_n(T_m))^2 I(|\Delta M_n(T_m)|>\varepsilon) = \Delta\sigma^\varepsilon[M_n](T_m) \;.$$

Consider now $m \in N_2$ and let $F(T_m-)$ denote the σ-field of events prior to T_m ; cf. [5], IV.54. Using [6], VI.(81.2), we obtain on $\{T_m \leq t\}$ with probability one

$$\Delta\overline{M}_n^\varepsilon(T_m)^2 = (\Delta A^\varepsilon[M_n](T_m) - \Delta\widetilde{A}^\varepsilon[M_n](T_m))^2$$

$$\leq 2\ \Delta A^\varepsilon[M_n](T_m) + 2\ E(\Delta A^\varepsilon[M_n](T_m)|F(T_m-))^2$$

$$\leq 2(\Delta M_n(T_m))^2 I(|\Delta M_n(T_m)| > \varepsilon) + 2E\{(\Delta M_n(T_m))^2 I(|\Delta M_n(T_m)| > \varepsilon)|F(T_m-)\}$$

$$= 2\ \Delta\sigma^\varepsilon[M_n](T_m) + 2\ E(\Delta\sigma^\varepsilon[M_n](T_m)|F(T_m-))$$

$$= 2\ \Delta\sigma^\varepsilon[M_n](T_m) + 2\ \Delta\widetilde{\sigma}^\varepsilon[M_n](T_m)\ .$$

Substituting these results into (3.13) yields for $0 \leq t < \infty$

$$[\overline{M}_n^\varepsilon](t) \leq 2\ \sigma^\varepsilon[M_n](t) + 2\ \widetilde{\sigma}^\varepsilon[M_n](t) \quad \text{a.s.}$$

Since the processes on both sides of this inequality are cadlag, we get $[\overline{M}_n^\varepsilon] \leq 2\ \sigma^\varepsilon[M_n] + 2\ \widetilde{\sigma}^\varepsilon[M_n]$, implying (3.5). Since $\langle \overline{M}_n^\varepsilon \rangle$ is the predictable compensator of $[\overline{M}_n^\varepsilon]$, cf. [6], VII.43, (3.6) follows from (3.5) and Proposition 3.2. The same kind of reasoning as in the proof of (3.5) shows that with probability one for $0 \leq t < \infty$

$$\Sigma_{s\leq t}|\Delta\overline{M}_n(s)| \leq \alpha^\varepsilon[M_n](t) + \widetilde{\alpha}^\varepsilon[M_n](t)\ . \tag{3.14}$$

From $[\underline{M}_n^\varepsilon,\overline{M}_n^\varepsilon](t) = \Sigma_{s\leq t}\Delta\underline{M}_n^\varepsilon(s)\Delta\overline{M}_n^\varepsilon(s)$ we have

$$|[\underline{M}_n^\varepsilon,\overline{M}_n^\varepsilon](t)| \leq \Sigma_{s\leq t}|\Delta\underline{M}_n^\varepsilon(s)||\Delta\overline{M}_n^\varepsilon(s)| \leq 2\varepsilon\ \Sigma_{s\leq t}|\Delta\overline{M}_n^\varepsilon(s)|$$

because of $|\Delta\underline{M}_n^\varepsilon| \leq 2\varepsilon$. Combining the last two estimates we get (3.7) immediately and infer (3.8) from $\langle \underline{M}_n^\varepsilon,\overline{M}_n^\varepsilon\rangle^* \leq [\underline{M}_n^\varepsilon,\overline{M}_n^\varepsilon]^{*,\sim} \leq 4\varepsilon\ \widetilde{\alpha}^\varepsilon[M_n]$. (3.9) and (3.10) are equivalent by Proposition 3.2. For the details of their proofs as well as for the proof of (3.11) the reader is referred to Lemma 5 and Lemma 7 in [36].

We are now prepared to verify the implications in (3.4). The proofs of (1) and (2) run exactly parallel to those of the corresponding implications in (2.18), employing part (ii) of Proposition 3.1 instead of (2.16). For the proof of implication (4) we first establish '$(CL_t^c) \Rightarrow (CL_{1,t}^c)$' : $\alpha^\varepsilon[M_n] \leq \varepsilon^{-1}\sigma^\varepsilon[M_n]$ is immediate from the definitions; $\widetilde{\alpha}^\varepsilon[M_n] \prec \alpha^\varepsilon[M_n]$ and $\sigma^\varepsilon[M_n] \prec \widetilde{\sigma}^\varepsilon[M_n]$ follow from Proposition 3.2. Combining we obtain $\widetilde{\alpha}^\varepsilon[M_n] \prec \varepsilon^{-1}\widetilde{\sigma}^\varepsilon[M_n]$, and part (i) of Proposition 3.1 yields the assertion.

We now show that (CL_t^c) and one of the conditions (N_t^c) or (R_t^c) imply

$$[M_n](t) - \langle M_n \rangle(t) \overset{P}{\to} 0 \quad \text{for each} \quad 0 \leq t < \infty . \tag{3.15}$$

This will prove particularly (4), whereas the implication '(CL_t^c) and $(R_t^c) \Rightarrow (N_t^c)$' will be needed in the proof of Theorem 3.4 below. For the proof of (3.15) assume first that only (CL_t^c) is satisfied. From $\tilde{\alpha}^{\varepsilon}[M_n] \prec \varepsilon^{-1} \tilde{\sigma}^{\varepsilon}[M_n]$ and (3.5) - (3.8) we obtain for each $\varepsilon > 0$ and $0 \leq t < \infty$

$$[M_n](t) - \langle M_n \rangle(t)$$

$$= [\underline{M}_n^{\varepsilon}](t) + 2[\underline{M}_n^{\varepsilon}, \overline{M}_n^{\varepsilon}](t) + [\overline{M}_n^{\varepsilon}](t) - \langle \underline{M}_n^{\varepsilon} \rangle(t) - 2\langle \underline{M}_n^{\varepsilon}, \overline{M}_n^{\varepsilon} \rangle(t) - \langle \overline{M}_n^{\varepsilon} \rangle(t)$$

$$= [\underline{M}_n^{\varepsilon}](t) - \langle \underline{M}_n^{\varepsilon} \rangle(t) + o_p(1) , \tag{3.16}$$

$$[M_n](t) = [\underline{M}_n^{\varepsilon}](t) + o_p(1) \tag{3.17}$$

and

$$\langle M_n \rangle(t) = \langle \underline{M}_n^{\varepsilon} \rangle(t) + o_p(1) . \tag{3.18}$$

For brevity, write $L_{n,\varepsilon} := [\underline{M}_n^{\varepsilon}] - \langle \underline{M}_n^{\varepsilon} \rangle$. If in addition to (CL_t^c) also (N_t^c) holds true, then one can use (3.10), (3.18) and the method of announcing stopping times for a predictable stopping time as employed already in the proof of part (i) of Proposition 3.1 to establish

$$\lim_{\varepsilon \downarrow 0} \limsup_{n\to\infty} P(L_{n,\varepsilon}^{*2}(t) > \delta) = 0 \tag{3.19}$$

for each $\delta > 0$ and $0 \leq t < \infty$. Combined with (3.16) this implies (3.15) and finishes the proof of (4). Next suppose that (CL_t^c) and (R_t^c) are satisfied. Then $[\underline{M}_n^{\varepsilon}](t) \overset{P}{\to} t$ in view of (3.17). Setting $R_n :=$ $\inf\{s : [\underline{M}_n^{\varepsilon}](s) \geq t+1\}$ we obtain $\mathbb{F}_n$-stopping times. For each $\delta > 0$ and $0 \leq t < \infty$ we find

$$P(L_{n,\varepsilon}^{*2}(t) > \delta)$$

$$\leq P(L_{n,\varepsilon}^{*2}(t) > \delta, R_n > t) + P(R_n \leq t)$$

$$\leq \delta^{-1} E(L_{n,\varepsilon}^{*2}(R_n \wedge t)) + P([\underline{M}_n^{\varepsilon}](t) \geq t+1)$$

$$\leq 16\ \delta^{-1} \varepsilon^2\ E([\underline{M}_n^{\varepsilon}](R_n \wedge t)) + P(|[\underline{M}_n^{\varepsilon}](t) - t| \geq 1)$$

$$\leq 16\ \delta^{-1} \varepsilon^2 (t + 1 + 4\varepsilon^2) + o(1)$$

which implies (3.19) again. Combined with (3.16) this yields (3.15) also under (CL^c_t) and (R^c_t) and concludes the proof.
For the proof of implication (5) in (3.4) observe that $\overline{M}^a_n$ is a martingale of finite variation whence for $0 < a < \infty$ and $0 \leq t < \infty$ in view of (3.14)

$$[\overline{M}^a_n](t) = \Sigma_{s\leq t}(\Delta\overline{M}^a_n(s))^2 \leq (\Sigma_{s\leq t}|\Delta\overline{M}^a_n(s)|)^2 \leq (\alpha^a[M_n](t) + \tilde{\alpha}^a[M_n](t))^2 .$$

Via Propositions 3.1 and 3.2 condition $(CL^c_{1,t})$ implies convergence to zero in probability of the rhs, and combined with (3.7) we obtain from $(CL^c_{1,t})$

$$[M_n](t) = [\overline{M}^a_n](t) + 2[\underline{M}^a_n, \overline{M}^a_n](t) + [\underline{M}^a_n](t) = [\underline{M}^a_n](t) + o_p(1) .$$

(R^c_t) therefore implies $(R^c_{a,t})$. Condition $(A_{a,t})$ follows from $(CL^c_{1,t})$; cf. (3.11) in Proposition 3.3. This concludes the proof of (5).

For local martingales the counterpart of Theorem 2.3 is

<u>Theorem 3.4.</u> Let $(M_n)_{n\in\mathbb{N}}$ be a sequence of local martingales and suppose that there exists a $0 < a < \infty$ such that $(R^c_{a,t})$ and $(A^c_{a,t})$ hold true. Then the sequence $(\underline{M}^a_n)_{n\in\mathbb{N}}$ of local martingales satisfies $(M^c_{2,t})$ and (N^c_t).

Observe that $(A^c_{a,t})$ in the present case just expresses the asymptotic equivalence of $(M_n)_{n\in\mathbb{N}}$ and $(\underline{M}^a_n)_{n\in\mathbb{N}}$ corresponding to (2.14). The proof of Theorem 3.4 is immediate from our preparatory considerations: As seen before, $(R^c_{a,t})$ implies $\max_{0\leq s\leq t}|\Delta\underline{M}^a_n(s)| \overset{P}{\to} 0$ for each $0 \leq t < \infty$, and combined with $|\Delta\underline{M}^a_n| \leq 2a$ a.s. we obtain $(M^c_{2,t})$ for $(\underline{M}^a_n)_{n\in\mathbb{N}}$ by dominated convergence. According to '$(M^c_{2,t}) \Rightarrow (CL^c_t)$' and equivalence of (N^c_t) and (R^c_t) under (CL^c_t) shown in the proof of (3.4) condition (N^c_t) for $(\underline{M}^a_n)_{n\in\mathbb{N}}$ is tantamount to $(R^c_{a,t})$. This concludes the proof.

To verify that all the conditions occurring in (3.4) imply $(FCLT^c)$, it now remains to show that this is the case for $(M^c_{1,t})$ and (R^c_t). This has been done by Helland in [18], but for the sake of completeness we include the main step here. For this, let $(M_n)_{n\in\mathbb{N}}$ be a sequence of martingales with $E(M^*_n(\infty)) < \infty$ for all n where $M^*_n(\infty) := \lim_{t\to\infty} M^*_n(t)$. As discussed before, it is enough to establish weak convergence to a standard Brownian motion B_1 on the time interval $[0,1]$. Define the rv's ξ_{mnk} by (3.1). Since the processes $(\Sigma_{k=0}^{[2^m t]} \xi_{mnk}^2)_{0\le t\le 1}$ and $[M_n]$ are non-decreasing, we have for all $m,n \ge 1$

$$\sup_{0\le t\le 1}|\Sigma_{k=0}^{[2^m t]} \xi_{mnk}^2 - t|$$

$$\le \max_{0\le i\le n}|\Sigma_{k=0}^{[2^m i/n]} \xi_{mnk}^2 - [M_n](\tfrac{i}{n})| +$$

$$\max_{1\le i\le n}|[M_n](\tfrac{i}{n}) - [M_n](\tfrac{i-1}{n})| + \sup_{0\le t\le 1}|[M_n](t) - t|$$

$$\le \max_{0\le i\le n}|\Sigma_{k=0}^{[2^m i/n]} \xi_{mnk}^2 - [M_n](\tfrac{i}{n})| + \frac{1}{n} + 3 \sup_{0\le t\le 1}|[M_n](t) - t| .$$

For fixed n the first summand on the rhs converges to zero in probability as $m \to \infty$ by (3.3). Consequently, for all sequences $(m(n))_{n\in\mathbb{N}}$ with $m(n) \to \infty$ fast enough we have

$$\sup_{0\le t\le 1}|\Sigma_{k=0}^{[2^{m(n)} t]} \xi_{m(n),n,k}^2 - t| \overset{P}{\to} 0 \quad \text{as} \quad n \to \infty .$$

By elementary analysis it is easy to verify that for each $f \in D[0,\infty)$

$$\max_{0\le t\le 1}|\Delta f(t)|$$

$$= \lim_{m\to\infty} \max_{0\le k\le 2^m} \sup\{|f(u) - f(v)| : k/2^m \le u,v \le (k+1)/2^m\} \tag{3.20}$$

$$= \lim_{m\to\infty} \max_{0\le k\le 2^m}|f(k/2^m) - f((k+1)/2^m)| .$$

Consequently, for each fixed n ,

$$\max_{0\le k\le 2^m}|\xi_{mnk}| \to \max_{0\le t\le 1}|\Delta M_n(t)| \quad \text{a.s. and in } L_1 \text{ as } m \to \infty ,$$

where convergence in L_1 follows by dominated convergence because of

$$\max_{0\le k\le 2^m}|\xi_{mnk}| \le 2 \sup_{0\le t\le 1}|M_n(t)| \le 2 M_n^*(\infty)$$

for all m and n . Combining with $(M_{1,t}^c)$ we obtain

$$E(\max_{0\le k\le 2^{m(n)}}|\xi_{m(n),n,k}|) \to 0 \quad \text{as} \quad n \to \infty$$

for all sequences $(m(n))_{n\in\mathbb{N}}$ with $m(n) \to \infty$ fast enough. Thus the mda $(\xi_{m(n),n,k}, F_{m(n),n,k}; n,k \in \mathbb{N})$ and the non-random time scales $\tau(n,t) := [2^{m(n)} t]$ satisfy (M_1) and (R_t) from Section 2 so that

$$(\Sigma_{k=0}^{[2^{m(n)}t]} \xi_{m(n),n,k})_{0\leq t\leq 1} \overset{\mathcal{L}}{\to} B_1 .$$

For each fixed n we have by (3.20)

$$\sup_{0\leq t\leq 1} |M_n(t) - S_{mn}(t)|$$

$$= \sup_{0\leq t\leq 1} |M_n(t) - M_n(([2^m t] + 1)/2^m)|$$

$$\leq \max_{0\leq k\leq 2^m} \sup\{|M_n(u) - M_n(v)| : k/2^m \leq u,v \leq (k+1)/2^m\}$$

$$\to \max_{0\leq t\leq 1} |\Delta M_n(t)| \quad \text{a.s. as } m \to \infty ,$$

and combined with $(M^c_{1,t})$ we obtain (3.2) for all sequences $(m(n))_{n\in\mathbb{N}}$ with $m(n) \to \infty$ fast enough which concludes the proof.

The reduction of the general case considered in Theorem 3.4 (i.e. local martingales) to the case discussed so far (i.e. martingales with $E(M^*_n(\infty)) < \infty$ for all n) may be carried out by familiar stopping time arguments according to the definition of local martingales and will not be presented here; see [18] for details and [6], VI.84 for necessary tools.

Theorem 3.4 leads to the following FCLT for continuous time local martingales.

<u>Theorem 3.5.</u> (i) Let $(M_n)_{n\in\mathbb{N}}$ be a sequence of local martingales. Then each set of conditions '$(M^c_{1,t})$ and (R^c_t)' , '$(CL^c_{1,t})$ and (R^c_t)' or '$(A^c_{a,t})$ and $(R^c_{a,t})$' implies $(FCLT^c)$.
(ii) If each M_n is locally square integrable, then '$(M^c_{2,t})$ and (N^c_t)' or '(CL^c_t) and (N^c_t)' also imply $(FCLT^c)$.

The reader may show that the other sufficient conditions for $(FCLT^c)$ introduced by Rebolledo [36] and not taken into account in (3.4) imply $(A^c_{a,t})$ and $(R^c_{a,t})$ so that these conditions may be incorporated into (3.4) and Theorem 3.5.

As a final remark, let us indicate that the approach presented here is likely to cover also considerable parts of the more general limit theory for semimartingales mentioned at the end of the introduction. On pp. 679 - 681 of [28] Liptser and Shiryayev reduce the proof of their main Theorem 1 to a situation considered in Theorem 3.5 above: Their sequence $M^n = (M^n_t, F^n_t)$, $t \leq T$, occurring in formula (55) is a sequence of locally square integrable local martingales which satisfies $(M^c_{2,t})$ and (N^c_t); cf. $|\Delta M^n_t| \leq 2\varepsilon_n \downarrow 0$ after (59), and (54).

Acknowledgement. The authors would like to express their gratitude to the Editors and the referee for their valuable comments and suggestions.

References

[1] P. BILLINGSLEY. Statistical Inference for Markov Processes. Institute of Mathematical Statistics - University of Chicago Statistical Research Monographs, University of Chicago Press, Chicago, 1961.

[2] P. BILLINGSLEY. Convergence of Probability Measures. Wiley, New York, 1968.

[3] E. BOLTHAUSEN. Exact convergence rates in some martingale central limit theorems. Ann. Probab. 10 (1982), 672 - 688.

[4] B. M. BROWN. Martingale central limit theorems. Ann. Math. Statist. 42 (1971), 59 - 66.

[5] C. DELLACHERIE and P. A. MEYER. Probabilities and Potential. North - Holland, Amsterdam, 1978.

[6] C. DELLACHERIE and P. A. MEYER. Probabilities and Potential B. North - Holland, Amsterdam, 1982.

[7] C. DOLÉANS. Variation quadratique des martingales continues à droite. Ann. Math. Statist. 40 (1969), 284 - 289.

[8] R. DROGIN. An invariance principle for martingales. Ann. Math. Statist. 43 (1972), 602 - 620.

[9] R. DURRETT and S. I. RESNICK. Functional limit theorems for dependent variables. Ann. Probab. 6 (1978), 829 - 846.

[10] A. DVORETZKY. Asymptotic normality for sums of dependent random variables. Proc. Sixth Berkeley Symp. Math. Statist. Probab., Vol. II, pp. 513 - 535. Univ. of California Press, Berkeley and Los Angeles, 1972.

[11] D. FREEDMAN. Brownian Motion and Diffusion. Holden - Day, San Francisco, 1971.

[12] P. GAENSSLER and E. HAEUSLER. Remarks on the functional central limit theorem for martingales. Z. Wahrsch. Verw. Gebiete 50 (1979), 237 - 243.

[13] B. V. GNEDENKO and A. N. KOLMOGOROV. Limit Distributions for Sums of Independent Random Variables. Addison - Wesley, Reading, Mass., 1954.

[14] E. HAEUSLER. An exact rate of convergence in the functional central limit theorem for special martingale difference arrays. Z. Wahrsch. Verw. Gebiete 65 (1984), 523 - 534.

[15] E. HAEUSLER. On the Lindeberg - Lévy - method for proving martingale central limit theorems. Statist. Decisions. To appear.

[16] P. HALL. On the Skorokhod representation approach to martingale invariance principles. Ann. Probab. 7 (1979), 371 - 376.

[17] P. HALL and C. C. HEYDE. Martingale Limit Theory and Its Application. Academic Press, New York, 1980.

[18] I. S. HELLAND. Central limit theorems for martingales with discrete or continuous time. Scand. J. Statist. 9 (1982), 79 - 94.

[19] I. S. HELLAND. Applications of central limit theorems for martingales with continuous time. Bull. Internat. Statist. Inst., Vol. L, pp. 346 - 360, 1983.

[20] C. C. HEYDE and B. M. BROWN. On the departure from normality of a certain class of martingales. Ann. Math. Statist. 41 (1970), 2161 - 2165.

[21] J. JACOD. Convergence en loi de semimartingales et variation quadratique. Sém. Probab. XV, pp. 547 - 560. Lecture Notes in Math. 850, Springer - Verlag, Berlin, Heidelberg, New York, 1981.

[22] P. JEGANATHAN. A solution of the martingale central limit problem, part I. Sankhyā Ser. A 44 (1982), 299 - 318.

[23] P. JEGANATHAN. A solution of the martingale central limit problem, part II. Sankhyā Ser. A 44 (1982), 319 - 340.

[24] E. LENGLART. Relation de domination entre deux processus. Ann. Inst. Poincaré, Sect. B (N. S.) 13 (1977), 171 - 179.

[25] P. LÉVY. Théorie de l'addition des variables aleatoires. Gauthier - Villars, Paris, 1937.

[26] J. W. LINDEBERG. Eine neue Herleitung des Exponentialgesetzes in der Wahrscheinlichkeitstheorie. Math. Z. 15 (1922),211 - 225.

[27] T. LINDVALL. Weak convergence of probability measures and random functions in the function space $D[0,\infty)$. J. Appl. Probab. 10 (1973), 109 - 121.

[28] R. SH. LIPTSER and A. N. SHIRYAYEV. A functional central limit theorem for semimartingales. Theory Probab. Appl. 25 (1980), 667 - 688.

[29] R. SH. LIPTSER and A. N. SHIRYAYEV. On a problem of necessary and sufficient conditions in the functional central limit theorem for local martingales. Z. Wahrsch. Verw. Gebiete 59 (1981), 311 - 318.

[30] R. SH. LIPTSER and A. N. SHIRYAYEV. On necessary and sufficient conditions in the functional central limit theorem for semimartingales. Theory Probab. Appl. 26 (1981), 130 - 135.

[31] R. SH. LIPTSER and A. N. SHIRYAYEV. On the rate of convergence in the central limit theorem for semimartingales. Theory Probab. Appl. 27 (1982), 1 - 13.

[32] D. L. MCLEISH. Dependent central limit theorems and invariance principles. Ann. Probab. 2 (1974), 620 - 628.

[33] P. A. MEYER. Le théorème fondamental sur les martingales locales. Sém. Probab. XI, pp. 463 - 464. Lecture Notes in Math. 581, Springer - Verlag, Berlin, Heidelberg, New York, 1977.

[34] D. POLLARD. Convergence of Stochastic Processes. Springer - Verlag, New York, 1984.

[35] R. REBOLLEDO. La méthode des martingales appliquée à l'étude de la convergence en loi des processus. Mémoirs de la Société Mathématique de France 62, 1979.

[36] R. REBOLLEDO. Central limit theorems for local martingales. Z. Wahrsch. Verw. Gebiete 51 (1980), 269 - 286.

[37] H. ROOTZÉN. On the functional central limit theorem for martingales. Z. Wahrsch. Verw. Gebiete 38 (1977), 199 - 210.

[38] H. ROOTZÉN. On the functional central limit theorem for martingales, II. Z. Wahrsch. Verw. Gebiete 51 (1980), 79 - 93.

[39] H. ROOTZÉN. Central limit theory for martingales via random change of time. Probability and Mathematical Statistics, Essays in Honour of Carl - Gustav Esséen, pp. 154 - 190. Uppsala University, 1983.

[40] D. J. SCOTT. Central limit theorems for martingales and for processes with stationary increments using a Skorokhod representation approach. Adv. in Appl. Probab. 5 (1973), 119 - 137.

[41] W. WHITT. Some useful functions for functional limit theorems. Math. Oper. Res. 5 (1980), 67 - 85.

University of Munich

Mathematical Institute

Theresienstrasse 39

D - 8000 Munich 2

West Germany

LIMIT LAWS FOR GENERALIZATIONS OF MARTINGALES

by
Ernst Eberlein

Given a sequence of random variables $(x_k)_{k\geq 1}$ one of the important problems in probability and statistics is the description of the asymptotic properties of the partial sum process $S(t) = \Sigma_{k\leq t} x_k$. Essentially three types of results are known: strong laws of large numbers (SLLNs), central limit theorems (CLTs) and laws of the iterated logarithm (LILs). All three of them as well as a number of refinements such as the functional versions of the last two or upper and lower class results can be derived from a strong or almost sure approximation of the partial sum process by a Brownian motion $(X(t))_{t\geq 0}$. Since as is well known Brownian motion satisfies all laws, the same holds for the partial sum process if the approximation is sharp enough, namely if

$$\sum_{k\leq t} x_k - X(t) \ll t^{1/2-\kappa} \quad \text{a.s.} \tag{1}$$

for some $\kappa > 0$. In other words formula (1) - once it is established - contains all information on the asymptotic behavior of the partial sums in a very compact form. Thus the question is to look for assumptions under which (1) holds.

Limit laws have been studied intensively for sequences of independent random variables. But independence is not necessary for (1) nor is it for any other of the classical laws. It is just the simplest case. From the point of view of the dependence structure of the sequence there are essentially two types of assumptions. First the so-called mixing conditions - initially introduced by Rosenblatt [22] in 1956 - which are defined directly in terms of the joint distributions of the given process. We shall not discuss mixing assumptions here. This is the topic of the articles of R. Bradley, M. Peligrad and W. Philipp in this volume. Let us only mention that there is no hope to characterize in this direction exactly the dependence structure such that (1) or any other limit law holds. The reason seems to be that by summing variables one looses most of the information on the joint distributions. An attempt to get the border line from the point of view of dependence using a mixing type assumption is [6] where (1) is established for very weak Bernoulli processes. This is a class of processes which goes beyond strong mixing sequences as has been shown in [4] .

The topic of this survey is the second type of dependence assumptions, those given in terms of conditional expectations. The simplest case here are martingale difference sequences. Strong approximations for martingales were already established by Strassen [24] but with a different error term. For an approximation with the order given in (1) see [21] and [16] . Again the class of martingales is too small for the validity of (1) .

Given a sequence $(x_k)_{k\geq 1}$ on a probability space $(\Omega,\mathfrak{F},P)$ denote

$$S_n(m) = \sum_{k=m+1}^{m+n} x_k$$

and write as usual $S_n = S_n(0)$. Let $\mathfrak{F}_m$ denote the σ-field generated by $x_1,\ldots,x_m$. For simplicity we shall not consider more general filtrations $(\mathfrak{F}_n)_{n\geq 1}$ here. Therefore sequences are always adapted. The following result is proved in [7] . We state the theorem for real-valued random variables only and in a somewhat streamlined form which reduces the number of parameters involved to a minimum.

Theorem. Let $(x_k)_{k\geq 1}$ be a sequence of random variables with uniformly bounded moments of order $2+\delta$ for some $\delta > 0$. Suppose there exists a $\theta > 0$ such that uniformly in m

$$\| E[S_n(m) | \mathfrak{F}_m] \|_1 \ll n^{1/2-\theta} \tag{2}$$

and

$$\| E[S_n^2(m) | \mathfrak{F}_m] - n \|_1 \ll n^{1-\theta} \quad . \tag{3}$$

Then without loss of generality there exists a Brownian motion $(X(t))_{t\geq 0}$ such that (1) holds.

As usual the phrase 'without loss of generality' here means that we can redefine the given sequence on a rich enough probability space on which a Brownian motion can be defined. The basic assumption on the dependence structure is (2). (3) is a second moment assumption which together with (2) also ensures that the variances of $(S_n)_{n\geq 1}$ grow asymptotically in a linear way.

What is the scope of assumption (2)? It holds trivially for martingales $(S_n)_{n\geq 1}$, since in this case $E[S_n(m) | \mathfrak{F}_m] = 0$ for all n, m. There is a hierarchy of successively larger classes of processes starting with the class of martingales. $(S_n)_{n\geq 1}$ is called a *quasimartingale* (Fisk [10]) if

$$\sum_{n\geq 1} \int | S_n - E[S_{n+1} | \mathfrak{F}_n] | dP < \infty \quad .$$

Let T denote the set of all bounded stopping times (with respect to $(\mathfrak{F}_n)_{n\geq 1}$). T is partially ordered with the definition $\sigma \leq \tau$ iff $\sigma(\omega) \leq \tau(\omega)$ for almost all $\omega \in \Omega$. A martingale $(S_n)_{n\geq 1}$ can be characterized by the fact that the net $(E[S_\tau])_{\tau \in T}$ is constant. If the same net converges, the process $(S_n)_{n\geq 1}$ is called an *amart* (for the name see Chacon and Sucheston [5] and Edgar and Sucheston [8]) and the underlying sequence an *amart difference sequence*. If the same net is only bounded, then we call $(S_n)_{n\geq 1}$ a *semiamart* ([8], [13]) and $(x_k)_{k\geq 1}$ a *semiamart difference sequence*. We have {martingales} $\subset$ {quasimartingales} $\subset$ {amarts} $\subset$ {semiamarts} . The first and the last inclusion are clear from the definitions, the second one is an easy exercise ([8], page 200). Examples for amarts and semiamarts arise naturally from the following facts. A sequence $(S_n)_{n\geq 1}$ is a semiamart if $\sup_{n\geq 1} |S_n|$ has finite expectation and a sequence with this property is an amart iff $(S_n)_{n\geq 1}$ converges almost surely ([8]). For an interesting application of amarts in the theory of sequential testing see [17].

To see that (2) is satisfied for these classes we use the Riesz decomposition theorem. According to Krengel and Sucheston [13] every semiamart $(S_n)_{n\geq 1}$ can be represented as

$$S_n = Y_n + Z_n \tag{4}$$

where $(Y_n)_{n\geq 1}$ is a martingale and $(Z_n)_{n\geq 1}$ is a L^1-bounded semiamart. Since the conditional expectation is a L^1-contraction we obtain from this decomposition

$$\begin{aligned} \| E[S_n(m) | \mathfrak{F}_m] \|_1 &= \| E[Z_{m+n} - Z_m | \mathfrak{F}_m] \|_1 \\ &\leq \| Z_{m+n} - Z_m \|_1 \quad . \end{aligned}$$

Thus we get for some constant $C > 0$

$$\| E[S_n(m) | \mathfrak{F}_m] \|_1 \leq C \tag{5}$$

for all n , m which is stronger than (2) . Let us mention that semiamarts can be characterized by such a decomposition. It was shown in [11] that $(S_n)_{n\geq 1}$ is a semiamart iff (4) holds for a martingale $(Y_n)_{n\geq 1}$ and a sequence $(Z_n)_{n\geq 1}$ which is dominated in absolute value by a positive supermartingale. Amarts $(S_n)_{n\geq 1}$ can be uniquely decomposed as in (4) in a martingale $(Y_n)_{n\geq 1}$ plus an amart $(Z_n)_{n\geq 1}$ converging to 0 in L^1 as well as almost surely ([8]) . Moreover the family $(Z_\tau)_{\tau\in T}$ is uniformly integrable. As for semiamarts the Riesz decomposition can be refined such that it characterizes amarts. $(S_n)_{n\geq 1}$ is an amart iff the summand $(Z_n)_{n\geq 1}$ in (4) is dominated in absolute value by a positive supermartingale converging to 0 in L^1 ([11]). For our purposes we need only the fact mentioned above that $(Z_n)_{n\geq 1}$ converges in L^1. With the same reasoning as for semiamarts we get here even

$$\lim_{m\to\infty} \| E[S_n(m) | \mathfrak{F}_m] \|_1 = 0 \tag{6}$$

uniformly in n.

Alloin [1] called $(S_n)_{n\geq 1}$ a *progressive martingale* iff $\{E[S_1(m-1)|\mathfrak{F}_{m-1}] = 0\} \subset \{E[S_1(m)|\mathfrak{F}_m] = 0\}$ for all m and

$$\lim_{m\to\infty} P[\ E[S_1(m)|\mathfrak{F}_m] = 0\ ] = 1 \quad .$$

Blake [2] defined a *game fairer with time* as a sequence $(S_n)_{n\geq 1}$ such that for all $\varepsilon > 0$

$$\lim_{m\to\infty} P[\ |E[S_n(m)|\mathfrak{F}_m]| > \varepsilon\ ] = 0$$

uniformly in n and Mucci [18] called $(S_n)_{n\geq 1}$ a *martingale in the limit* if in this definition convergence in probability is replaced by almost sure convergence, i.e. if

$$\lim_{m\to\infty} E[S_n(m)|\mathfrak{F}_m] = 0 \quad \text{a.s.}$$

uniformly in n . Every progressive martingale is a martingale in the limit ([27]). According to Edgar and Sucheston ([9]) and Blake ([3]) every amart is a martingale in the limit. Therefore we have {amarts} $\subset$ {martingales in the limit} $\subset$ {games fairer with time} .

What can be said about the largest class here? Mucci ([18]) and Subramanian ([25]) showed that a uniformly integrable fairer with time game converges in L^1. Again by the L^1-contraction property this immediately implies (6). If even $(|S_n|^p)_{n\geq 1}$ is uniformly integrable for some $p > 1$, then $(S_n)_{n\geq 1}$ converges in L^p ([25]). Therefore (6) holds with 1 replaced by p.

A tool that seems to be attractive to describe dependence structures is the *associated measure* or *Doléans measure* λ of the process (see e.g. [15]). Let $m < n$ be positive integers and $F \in \mathfrak{F}_m$. Then the subset $]m,n] \times F$ of the product space $\mathbb{R}^+ \times \Omega$ is called a *predictable rectangle*. Suppose the x_k have finite expectation then λ is defined by

$$\lambda(]m,n] \times F) = E[1_F(S_n - S_m)] .$$

Consider for example the assumption that there exists a constant $C > 0$ such that

$$|\lambda(]m,n] \times F)| \leq C\, P(F) \tag{7}$$

for all predictable rectangles. (7) is equivalent to

$$|E[S_n(m)|\mathfrak{F}_m]| \leq C \quad \text{a.s.}$$

for all m, $n \geq 1$. Therefore (7) implies (5).

Now let us discuss (2) with L^2-norm instead of L^1-norm. For the following definition let $(\mathfrak{F}_k)_{k\geq 1}$ be an arbitrary filtration. A sequence $(x_k, \mathfrak{F}_k)_{k\geq 1}$ is called a *mixingale* (McLeish [14]) if there exists a sequence of positive numbers $\psi_k \to 0$ such that for all $i \geq 1$, $k \geq 0$

$$\| E[x_i|\mathfrak{F}_{i-k}] \|_2 \leq \psi_k$$

and

$$\| x_i - E[x_i|\mathfrak{F}_{i+k}] \|_2 \leq \psi_{k+1} .$$

Since we are considering in this article only the filtration generated by $(x_k)_{k\geq1}$, i.e. sequences which are adapted, the second requirement in this definition is not relevant. From Minkowski's inequality we see that the first requirement implies

$$\| E[S_n(m)|\mathfrak{F}_m] \|_2 \leq C \tag{8}$$

for all $m, n \geq 1$ if $\Sigma_{k\geq1} \psi_k < \infty$. In short, adapted mixingales with summable coefficients satisfy (8).

In a paper where weak invariance principles for dependent random variables are studied, Peligrad [20] defined the following class. Assume that for each fixed m the sequence $(E[S_n(m)|\mathfrak{F}_m])_{n\geq1}$ converges in L^2 to a function U_m such that the sequence $(U_m^2)_{m\geq1}$ is uniformly integrable. From the first part of this definition one easily gets a decomposition $S_n = Z_n - U_n$ where $(Z_n)_{n\geq1}$ is a martingale. Thus

$$\| E[S_n(m)|\mathfrak{F}_m] \|_2 \leq \| U_{m+n} - U_m \|_2 .$$

Uniform integrability of $(U_m^2)_{m\geq1}$ now implies (8) for some constant $C > 0$. Peligrad [19] called $(S_n)_{n\geq1}$ a *martingale in the L^2-limit* if

$$\lim_{m\to\infty} \| E[S_n(m)|\mathfrak{F}_m] \|_2 = 0 \tag{9}$$

uniformly in n . In other words martingales in the L^2-limit constitute the subclass where $U_m = 0$ for all m . Let us conclude this discussion of the basic assumption of the Theorem with a remark on amarts again. The amart $(Z_n)_{n\geq1}$ in the Riesz decomposition of $(S_n)_{n\geq1}$ converges to 0 almost surely. Thus if $(Z_n)_{n\geq1}$ is L^2-dominated, i.e. if there exists a variable Z in L^2 such that $|Z_n| \leq Z$, then we get (9).

The result discussed here refers to the case where the variances are asymptotically linear, i.e. where $Var(S_n)/n$ converges to a finite positive limit as $n \to \infty$ (in our statement this value is 1). If this is not the case, the corresponding partial sum process to be approximated by a Brownian motion is

$$S(t) = \sum_{n\geq 1} x_n 1_{\{V_n \leq t\}}$$

where V_n is for mean zero variables the conditional variance given by

$$V_n = \sum_{k\leq n} E[x_k^2 | \mathfrak{F}_{k-1}] \quad .$$

Let us make an outlook on this situation. Certainly one has to assume that $V_n \to \infty$ a.s. . As an assumption to replace (2) one could require

$$\| E[S_\tau - S_\sigma | \mathfrak{F}_\sigma] \|_1 \leq C \tag{10}$$

for all bounded stopping times σ , τ such that $\sigma \leq \tau$. Amarts are covered by (10). This can easily be seen from the amart Riesz decomposition and the optional sampling theorem for martingales. For the strong approximation instead of a constant C on the right side, a suitable growth rate depending on the difference $\tau - \sigma$ of the stopping times would do as well.

In this paper only references are given which are of interest for the strong approximation problem. A central limit theorem under assumptions on the conditional expectations was already proved in 1968 (Serfling [23]). There is quite a number of papers starting about 1970 where the preservation of martingale properties such as optimal stopping, optional sampling, Riesz decomposition and the martingale convergence theorem is investigated for classes generalizing martingales. A strong law of large numbers is proved in several of these papers. In the recent paper of Tomkins [28] a systematic discussion of seven martingale properties can be found. A bibliography of the literature on amarts, including papers which led to or were inspired by amart theory, is given at the end of [12]. A recent review of [12] (Sucheston [26]) contains a general evaluation of certain aspects of amart theory.

References

[1] Alloin, C. Martingales progressives. *Cahiers Centre Etudes Rech. Opér. 12* , 201-210 (1970)

[2] Blake, L. A generalization of martingales and two consequent convergence theorems. *Pacific J. Math. 35* , 279-283 (1970)

[3] Blake, L. Every amart is a martingale in the limit. *J. London Math. Soc. 18* , 381-384 (1978)

[4] Bradley, R. On a very weak Bernoulli condition. *Stochastics 13* , 61-81 (1984)

[5] Chacon, R.V. and Sucheston, L. On convergence of vector-valued asymptotic martingales. *Z. Wahrscheinlichkeitstheorie verw. Geb. 33* , 55-59 (1975)

[6] Eberlein, E. Strong approximation of very weak Bernoulli processes. *Z. Wahrscheinlichkeitstheorie verw. Geb. 62* , 17-37 (1983)

[7] Eberlein, E. On strong invariance principles under dependence assumptions. *Ann. Probability* (to appear Jan. 1986)

[8] Edgar, G. and Sucheston, L. Amarts: A class of asymptotic martingales. I, II *J. Multivariate Anal. 6* , 193-221, 572-591 (1976)

[9] Edgar, G. and Sucheston, L. Martingales in the limit and amarts. *Proc. Amer. Math. Soc. 67* , 315-320 (1977)

[10] Fisk, D.L. Quasimartingales. *Trans. Amer. Math. Soc. 120* , 369-389 (1965)

[11] Ghoussoub, N. and Sucheston, L. A refinement of the Riesz decomposition for amarts and semiamarts. *J. Multivariate Anal. 8* , 146-150 (1978)

[12] Gut, A. and Schmidt, K. *Amarts and set function processes.* Lecture Notes in Math. 1042 , Springer , Berlin (1983)

[13] Krengel, U. and Sucheston, L. On semiamarts, amarts, and processes with finite value. In *Probability on Banach Spaces* , Ed., J. Kuelbs , *Advances in Probability and Related Topics 4* , 197-266. Marcel Dekker , New York (1978)

[14] McLeish, D.L. Invariance principles for dependent variables. *Z. Wahrscheinlichkeitstheorie verw. Geb. 32* , 165-178 (1975)

[15] Métivier, M. *Semimartingales.* de Gruyter , Berlin (1982)

[16] Morrow, G. and Philipp, W. An almost sure invariance principle for Hilbert space valued martingales. *Trans. Amer. Math. Soc. 273* , 231-251 (1982)

[17] Müller-Funk, U. A weak compactness theorem for sequential tests. Preprint Freiburg (1985)

[18] Mucci, A.G. Limits for martingale-like sequences. *Pacific J. Math. 48* , 197-202 (1973)

[19] Peligrad, M. Limit theorems and the law of large numbers for martingale-like sequences. *Math. Nachr. 99* , 211-216 (1980)

[20] Peligrad, M. An invariance principle for dependent random variables. *Z. Wahrscheinlichkeitstheorie verw. Geb. 57* , 495-507 (1981)

[21] Philipp, W. and Stout, W. Almost sure invariance principles for partial sums of weakly dependent random variables. *Mem. Amer. Math. Soc. 161* , 1-140 (1975)

[22] Rosenblatt, M. A central limit theorem and a strong mixing condition. *Proc. Nat. Acad. Sci. USA 42* , 43-47 (1956)

[23] Serfling, R.J. Contributions to central limit theory for dependent variables. *Ann. Math. Statist. 39* , 1158-1175 (1968)

[24] Strassen, V. Almost sure behavior of sums of independent random variables and martingales. *Proc. Fifth Berkeley Symp. Math. Statist. Prob. 2* , 315-343 (1967)

[25] Subramanian, R. On a generalization of martingales due to Blake. *Pacific J. Math. 48* , 275-278 (1973)

[26] Sucheston, L. Review of *Amarts and set function processes* by A. Gut and K. Schmidt. *Bulletin Amer. Math. Soc. 12* , 306-312 (1985)

[27] Tomkins, R.J. Properties of martingale-like sequences. *Pacific J. Math. 61* , 521-525 (1975)

[28] Tomkins, R.J. Martingale generalizations and preservation of martingale properties. *Canadian J. Statistics 12* , 99-106 (1984)

Ernst EBERLEIN
Institut f. Math. Stochastik
Universität Freiburg
Hebelstrasse 27
7800 FREIBURG I. BR.
WEST GERMANY

SECTION 4

OTHER DEPENDENCE STRUCTURES

Joint Characteristic Functions and Associated Sequences

André Robert Dabrowski

The proof of a limit theorem for independent and identically distributed random variables usually uses independence in a fundamental way. For dependent sequences, the dependence structure is frequently used to approximate the classical tools of independence, i.e. the expression of joint probabilities as a product of marginal probabilities, and/or the factorization of joint characteristic functions. Here, we wish to review the connection between characteristic functions, associated sequences and two fundamental limit theorems.

1. Joint characteristic functions

Let $\{X_i: i \geq 1\}$ be a sequence of independent, identically distributed real random variables such that $E\, X_1 = 0$ and $E\, X_1{}^2 = 1$. Define $S_n(t)$ to be the random function in $t \in [0,1]$ which is linear in every interval of form

$$[(k-1)/n,\ k/n] \qquad k = 1,2,\ldots,n$$

and for $t = k/n$, $k = 1,2,\ldots,n$

$$S_n(t) = \sum_{1 \leq j \leq k} X_j \qquad (S_n(0) \equiv 0).$$

Further, set $f_n(t) = n^{-1/2}\, S_n(t)$ and $g_n(t) = (2n\log\log n)^{-1/2}\, S_n(t)$. By well-known theorems of Donsker [8] and Strassen [27], the sequence $\{X_i: i \geq 1\}$ satisfies both the functional central limit theorem (FCLT), and the functional form of the law of the iterated logarithm (FLIL).

FCLT: With respect to the supnorm topology on C[0,1], as $n \to \infty$

$$f_n \to W,$$

where W is a standard Brownian motion.

FLIL: With probability one, the sequence $\{g_n : n \geq 1\}$ is equicontinuous, and the set of its norm-limit points (in the supnorm topology on [0,1]) coincides with the set,

$$K = \{x: x \text{ is absolutely continuous on } [0,1],\ x(0) = 0, \text{ and } \int_0^1 (\dot{x}(s))^2\, ds \leq 1\}.$$

The FCLT makes a statement over the probability space on the distribution of values of f_n about zero. The FLIL describes the extreme fluctuations of g_n about zero for almost every point in the probability space. Applications of the FCLT to estimation problems, hypothesis testing, and other procedures are widespread throughout the statistical literature. Applications of the FLIL are, for example, those introduced by Strassen [27], and the concept of 100% confidence intervals (cf. Robbins [24], Robbins and Siegmud [25] and Lai [13]).

Now consider a sequence of real, possibly dependent, random variables, $\{X_i: i \geq 1\}$ with mean values zero and unit variances. By a well-known result of Prohorov, this sequence has the FCLT if and only if

a) the finite-dimensional distributions of f_n converge weakly, as $n \to \infty$, to the corresponding finite-dimensional distributions of W, and

... (1)

b) the distributions of $\{f_n;\ n \geq 1\}$ are tight.

An analagous result for the FLIL is developed in Oodaira [21]. Here $\{X_i: i \geq 1\}$ satisfies the FLIL if and only if the following two properties hold.

a) Let $T = T_m = \{t_1, t_2, \ldots, t_m\}$ be a finite subset of $[0,1]$. Denote by $g^T = [g(t_1), g(t_2), \ldots, g(t_m)]$ the restriction of g to T. Let $A^T = [g^T: g \in A]$ for a class of functions A. Let $\{T_m\}$ be an increasing sequence of sets T_m such that $\cup T_m$ is dense in $[0,1]$. Then for each $T \in \{T_m\}$, the set of limit points of the random vectors $\{g_n^T\}$ is K^T with probability one.

... (2)

b) The sequence $\{g_n\}$ is with probability one relatively compact.

These results describe clearly the content of the FCLT or the FLIL for a sequence $\{X_i: i \geq 1\}$ from a computational viewpoint. In most applications, these rather theoretical requirements may be replaced by more readily computable sufficient conditions. Here we wish to consider one set of conditions stated in terms of characteristic functions. We follow the approach of Berkes [2].

For fixed $0 \leq t_1 < t_1' \leq \ldots \leq t_r < t_r' \leq 1$ $(r = 1,2,\ldots)$ let $\{t\}$ denote $\{t_1, t_1', t_2, \ldots, t_r, t_r'\}$, and define $\phi_n^{\{t\}}[\lambda_1, \lambda_2, \ldots, \lambda_r]$ as the characteristic function of the random vector

$$\left[\sum_{j>nt_i}^{nt_i'} X_j/[nt_i' - nt_i]^{1/2} : i = 1,2, \ldots, r\right]. \quad \ldots (3)$$

Let $\psi_n^{\{t\}}[\lambda_1, \lambda_2, \ldots, \lambda_r]$ be the characteristic function of

$$[[W(t_i') - W(t_i)] / [t_i'-t_i]^{1/2}: i = 1,2,\ldots, r],$$

where W is a standard Brownian motion. Consider the following versions of (1) and (2).

a) For any fixed choice of {t} and $\lambda \in \mathbb{R}^r$, as $n \to \infty$

$$|\phi_n^{\{t\}}(\lambda) - \psi_n^{\{t\}}(\lambda)| \to 0, \text{ and} \qquad \ldots (4)$$

b) for any $\epsilon > 0$

$$\lim_{h\to 0} \limsup_{n\to\infty} P[\sup\{|\sum_{j>nt}^{nt'} X_j|;\ 0 \le t < t' \le t + h \le 1\} > \epsilon\, n^{1/2}] = 0.$$

For {t} as above define $\Delta = \min\,[t_i' - t_i;\ i = 1,2,\ldots,r]$.

For each choice of {t} as in (3), for every $\lambda \in \mathbb{R}^{2r}$ satisfying

a) $||\lambda|| \le C_1\ (n\Delta)^{\gamma}$, we have

$$|\phi_n^{\{t\}}(\lambda) - \psi_n^{\{t\}}(\lambda)| < C_2\ (n\Delta)^{-\delta},$$

where C_1, C_2, γ and δ are positive constants depending only on r and the distribution of the sequence $\{X_i: i \ge 1\}$, and ... (5)

b) $E|X_n|^{2+\epsilon} \le C_3 < +\infty$ for n = 1,2,3,..., and for any positive integers m, n

$$P[\sup\{|\sum_{j=m+1}^{m+v} X_j|: 1 \le v \le n\} \ge C_4\ (n\log\log n)^{1/2}] \le C_5 \exp(-2\log\log n),$$

where C_3, C_4, C_5 and ϵ are positive constants depending only on the distribution of $\{X_i: i \ge 1\}$.

Condition (4) is equivalent to the FCLT by well-known arguments (eg. Billingsley [5]). Condition (5) implies the conditions of Theorem 3 of Berkes [2], and consequently implies the FLIL holds for $\{X_i: i \ge 1\}$. However, (5) is clearly not a necessary condition. Even for independent and identically distributed sequences, the requirement of finite $(2+\epsilon)^{th}$ moments is superfluous. By considering exchangeable sequences, one may construct simple examples where the FLIL holds, but the rates of (5a) or (5b) are not maintained.

Conditions (4a) and (5a) are usually interpreted as some form of asymptotic independence for the sequence $\{X_i: i \geq 1\}$. The following example shows that a high degree of asymptotic dependence may be present in $\{X_i: i \geq 1\}$, and yet the FCLT and FLIL are still valid. This is an "embedding" of a triangular array into a sequence.

Let $\{Z_v: v \geq 0\}$ be a sequence of independent standard normal variables. For each $j \geq 1$, let $Y(j,1), Y(j,2), \ldots, Y(j,n_j)$ be a finite sequence of independent standard normal variables such that

$$Z_0 = \sum_{i=1}^{n_j} Y(j,i)/n_j^{1/2},$$

and such that $\{Y(j,i): j \geq 1, 1 \leq i \leq n_j\}$ is independent of $\{Z_v: v \geq 1\}$. Set $n_1 = 1 = m_1$, $m_{k+1} = \exp[n_k+m_k]$, and $n_k=m_k$. Define

$$X_v = \begin{cases} Z_v & \text{for } m_k + n_k < v \leq m_{k+1} \\ Z_0 & \text{for } v = 1 \\ Y(k,v-n_k) & \text{for } m_k < v \leq m_k + n_k. \end{cases}$$

The sequence $\{X_v: v \geq 1\}$ is not asymptotically independent in any sense since for all $k \geq 1$

$$X_1 = Z_0 = \sum_{m_k+1}^{m_k+n_k} X_v/n_k^{1/2} .$$

Yet the FCLT and FLIL for this sequence follow from the usual results on independent sequences and the fact that

$$\sup\{m_{k+1}^{-1/2} \sum_{j \le v} X_j : v \le m_k\} \to 0 \quad \text{a.s. as } k \to \infty. \quad \ldots (6)$$

The very weak form of stationarity inherent in (5) fails for this example. We may incorporate (6) into a slightly weaker form of (5) which still is sufficient for the FLIL.

a) There is a sequence of constants $\{\tau_n\}$ decreasing to zero such that for each n(5a) holds uniformly over all $\{t\}$ chosen such that $t_1 \ge \tau_n$,

b) $\sup\{|n^{-1/2} \sum_{i \le nt} X_i| : 0 \le t \le \tau_n\} \to 0$ a.s., and ... (5')

c) (5b) holds.

A well-established strengthening of both the FCLT and the FLIL is the almost sure invariance principle with error $o(n^{1/2})$, e.g. Philipp and Stout [22]. The above example readily yields that a strong invariance principle of that type, or even an invariance principle in probability, does not follow from the FCLT and FLIL alone. A FLIL as in Strassen [27] with error rate $o((n\log\log n)^{1/2})$ may yet hold.

2. Associated Sequences

As is evident from the majority of articles in this book, there is a well-developed theory concerning the central limit theorem and law of the iterated logarithm for mixing sequences and martingales. A number of excellent techniques have led to precise weak convergence and almost sure results for these, and related dependent sequences. The demonstrations of such results naturally rely heavily on the exact dependence structure. The methods frequently rely on factorizations of joint probabilities to within a reasonable degree of approximation. The advantage of

(4) and (5) is that they may be applied to a disparate collection of dependent sequences. Berkes ([3], [4]) applied his results on (5) to lacunary sequences, Walsh functions, and strongly multiplicative systems. Of more recent interest are the associated sequences of Esary, Proschan and Walkup [9].

A finite family of finite variance real random variables $\{X_1, X_2, \ldots, X_n\}$ is said to be associated if, for any real co-ordinatewise non-decreasing functions f_1, and f_2 on $\mathbb{R}^n$ with $\mathrm{Var}[f_i(X_1,\ldots,X_n)] < +\infty$,

$$\mathrm{Cov}\,[f_1(X_1,\ldots,X_n),\ f_2(X_1,\ldots,X_n)] \geq 0.$$

An infinite sequence is associated if every finite subfamily is associated.

For the case of strong-mixing sequences, the mathematical definition of independence is relaxed (cf. Rosenblatt [26]). In contrast, the structure of associated sequences arises from several separate models. The book of Barlow and Proschan [1] clearly indicates the usefulness of these sequences to reliability theory. The application of associated variables to percolation models and statistical mechanics may be traced to Harris [11] and Fortuin, Kastelyn and Ginibre [10]. This system of dependence may also be viewed as a form of the positive quadrant dependence introduced by Lehmann [15]. An excellent presentation of these concepts, theorems, and related references is found in Newman [18].

Pitt [33] has established that positively correlated Gaussian sequences are associated. Under mild conditions, Lebowitz [14] has that associated sequences are mixing in the ergodic sense. Wood [28] has examples of associated sequences which are not strong-mixing, and yet whose normalized partial sums converge quickly to a normal distribution. Clearly strong-mixing sequences are not necessarily associated. Association thus defines a class of dependent sequences distinct from strong-mixing sequences. The dependence structure of associated sequences is reflected in the factorization of the characteristic functions of partial sums, and is controlled by its covariance structure. In particular, most results on stationary associated sequences require a finite summability condition, i.e.

$$0 < E X_1^2 + 2 \sum_{i=2}^{\infty} \text{Cov}[X_1, X_i] < +\infty \qquad \ldots (7)$$

An example of Herrndorff [12] shows that this condition cannot realistically be relaxed for stationary sequences. The rate at which the value of

$$\text{Var}[n^{-1/2} \sum_1^n X_i]$$

approaches the limit of (7) may be used for almost sure results as the mixing rate is used for mixing sequences (cf. Dabrowski [7], Philipp and Stout [22]). In view of these remarks, the example given previously is a non-stationary associated sequence possessing both the FCLT and FLIL, and which is not strong-mixing. However, the sum of (7) for this sequence is infinite.

Key tools for the study of associated sequences are a characteristic function inequality of Newman [16] and a maximal inequality of Newman and Wright [19].

Newman's Inequality: Suppose $Y_1, Y_2, \ldots, Y_m$ are associated finite variance random variables with joint and marginal characteristic functions $\phi[\lambda_1, \lambda_2, \ldots, \lambda_m]$ and $\phi_j[\lambda_j]$, respectively. Then

$$|\phi[\lambda_1, \lambda_2, \ldots, \lambda_m] - \prod_{j=1}^{m} \phi_j[\lambda_j]| \le \frac{1}{2} \sum_{1< k\neq j\le m} |\lambda_j \lambda_k| \text{Cov}[Y_j, Y_k] . \qquad \ldots (8)$$

Maximal Inequality: Suppose $Y_1, Y_2, \ldots, Y_m$ are associated mean zero, finite variance real random variables. Define

$M_m = \max \{ \sum_{j \in v} Y_i : v = 1,2,\ldots,m\}$. Then

$$E\, M_m^2 \le \mathrm{Var}\,(S_m). \qquad \ldots (9)$$

Newman and Wright [19] exploit (4), (8), and (9) to obtain the FCLT for associated sequences satisfying (7). These inequalities lead to additional results similar to those found for independent variables. Cox and Grimmett [6] extend the Newman-Wright FCLT to non-stationary associated sequences with application to the voter model and the percolation model. Wood [28] establishes a Berry-Esseen theorem for $\{X_i : i \ge 1\}$ under the additional assumption that $E\, X_1^3 < +\infty$. Dabrowski [7] uses (8), (9) and a condition on the growth of the third moments of partial sums to establish (5), and hence the FLIL for a class of associated variables. Similar conditions and Lemma 3 of Berkes [2] yield a multidimensional analogue to the Berry-Esseen theorem for associated sequences. Wood [29] has a local limit theorem for associated sequences.

3. Problems

The inequalities (7), (8) and (9) are key to the description of associated sequences. An additional result of importance is the demimartingale formulation of Newman and Wright [20]. With these tools, one may explore to which extent the properties of independent sequences carry over to associated sequences.

Newman [16] shows that (8) holds for dependent sequences far more general than stationary associated sequences. Under (7), he obtains the central limit theorem. The probabilistic tools required to prove a central limit theorem may yield a law of the iterated logarithm as well. A rate of approximation in the central limit theorem is frequently the additional step. A law of the iterated logarithm for the more general structures described in Newman [18] seems attainable. In addition, Newman [18] describes a number of positive dependence conditions which lead to inequalities on distribution functions instead of characteristic functions. Berkes [2] has sufficient conditions [Condition B_2] for the FLIL under distribution functions inequalities. The events over which these two sets of inequalities hold are, however, not the same.

The results of Cox and Grimmett [6], Wood [28] and Dabrowski [7] assume some control over third moments of an associated sequence. A most attractive feature of the FCLT demonstrated by Newman and Wright [19] is that the conditions on the sequence are stated only in terms of second order moments. Since (7) apparently holds the key to asymptotic independence, it would be preferable that the third moment conditions be replaced by second order moment conditions, in particular in Wood [28] and Dabrowski [7]. From a practical point of view, it would be preferable to test the covariance structure of an associated sequence, rather than the structure of higher moments.

The author expresses his thanks to the referee for his valuable comments and to the Natural Sciences and Engineering Research Council of Canada for its support of this research.

4. References

[1] Barlow, R. and F. Proschan (1975). Statistical Theory of Reliability and Life Testing. New York. Holt, Rinehard and Winston.

[2] Berkes, I. (1973). The Functional Low of the Iterated Logarithm for Dependent Random Variables. Z. Wahrsch. verw. Gebiete 26; 245-258.

[3] Berkes, I. (1973). On Strassen's Version of the Loglog Law for Multiplicative Systems. Studia Sci. Math. Hung. 8, 425-431.

[4] Berkes, I. (1973). Functional Limit Theorems for Lacunary Trigonnetric and Walsh Series. Studia Sci. Math. Hung. 8, 411-423.

[5] Billingsley, P. (1968). Convergence of Probability Measures. Wiley. New York.

[6] Cox, J.T. and G. Grimmett (1984). Central Limit Theorems for Associated Random Variables and the Percolation Model. A. Probability 12, 514-528.

[7] Dabrowski, A.R. (1985). A Functional Law of the Iterated Logarithm for Associated Sequences. To appear in Statistics and Probability Letters.

[8] Donsker, M. (1951). An Invariance Principle for Certain Probability Limit Theorems. Mem. Amer. Math. Soc. 6.

[9] Esary, J., F. Proschan and D. Walkup (1967). Association of Random Variables with Applications. Ann. Math Statist. 38, 1466-1474.

[10] Fortuin, C.P. Kastelyn and J. Ginibre (1971). Correlation Inequalities on some Partially Ordered Sets. Comm. Math. Phys. 22, 89-103.

[11] Harris, T.E. (1960). A Lower Bound for the Critical Probability in a Certain Percolation Process. Proc. Camb. Phil. Soc. 59, 13-20.

[12] Herrndorf, N. (1984). An Example on the Central Limit Theorem for Associated Sequences. A. Probability 12, 912-917.

[13] Lai, T.L. (1977), Power-one Tests Based on Sample Sums. A. Statistics, 5, 866-880.

[14] Lebowitz, J. (1972). Rounds on the Correlations and Analyticity Properties of Ferromagnetic Ising Spin Systems. Comm. Math. Phys. 28, 313-321.

[15] Lehmann, E.L. (1966). Some Concepts of Dependence. Ann. Math. Statist. 37, 1137-1153.

[16] Newman, C.M. (1980). Normal Functuations and the FKG Inequalities. Comm. Math. Phys. 74, 119-128.

[17] Newman. C.M. (1983). A General Central Limit Theorem for FKG Systems. Comm. Math. Phys. 91, 75-80.

[18] Newman, C.M. (1984). Asymptotic Independence and Limit Theorems for Positively and Negatively Dependent Random Variables. Inequalities in Statistics and Probability. Volume 5 of the IMS Notes-Monograph Series, Y.L. Tong, editor.

[19] Newman, C.M. and A.L. Wright (1981). An Invariance Principle for Certain Dependent Sequences. A. Probability 9, 671-675.

[20] Newman, C.M. and A.L. Wright (1982). Associated Random Variables and Martingale Inequalities. Z. Wahrsch. verw. Gebiete. 59, 361-371.

[21] Oodaira, H. (1975). Some Functional Laws of the

Iterated Logarithm for Dependent Random Variables. Collo. Math. Soc. Janos Bolyai 11, 289-299.

[22] Philipp, W. and W. Stout (1975). Almost Sure Invariance Principles for Partial sums of Weakly Dependent random variables. Mem. Amer. Math. Soc. 161.

[23] Pitt, L. (1982). Positively Correlated Normal Variables are Associated. A. Probability, 10,496-499.

[24] Robbins, H. (1970). Statistical Methods Related to the Law of Iterated Logarithm. Ann. Math. Statist. 41, 1397-1409.

[25] Robbins, H. and D. Siegmund (1974). Expected Sample Size of some Tests of Power One. A. Statistics 2, 415-436.

[26] Rosenblatt, M. (1956). A Central Limit Theorem and a Strong Mixing Condition. Proc. Natl. Acad. Sci. USA 42, 43-47.

[27] Strassen, V. (1964). An Invariance Principle for the Law of the Iterated Logarithm. Z. Wahrsch. verw. Gebiete 3, 211-226.

[28] Wood, T.E. (1983). A Berry-Esseen Theorem for Associated Random Variables. A. Probability 11, 1042-1047.

[29] Wood, T.E. (1985). A Local Limit Theorem for Associated Sequences. A Probability 13, Vol. 2.

A.R. Dabrowski
Department of Mathematics
University of Ottawa
Ottawa, Ontario, K1N 6N5
Canada.

INFINITELY DIVISIBLE DISTRIBUTIONS; GIBBS STATES AND CORRELATIONS

by

Ed Waymire

1. *Introduction and Preliminaries*

Let S be a non-empty set. S may be either finite or countably infinite. Elements of S are referred to as *sites*. We wish to consider random distributions of ± 1 values on S. The *configuration space* is the product space $\Omega = \{-1,1\}^S$ equipped with the Borel sigma-field $\mathfrak{F}$ for the product topology on Ω when $\{-1,1\}$ has the discrete topology, Equivalently, $\mathfrak{F}$ is the sigma-field generated by the finite-dimensional events. Let $\mathcal{L}$ denote the collection of finite subsets of S.

Borrowing from the terminology of statistical mechanics, the values of the variables associated with sites of S are referred to as *spins*. The *spin variables* $\{X_n : n \in S\}$ are represented by the coordinate projections, $X_n(w) = w_n$, $n \in S$, $w \in \Omega$. The distribution of the random field $X = \{X_n : n \in S\}$ is a probability measure P on $(\Omega, \mathfrak{F})$. For $T \subset S$, the sigma-field of events in $\mathfrak{F}$ which depend only on coordinates in T shall be denoted $\sigma\{X_n : n \in T\}$.

The *block correlations* of the random field X (or P) associated with finite sets $M \in \mathcal{L}$ are defined by

$$\phi_X(M) = E\{\prod_{m \in M} X_m\}, \quad M \in \mathcal{L}. \tag{1.1}$$

Observe that under coordinatewise multiplication Ω is a compact abelian topological group. The continuous characters (i.e. continuous homomorphisms of Ω into the circle group) are given by γ_M, $M \in \mathcal{L}$, where

$$\gamma_M(w) = \prod_{m \in M} w_m , \quad M \in \mathcal{L} . \tag{1.2}$$

The dual group (or character group) of Ω can be identified with $\mathcal{L}$ under symmetric difference Δ and endowed with the discrete topology; note that $\gamma_M \cdot \gamma_N = \gamma_{M \Delta N}$. Moreover, the block correlations furnish the Fourier transform of the distribution P of X via

$$\phi_X(M) = \int_\Omega \gamma_M(w) P(dw), \quad M \in \mathcal{L} . \tag{1.3}$$

In particular it follows from the uniqueness theorem for Fourier transforms that the block correlations uniquely determine P. We shall take the liberty of writing ϕ_X or ϕ_P interchangeably according to convenience.

The central questions to modern statistical physics coincide with those of probability theory to the extent that the interest is in the large scale behavior of sums (large scale magnetization) of the form

$$T_M = \sum_{m \in M} X_m, \quad M \in \mathcal{L} , \tag{1.4}$$

as M varies through the directed set $\mathcal{L}$. However, owing to the nature of the dependencies in the distributions of interest in statistical physics, extensions of the ideas underlying the classical limit theorems in probability are generally required in the applications; e.g. the law of large numbers, central limit theorem, law of the iterated logarithm. The most dramatic departures from classical limit theorems, e.g. non-central limit theorems, are anticipated in the physics at certain critical values of the parameters of the distribution. The nature of dependencies in P for various ranges of the parameter values is often reflected most explicitly in the form of the correlations;

e.g. power law vs. exponential decay in the two-point correlations.

Described below are two classes of distributions for the random field $X = \{X_n : n \in S\}$ which will be the object of focus in the present paper. The first case, that of the ferromagnetic Ising model, is well-studied in statistical physics. In particular the block correlations are objects of considerable interest and a variety of results are known in this connection. The second case, that of a special formulation of infinite divisibility, is natural from a probabilistic point of view. In this case the most celebrated result for the block correlations (i.e. Fourier transform) is the Levy-Khinchine representation. It is not at all obvious, from the types of results known in the two traditions of physics and probability, that there could be a connection between these two cases. However, some connections between the two classes of distributions have recently been discovered in [25] and in [5]. While the precise nature of the overlap between the two cases is unknown, the investigation has turned up new results for the block correlations in each of the two cases. For example, certain correlation inequalities which were previously known to hold for ferromagnetic Ising models are shown to hold for the infinitely divisible laws, and, likewise, the Levy-Khinchine representation is shown to apply to certain Ising models.

2. *Two Random Field Distributions*

In this section the two classes of distributions discussed in the introduction are described.

2.1 *The General Ising Ferromagnet*

Let P be a probability measure on $(\Omega, \mathfrak{F})$. Versions of the conditional distributions $P(\cdot | X_m, m \in S \setminus F)$, $\phi \neq F \in \mathcal{L}$, are collectively referred to as the *local structure* of P. The general Ising models (or Gibbs distributions) are probability measures on $(\Omega, \mathfrak{F})$ having certain prescribed local structure as follows. The parameters are furnished by a real-valued function $J = \{J_M : M \in \mathcal{L}\}$

on $\mathcal{L}$. We take $J_\phi = 1$, throughout, without any loss of generality. J_M is referred to as a *coupling constant* when $\#M > 2$. For $M = \{n\}$ $J_M = J_n$ is called the *external field* at site $n \in S$. We shall assume that in any prescription of the parameters $J = \{J_M : M \in \mathcal{L}\}$ we have

$$\sum_{m \in M \in \mathcal{L}} |J_M| < \infty \quad \text{for each} \quad m \in S. \tag{2.1}$$

The limit in (2.1) is the limit of the corresponding partial sums $m \in M \subset F$ as F varies over the directed set $\mathcal{L}$. The condition (2.1) makes the following definition non-vaccuous; see [18].

Definition (2.2). A general Ising model for the parameters $J = \{J_M : M \in \mathcal{L}\}$ is a probability measure P on $(\Omega, \mathfrak{F})$ having local structure of the form

$$g_F(\epsilon, w) = Z_F^{-1}(w) \left\{ \exp \sum_{\substack{M \subset F \\ M \neq \phi}} \sum_{\substack{N \subset F^c \\ N \in \mathcal{L}}} J_{M \cup N} \prod_{m \in M} \epsilon_m \cdot \prod_{n \in N} w_n \right\}, \tag{2.3}$$

$$w \in \Omega ,$$

for each function $\epsilon = \{\epsilon_n : n \in F\}$ on F taking values in $\{-1,1\}$, where $Z_F(w)$ is the normalization selected so that

$$\sum_{\epsilon \in \{-1,1\}^F} g_F(\epsilon, w) = 1, \quad w \in \Omega . \tag{2.4}$$

$Z_F(w)$ is referred to as a *partition function* in this context. P is called *ferromagnetic* if

$$J_M \geq 0 \quad \text{for all} \quad M \in \mathcal{L} . \tag{2.5}$$

We adopt the convention that sums over the empty set are zero and products over the empty set are one. Notice that by definition the local structure corresponding to

$F \in \mathcal{L}$ and $w \in \Omega$ is concentrated on a finite set and can therefore be prescribed in terms of a probability mass function as is done in (2.3).

The condition (2.1) on the parameters makes g_F a continuous function of $w \in \Omega$. Observe that in the case S is finite, the general Ising model for the parameters $J = \{J_M : M \in \mathcal{L}\}$ is the probability measure P on $(\Omega, \mathfrak{F})$ determined by

$$P(\{w\}) = Z^{-1} \exp\left\{ \sum_{M \subset S} J_M \cdot \prod_{m \in M} w_m \right\}, \quad w \in \Omega, \tag{2.6}$$

where

$$Z = \sum_{w \in \Omega} \exp\left\{ \sum_{M \subset S} J_M \cdot \prod_{m \in M} w_m \right\}. \tag{2.7}$$

In particular P exists and is uniquely determined according to (2.2) when S is finite. In general, it is well-known that under (2.1) there is always at least one probability measure P for the given parameters. In fact, letting $\mathcal{G} = \mathcal{G}(J = \{J_m : M \in \mathcal{L}\})$ denote the collection of general Ising models for the given parameters $J = \{J_M : M \in \mathcal{L}\}$, $\mathcal{G}$ is non-empty, convex, and compact in the weak* topology; see [18]. Moreover, letting ε denote the collection of extreme points of $\mathcal{G}$, for each $P \in \mathcal{G}$ there is a probability measure ρ concentrated on ε such that

$$P = \int_{\varepsilon} \nu \rho(d\nu) . \tag{2.8}$$

The case $\#\mathcal{G} > 1$ is said to correspond to the phenomena of *phase transition* and the distributions belonging to ε are referred to as the *pure phases*. In the present context the block sums T_M given in (1.4) refer to the *net magnetization* for the block of sites in M.

To get specific structural results of the type desired here it seems necessary that we keep in mind certain graph structures on S; see [17]. For such S the coupling constants $J = \{J_M : M \in \mathcal{L}\}$ of a general Ising model are called *nearest neighbor* if $J_M = 0$ unless M is a singleton or $M = \{m,n\}$ with n and m neighbors; i.e. connected by an edge. The corresponding probability measures for such parameters are called *nearest neighbor distributions*. The *classical Ising model* refers to the case of nearest neighbor coupling constants with $J_{\{n,m\}} = J$ for all pairs of neighbors $n,m \in S$ and $J_{\{n\}} = H$ for all $n \in S$. That the class of *Markov random fields* coincides with the class of nearest neighbor Ising models has appeared in several places in the literature under various formulations; cf. [8], [16], [20], [21], [23]. In the case S is the one-dimensional integer lattice, the classical Ising models coincide with the class of stationary Markov chains.

2.2 *Infinitely Divisible Distributions*

The structure of the class of distributions given here is intimately connected with the group structure of the configuration space Ω regarded as the direct product of cyclic groups $\{-1,1\}$ of order two as described in the introduction. For independent random fields X and Y with distributions P_1 and P_2, respectively, the convolution $P_1 * P_2$ is the distribution of the coordinatewise product field $Z = \{Z_n = X_n \cdot Y_n : n \in S\}$.

Definition (2.9). A probability measure P on $(\Omega,\mathfrak{F})$ is called *infinitely divisible* if for each integer $n \geq 1$ there is a probability measure P_n on $(\Omega,\mathfrak{F})$ such that

$$P = P_n^{*n}, \tag{2.10}$$

where P_n^{*n} denotes the n-fold convolution $P_n * P_n * \cdots * P_n$ in the factorization property (2.10). The collection of infinitely divisible distributions on Ω will be denoted by $\mathfrak{D}$.

An equivalent formulation of infinite divisibility in terms of characteristic functions (i.e. Fourier transforms) is that for each $k \geq 1$ there is a characteristic function ϕ_k of a probability distribution on $(\Omega,\mathfrak{F})$ such that

$$\phi_p = \phi_k^k \; . \tag{2.11}$$

The characterization of Fourier transforms of probability distributions on $(\Omega,\mathfrak{Z})$ is by Bochner's Theorem; see [19].

For the infinitely divisible distributions the Levy-Khinchine Theorem provides formulae for the block correlations; see [15].

Theorem (2.12). Let P be an infinitely divisible distribution on $(\Omega,\mathfrak{Z})$. Then there is a compact subgroup H of Ω and a σ-finite measure F, which is finite valued outside every neighborhood of the identity, such that

$$\phi_p(M) = \lambda_H(M) \exp\left\{\int_\Omega [\gamma_M(w) - 1](dw)\right\} , \quad M \in \mathcal{L} ,$$

where λ_H is the Fourier transform of the normalized Haar measure on H. Moreover, F, so determined, is unique.

The three subgroups $H = \Omega$, $H = \{\sigma^+\}$ and $H = \{\sigma^+,\sigma^-\}$ are of special importance here. The collection $\mathfrak{D}_o \subset \mathfrak{D}$ denotes infinitely divisible laws corresponding to such subgroups under the Levy-Khinchine formula, where $\sigma_n^+ = +1$, $n \in S$ and $\sigma_n^- = -1$, $n \in S$.

3. *Examples*

An important class of examples of infinitely divisible distributions on $(\Omega,\mathfrak{Z})$ is given by certain Bernoulli product measures. Let $\underline{\mu} = \{\mu_n : n \in S\}$ where each μ_n is a probability measure on $\{-1,1\}$. Define $Q_{\underline{\mu}}$ on $(\Omega,\mathfrak{Z})$ by

$$Q_{\underline{\mu}}([M,N]) = \prod_{m\in M} \mu_m(-1) \cdot \prod_{n\in N\setminus M} \mu_n(1), \qquad M \subset N \in \mathcal{L} , \tag{3.1}$$

where $[M,N]$ is the finite dimensional event given by

$$[M,N] = \left\{ X_m = -1,\ m \in M,\ X_n = +1,\ n \in N \setminus M \right\}, \qquad (3.2)$$
$$M \subset N \in \mathcal{L} .$$

Then $Q_{\underline{\mu}}$ is infinitely divisible provided that $p_n = \mu_n(1) \geq \mu_n(-1) = q_n$ for each $n \in S$. In the case $p_n = p$ for all $n \in S$ we write Q_p in place of $Q_{\underline{\mu}}$. Observe that each Bernoulli product measure $Q_{\underline{\mu}}$ is a nearest neighbor Ising model with

$$J_M = \begin{cases} 0 & \text{if} \quad \#M \geq 2 \\ 1/2 \log\left[\dfrac{\mu_m(1)}{\mu_m(-1)}\right] & \text{if} \quad M = \{m\} . \end{cases} \qquad (3.3)$$

In particular, $Q_{\underline{\mu}}$ is infinitely divisible if and only if

$$J_m \geq 0 \quad \text{for each} \quad m \in S . \qquad (3.4)$$

That is, as an Ising model, $Q_{\underline{\mu}}$ must be ferromagnetic; see (2.5).

Unlike the case of infinitely divisible distributions on $\mathbb{R}^d$, it is possible for the characteristic function of an infinitely divisible distribution on $(\Omega, \mathfrak{F})$ to have zero values due to idempotent elements of the group Ω. In particular, for the Bernoulli measure $Q_{1/2}$ we have

$$\phi_{1/2}(M) = \begin{cases} 0 & \text{if} \quad M \neq \phi \\ 1 & \text{if} \quad M = \phi . \end{cases} \qquad (3.5)$$

Observe that since translations on Ω correspond to spin flips at certain sites of S, $Q_{1/2}$ is the Haar measure on Ω.

The above observations are conveniently summarized in the form of the following simple proposition.

Proposition (3.6). Let $J = \{J_M : M \in \mathcal{L}\}$ denote the no-interaction coupling constants defined by

$$J_M = \begin{cases} J_m , & M = \{m\} \\ 0 , & \#M \geq 2 . \end{cases}$$

Then

$$\mathcal{S}(J) \subset \mathcal{S}_o$$

if and only if J is ferromagnetic.

It is well-known and easy to check that in the case of no interaction (i.e. infinite temperature) as described in Proposition (3.6) we have that $\mathcal{S}(J)$ is the singleton consisting of the corresponding independent Bernoulli distribution.

Slightly less trivial classes of examples illustrating results of the type in Proposition (3.6) are given in [25] and in [5]. In particular, examples for which phase transition is possible (as described in section two) are now known. However, the physical significance of these examples is limited by the choice of the graph structure. The main features of this class of examples are described below. The reader may consult the references cited above for the details of proofs.

For S we take the vertices of the infinite tree graph such that each vertex is adjacent to precisely $N + 1$ distinct vertices. S is referred to as the Bethe lattice in the statistical mechanics literature. In the case $N = 1$ S is the 1-dimensional integer lattice. Note that for $N = 2d - 1$ each site of S has the same number of neighbors as each site of the d-dimensional integer lattice with the usual cubic graph structure. It is known that for $N \geq 2$, S is a phase transition graph for the classical Ising model on S; see [17], [22].

For $0 < p < 1$, $q = 1 - p$, let P_p denote the unique probability measure on $(\Omega,\mathfrak{F})$ determined by the following specification of finite dimensional distributions,

$$P_p(X_n = \epsilon_n, n\in D) = \frac{1}{2}\, p^{n_e(\varepsilon,D)} q^{n_o(\varepsilon,D)}, \quad \epsilon_n = \pm 1, \tag{3.7}$$

where D is a finite connected set of sites in S, and $n_e(\epsilon,D)n_o(\epsilon,D)$ denote the numbers of pairs of even and odd bonds, respectively, in the configuration $\epsilon = \{\epsilon_n : n\in D\}$. An even bond is an edge connecting neighboring sites n,m such that $\epsilon_n = \epsilon_m$, and likewise for an odd bond with $\epsilon_n \neq \epsilon_m$.

The local structure of P_p is determined by

$$\begin{aligned} &P(X_n = -1 \mid X_m = -1 \text{ at } k \text{ neighbors } m \text{ of } n) \\ &\quad = [1 + (q/p)^{2k-N-1}]^{-1} \end{aligned} \tag{3.8}$$

for $k = 0,1,\ldots,N+1$. In particular this corresponds to the local structure of the classical Ising model with

$$J_M = \begin{cases} \frac{1}{2}\log[p/q] & \text{if } M = \{m,n\} \text{ with } m\in\partial(n),\ n\in S \\ 0 & \text{otherwise,} \end{cases} \tag{3.9}$$

where $\partial(n)$ denotes the set of neighbors of the site n.

So the ferromagnetic case corresponds to $p \geq q$. In view of the phase transition property quoted above, for $N \geq 2$ there is a critical value of the parameter $J = \frac{1}{2}\log[p/q]$, above which there are additional Ising models with the same local structure as P_p. The following result is proved in [5] for $N \geq 2$ and in [25] for $N = 1$.

Theorem (3.10). Let S denote the valence $N + 1$ tree graph and let $J = \{J_M : M\in\mathcal{L}\}$ denote the coupling constants defined in (3.9) with $p > q$. Then

(i). $\mathcal{S}(J) \subset \mathcal{S}_o$ if $N = 1$

and

(ii). $\mathcal{S}(J) \cap \mathcal{S}_o \neq \emptyset$ if $N \geq 2$.

In the case $N = 1$ it is well known that $\mathcal{S}(J)$ is the singleton consisting of the corresponding Markov chain; this was the case first introduced and studied by Ising in [11].

Because of infinite divisibility, the Levy-Khinchine Theorem provides a representation of the block correlations. Complete descriptions of block correlations are unknown in the case of the d-dimensional integer lattice for $d \geq 2$ with the usual lattice graph structure, although partial results are known in $d = 2$ dimensions.

4. *Correlation Inequalities*

The study of correlation inequalities has occupied a large part of the statistical physics literature. In this section we shall present some of the standard sets of correlation inequalities for the Ising model, and then give some new companion results for infinitely divisible distributions. The discussion of correlation inequalities presented here is by no means exhaustive in the case of Ising models, nor have all of the companion results been explored for infinitely divisible distributions.

4.1 *Ising Model and Correlation Inequalities*

The earliest correlation inequalities in statistical physics are generally attributed to Griffiths; cf. [6]. These refer to positivity of block correlations in the case of Ising ferromagnets in a sense to be described below. This particular set of inequalities were later generalized by Kelly and Sherman in [12] and are now most commonly referred to as the *generalized Griffiths inequalities.*

Theorem (4.1). Let P be a ferromagnetic Ising model on a finite set. Then

(i). $E\gamma_M \geq 0, \quad M \in \mathcal{L}$

(ii). $\text{cov}(\gamma_M, \gamma_N) = E\gamma_{M\Delta N} - E\gamma_M E\gamma_N \geq 0, \quad M,N \in \mathcal{L}.$

The first generalized Griffith inequality (i) is an expression of the tendency of spins to align in ferromagnetic materials. To appreciate the general flavor of the second generalized Griffith inequality (ii) in the present context observe that increasing the region S may be thought of in terms of the introduction of additional coupling constants in the formula for P. In particular it will follow that the average net magnetization is a nondecreasing function of $H > 0$ and in fact concave in the ferromagnetic case; see [7].

The next set of inequalities are the so-called *FKG-inequalities* in the statistical mechanics literature after Fortuin, Kasteleyn, and Ginibre [4]. This same set of inequalities were independently investigated for distributions arising in statistics and reliability theory by Esary, Proschan, and Walkup [3]. In this latter reference the term *associated dependence* was introduced to describe random variables satisfying the FKG-inequalities.

Theorem (4.3). Let P be a general Ising model for the parameters $J = \{J_M : M \in \mathcal{L}\}$ for finite S. Assume that for all $m,n \in S$, $m \neq n$,

$$\sum_{M \not\ni m,n} J_{M \cup \{m,n\}} \cdot \gamma_M(w) \geq 0, \quad w \in \Omega. \tag{4.4}$$

Then for any (coordinatewise) nondecreasing functions f,g on Ω

$$\text{(FKG)} : \operatorname{cov}(f,g) \geq 0 . \tag{4.5}$$

Random variables satisfying the FKG-inequalities (4.5) are said to be *associated*. Notice that the nearest neighbor Ising model satisfies the FKG-inequalities provided that

$$J_{\{m,n\}} \geq 0, \quad m \neq n, \quad m,n \in S. \tag{4.6}$$

In fact ferromagnetism is sufficient whenever $J_M = 0$ for $\#M \geq 3$.

There are many interesting applications of the FKG-inequalities in the literature, including results on asymptotic independence as well as fluctuation laws; see [14] for a recent survey of consequences of these as well as other important inequalities. The condition (4.4) is a reformulation of the *strong FKG-inequalities* for the special case of the Ising model. It is important to note that the condition for association is not merely a consequence of ferromagnetism when three point or higher interactions are incorporated; see example (4.7) below. However, in the case of pairwise interactions ferromagnetism is sufficient for the strong FKG-inequalities and therefore association; see [4].

Example (4.7). Take $S = \{a,b,c\}$, $J_{ab} = J_{ac} = 0$, and $J_a = J_b = J_c = J_{bc} = J_{abc} = \log(1-\varepsilon)^{-1} > 0$ for $\varepsilon > 0$. Then $P \in \mathcal{G}(J)$ is ferromagnetic but is neither strong FKG nor associated for ε sufficiently small.

The final set of inequalities to be discussed were originally introduced for S finite by Kelly and Sherman in [12] and will be referred to as the *Kelly-Sherman inequalities*. Namely, $X = \{X_n : n \in S\}$ is said to satisfy the Kelly-Sherman inequalities provided that

$$\prod_{M:\#M\cap N\text{even}} E\{\prod_{m\in M} X_m\} \geq \prod_{M:\#M\cap N\text{odd}} E\{\prod_{m\in M} X_m\}, \qquad N \subset S. \tag{4.8}$$

As noted by Kelly and Sherman in [12], via Example (4.7), the inequalities (4.8) are not merely a consequence of ferromagnetism.

4.2 *Infinite Divisibility and Correlation Inequalities*

For the first result we get that the Griffiths inequalities, which are purely a consequence of ferromagnetism in the Ising model case, hold for the infinitely divisible laws as well.

Theorem (4.9). Let P be an infinitely divisible distribution on $(\Omega, \mathfrak{F})$ without idempotent factors. Then

(i). $E\gamma_M \geq 0$ for all $M \in \mathcal{L}$

(ii). $\text{cov}(\gamma_M, \gamma_N) \geq 0$ for $M,N \in \mathcal{L}$.

Proof. In view of the Levy-Khinchine representation we have

$$E\gamma_M = \phi_p(M) = \exp\left\{\int_\Omega [\gamma_M(w) - 1]F(dw)\right\} \geq 0, \quad M \in \mathcal{L}.$$

Moreover, by nonnegativity and monotonicity of F as a measure, together with finiteness outside neighborhoods of the identity, we have

$$E\gamma_{M\Delta N} - E\gamma_M E\gamma_N >$$
$$\exp\{-2[F(\gamma_{M\Delta N}=-1) - F(\gamma_M=-1) - F(\gamma_N=-1)]\} > 0 ,$$

since

$$F(\gamma_{M\Delta N}=-1) = F(\gamma_M=-1), \gamma_N=1) + F(\gamma_M=1, \gamma_N=-1)$$
$$< F(\gamma_M=-1) + F(\gamma_N=-1) . \qquad \text{QED.}$$

Theorem (4.9) provides the Griffiths inequalities for $P \in \mathcal{D}_o$ corresponding to $H = \{\sigma_+\}$. One can easily check the case $H = \Omega$ directly. So the following Corollary is immediate from (4.9).

Corollary (4.10). The Griffiths inequalities are valid for all $P \in \mathcal{D}_o$.

In the case of the FKG-inequalities note that the Bethe lattice Ising model consists of pairwise interactions and therefore, in the ferromagnetic case, is both infinitely divisible and by (4.6) is associated. Another class of examples in this connection is obtained as follows. Let $X^{(1)}, X^{(2)},\ldots$ be iid with Bernoulli product distribution Q_p, $p > \frac{1}{2}$, as defined in section 3. Let N_λ be a Poisson random variable, independent of $X^{(1)}, X^{(2)},\ldots$, having parameter $\lambda > 0$. Let P_λ denote the distribution of the random product $Z = Z(\lambda) = X^{(1)} \ldots X^{(N_\lambda)}$ (coordinatewise). Then we have

Proposition (4.11). $P_\lambda \in \mathcal{D}_o$ and P_λ is associated.

The proof of Proposition (4.11) follows from applications of a few well-known tools. We shall give the proof by a sequence of lemmas.

Lemma (4.12). $P_n = Q_p^{*n}$ is associated for each integer $n \geq 1$, $p > \frac{1}{2}$.

Proof. Simply observe that $Q_p^{*n} = Q_{p_n}$ is the distribution of independent variables, and therefore is associated, where $p_n = \frac{1 + (1-2q)^n}{2}$. QED.

Notice that with $p > \frac{1}{2}$, the density $q_n = 1 - p_n$ of -1's increases as a function of n under n-fold convolution.

Definition (4.13). Let P_1, P_2 be probability measures on $(\Omega, \mathfrak{F})$. P_1 is said to be stochastically larger than P_2, denoted $P_1 \underset{s}{\geq} P_2$, if and only if

$$\int_\Omega f dP_1 \leq \int_\Omega f dP_2$$

for all (coordinatewise) non-decreasing continuous functions f on Ω.

Lemma (4.14). For $p > \frac{1}{2}$,

$$\delta_{\sigma+} = Q_p^{*0} \underset{s}{\geq} Q_p^{*1} \underset{s}{\geq} Q_p^{*2} \underset{s}{\geq} \cdots \underset{s}{\geq} Q_p^{*n} \underset{s}{\geq} \cdots$$

Proof. The lemma follows by a direct application of Holley's inequality; see [10]. QED.

Proof of (4.9). Let f and g be coordinatewise non-decreasing functions on Ω. Then we may take $f \geq 0$, $g \geq 0$ by adding constants if necessary.

$$\mathrm{Cov}_\lambda(f,g) = E_{P_\lambda}(f,g) - E_{P_\lambda}(f)E_{P_\lambda}(g)$$

$$= \sum_{n=0}^{\infty} E_{Q^{*n}}(f \cdot g)\frac{\lambda^n}{n!}e^{-\lambda} - \sum_{n=0}^{\infty} E_{Q^{*n}}(f)\frac{\lambda^n}{n!}e^{-\lambda} \sum_{n=0}^{\infty} E_{Q^{*n}}(g)\frac{\lambda^n}{n!}e^{-\lambda} .$$

Therefore,

$$\mathrm{Cov}_\lambda(f,g) \geq \sum_{n=0}^{\infty} E_{Q^{*n}}(f)E_{Q^{*n}}(g)\frac{\lambda^n}{n!}e^{-\lambda} - \sum_{n=0}^{\infty} E_{Q^{*n}}(f)\frac{\lambda^n}{n!}e^{-\lambda} \sum_{n=0}^{\infty} E_{Q^{*n}}(g)\frac{\lambda^n}{n!}e^{-\lambda}$$

by Lemma (4.12). Let $F_n = E_{Q^{*n}}(f)$, $G_n = E_{Q^{*n}}(g)$. Then, by Lemma (4.18), we have $F_n \leq F_{n+1}$, $G_n \leq G_{n+1}$, and $F_nG_n \leq F_{n+1}\, G_{n+1}$ for $n = 0,1,2,\ldots$. Define functions $\tilde{f}$ and $\tilde{g}$ on the set of non-negative integers by $\tilde{f}(n) = F_n$, $\tilde{g}(n) = G_n$, $n = 0,1,2,\ldots$. Then, we have

$$\mathrm{cov}_\lambda(f,g) \geq \mathrm{cov}(\tilde{f}(N_\lambda), \tilde{g}(N_\lambda)) \geq 0$$

since a single random variable, N_λ is this case, is associated. QED.

The general problem of finding further conditions on $\mathcal{B}_o$ for which the infinitely divisible laws are associated is presently unsettled. However, the following proposition indicates a step in that direction.

Proposition (4.15). Let P_λ be an infinitely divisible distribution on $(\Omega,\mathfrak{J})$ for finite S without idempotent factors and having Levy-Khinchine measure $F_\lambda = \lambda \cdot Q$, $\lambda > 0$, where Q is any probability measure on $(\Omega,\mathfrak{J})$ such that $Q([M,N]) > 0$ for all $M \subset N \in \mathcal{L}$. Let f,g be coordinatewise non-decreasing functions on Ω. Then there is a $\Delta = \Delta(f,g) > 0$ such that

$$p(\lambda) = \mathrm{cov}_\lambda(f,g) \geq 0 \quad \text{for} \quad \lambda < \Delta .$$

Proof. It is convenient to identify Ω with $\mathcal{L}$ and suppose that f,g are non-decreasing with respect to partial order on $\mathcal{L}$ by set inclusion. Also we write

$$p_\lambda(A) = P_\lambda([A,S]),$$

$$q(A) = Q([A,S]) , \quad \text{for} \quad A \in \mathcal{L}.$$

Then

$$p(\lambda) = \mathrm{cov}_\lambda(f,g) = \frac{1}{2} e^{-2\lambda} \sum_{n=0}^{\infty} \sum_{k=0}^{m} \frac{\lambda^n}{n!} \binom{n}{k} \cdot$$

$$\{E_{n-k}(fg) + E_k(fg) - E_k f E_{n-k} g - E_k g E_{n-k} f\}$$

where

$$E_j f = \sum_A f(A) q^{*j}(A), \quad j = 1,2,\ldots, \quad E_o f = f(\phi).$$

So, $p(\lambda) = e^{-2\lambda} \sum_{n=0}^{\infty} \frac{\lambda^n}{n!} c_n$, where

$$c_n = \sum_{k=0}^{n} \binom{n}{k} \{E_k(fg) - E_k f E_{n-k} g\}$$

is an umbral product; cf. [24] where such products also arise in a different context.

Now observe that

$$p(0) = c_o = E_o(f\cdot g) - E_o f E_o g = f(\phi)g(\phi) - f(\phi)g(\phi) = 0 .$$

and

$$c_1 = \sum_A \{f(A) - f(\phi)\}\{g(A) - g(\phi)\} \cdot$$

If one of f or g is constant then $p(\lambda) = 0$ for all $\lambda > 0$, so take $\Delta(f,g) = +\infty$. If neither f nor g is constant then since $q(A) > 0$ for all $A \in \mathcal{L}$ we have $c_1 > 0$. In particular, $\rho(0) = 0$ and $\rho'(0) > 0$. The statement in the proposition now follows. QED.

The third set of correlation inequalities, namely the Kelly-Sherman inequalities, are known to be a consequence of infinite divisibility under the positive mass condition on the Levy-Khinchine measure; see [25]. The result is given as follows.

Theorem (4.16). Let $P_\lambda \in \mathfrak{D}_o$ be of the form given in (4.15) with $Q([M,N]) > 0$ for all $M \subset N \in \mathcal{L}$. Then P_λ satisfies the Kelly-Sherman inequalities (4.8).

Note that up to the requirement that the Levy-Khinchine measure assign positive mass to all events of the form $[M,N]$, the form of P_λ given in Proposition (4.15) and Theorem (4.16) is quite general for infinitely divisible distributions with finite S by the Levy-Khinchine formula. The positive mass condition is essential for these inequalities; see [25].

5. *Some Open Problems*

A. The general problem of determining conditions on the potential which makes an Ising model (Gibbs state) infinitely divisible is unsolved; see [25] for results in this direction. Important examples in this connection for which results are not known are the classical Ising model on the 2-dimensional integer lattice and/or one-dimensional models with long range interactions.

B. While the Kelly-Sherman inequalities are in general not a consequence of ferromagnetism alone, see Example (4.7), unlike the case of the FKG inequalities it is not known whether or not this is still true for pairwise ferromagnetic interactions. Example (4.7) is the example introduced by Kelly and Sherman in [12] to illustrate that ferromagnetism is not sufficient and it was observed by the present author that neither do the FKG-inequalities hold for this example.

C. Association of infinitely divisible laws for "group valued" random fields on S is unsettled beyond the point discussed here. This is not the case for point random fields on $\mathbb{R}^d$, where the issue has been completely resolved in [1] and [2]. In this latter case the mod two cancellations are not present in the superposition of points.

D. The random field of the form $Z(\lambda)$, $\lambda \geq 0$, defined by P_λ may be regarded as a process with independent increments indexed by the parameter λ, representing the total mass of the Levy-Khinchine measure, taking values in the partially ordered space $\Omega = \{-1,1\}^S$. In particular $Z(\lambda)$, $\lambda \geq 0$ is a Markov process starting at $Z(0) = \sigma_+$. In [9] Harris has given necessary and sufficient conditions on monotone Markov evolutions $X(t)$, $t \geq 0$, the so-called up-down condition, such that $X(t)$ is associated for each $t \geq 0$ whenever $X(0)$ has an associated distribution; see [13] for a complete discussion and simple proofs. In the present context the problem of establishing association for $Z(\lambda)$, $\lambda \geq 0$ is to preserve association of $Z(0) = \sigma_+$ and the Haar measures of certain subgroups of Ω; e.g. those defining $\mathcal{S}_0$. However, the up-down condition is inappropriate for $Z(\lambda)$, $\lambda \geq 0$. Proposition (4.11) can be viewed as a special case of the more general result which is anticipated in this connection. It is interesting to note that the positive mass condition is enough for (4.16), though this may not be enough for association.

6. *References*

[1] Burton, R. and E. Waymire, Scaling limits for associated random measures. Ann. of Prob., in press, (1985).

[2] Burton, R. and E. Waymire, The central limit problem for infinitely divisible random measures, This edition, (1985).

[3] Esary, J., Proschan F., and D. Walkup, Association of random variables with applications. Ann. Math. Stat., *58* (1967), 1466-1474.

[4] Fortuin, C., Kasteleyn, P., and J. Ginibre, Correlation inequalities on some partially ordered sets. Comm. Mth. Phys., *22* (1971), 89-103.

[5] Glaffig, C. and E. Waymire, Infinite divisibility of the Bethe lattice Ising model. Preprint (1985).

[6] Griffiths, R.B., Correlations in Ising ferromagnets. J. Math. Phys., *8* (1967), 478-489.

[7] Griffiths, R.B., Hurst, C.A., and S. Sherman, Concavity of magnetization of an Ising ferromagnet in a positive external field. J. Math. Phys., *11* (1970), 790-795.

[8] Grimmet, G.R., A theorem about random fields. Bull. London Math. Soc., *5* (1973), 81-84.

[9] Harris, T.E., A correlation inequality for Markov processes in partially ordered spaces. Ann. of Prob., *5* (1977), 451-454.

[10] Holley, R., Remarks on the FKG inequalities. Comm. Math. Phys., *36* (1974), 227-231.

[11] Ising, E., Beitrag sur theorie des ferromagnetismus. Zeit. fur Physik, *31* (1925), 253-258.

[12] Kelly, D.G. and S. Sherman, General Griffiths inequalities on correlations in Ising ferromagnets. J. Math. Phys., *9* (1968), 466-484.

[13] Liggett, T.M., *Interacting Particle Systems*. Springer-Verlag, New York, 1985.

[14] Newman, C.M., Asymptotic independence and limit theorems for positively and negatively dependent random variables. *Inequalities in Statistics and Probability*, IMS Lecture Notes, 1984.

[15] Parthasarathy, K.R., *Probability measures on metric spaces*. Academic Press, New York, 1967.

[16] Preston, C., Generalized Gibbs states and Markov random fields. Adv. in Appld. Prob., *5* (1973), 242-261.

[17] Preston, C., *Gibbs states on countable sets*. Cambridge Univ. Press, 1974.

[18] Preston, C., *Random Fields*. Springer-Verlag, No. 534, New York, 1976.

[19] Rudin, W., *Fourier analysis on groups*. Wiley, New York, 1962.

[20] Sherman, S., Markov random fields and Gibbs random fields. Israel J. Math., *14* (1973), 92-103.

[21] Spitzer, F., Random fields and interacting particle systems. M.A.A. Summer Seminar Notes, 1971.

[22] Sptizer, F., Markov random fields on an infinite tree. Ann. of Prob., *3* (1975), 387-398.

[23] Sullivan, W.G., Potentials for almost Markovian random fields. Comm. Math. Phys., *133* (1973), 61-74.

[24] Taqqu, M., Self-similar processes and long-range dependence: a bibliographical survey. This edition, 1985.

[25] Waymire, E., Infinitely divisible Gibbs states. Rcky. Mtn. J. Math., *14*(3) (1984), 665-678.

Ed Waymire
Department of Mathematics
Oregon State University
Corvallis, OR 97331

THE CENTRAL LIMIT PROBLEM FOR INFINITELY DIVISIBLE RANDOM MEASURES

by

Robert M. Burton and Ed Waymire

1. *Introduction and Preliminaries*

A good reference to the technical preliminaries described below is Kallenberg's book, [10]. We are taking the liberty of stating facts without proof as our intention is to provide the logical development which is used to get the main results.

Let S be a locally compact Polish space with Borel σ-field $\mathcal{B}$ and let $\mathcal{B}_0$ denote the ring of bounded Borel sets. $\mathcal{F}_0$ will denote the cone of nonnegative continuous functions $f : S \to \mathbb{R}$ with compact support. Let M denote the cone of Radon measures of $(S,\mathcal{B})$. Equip M with the σ-field $\mathcal{M}$ generated by sets of the form $\{\mu \in M \mid \mu(B) < x\}$ for $B \in \mathcal{B}_0$, $x \geq 0$. M is a Polish space when equipped with the vague topology and $\mathcal{M}$ is the Borel σ-field for this topology.

We distinguish the set of counting measures by $N = \{\mu \in M \mid \mu(B) \in Z^+ = \{0,1,2,\ldots\}$ for every $B \in \mathcal{B}_0\}$. Each $\mu \in N$ may be identified with an unordered countable sequence $\{x_i\}$ of points in S that have no limit points, but may have multiplicities. N has the σ-field $\mathcal{N}$ generated by sets of the form $\{\mu \in N \mid \mu(B) = k\}$ for $B \in \mathcal{B}_0$ and $k \in Z^+$. N is a closed subspace of M. A *random measure* X is a measurable mapping $X : (\Omega,\mathcal{A},P) \to (M,\mathcal{M})$ where $(\Omega,\mathcal{A},P)$ is a probability space. In the case $P(X \in N) = 1$ the random measure X is called a *point random field*. If $B \in \mathcal{B}_0$ and $f \in \mathcal{F}_0$ then $X(B)$

is the random variable $(X(w))(B)$ and $X(f)$ is the random variable $\int_S f(x)(X(w))(dx)$. The distribution of X is the probability measure $P_X = P \circ X^{-1}$ on $(M,\mathfrak{m})$. The symbol $\doteq$ is used to denote equality in distribution.

If X is a random measure the *Laplace functional* of X is $L_X(f) = E[\exp(-X(f))]$ for $f \in \mathfrak{F}_0$. In the case of point random fields it is sometimes more convenient to use the probability generating functional $G_X(f) = E[\exp(X(\log(f)))] = L_X(-\log(f))$ for $1 - f \in \mathfrak{F}_0$, $0 < f < 1$. The following theorem provides a summary of standard facts; see [10].

Theorem 1.1. Let X and Y be random measures. The following are equivalent:

(1) $X \doteq Y$

(2) $X(f) \doteq Y(f)$ for all $f \in \mathfrak{F}_0$

(3) $L_X(f) = L_y(f)$ for all $f \in \mathfrak{F}_0$

(4) $(X(B_1),\ldots,X(B_n)) \doteq (Y(B_1),\ldots,Y(B_n))$ for all $B_1,\ldots,B_n \in \mathfrak{B}_0$.

All the statements that are made about random measures are statements about their distributions. Let $\mathbb{P}$ be the set of distributions on $(M,\mathfrak{m})$ and let $\mathbb{P}$ have the smallest σ-field that allows the mappings $P \mapsto P(M_0)$, for $P \in \mathbb{P}$ and $M_0 \in \mathfrak{m}$, to be measurable.

We say that random measures X_n , $n \geq 1$, converge in distribution to X , denoted $X_n \to X$, if their distributions converge weakly. The following theorem is well-known; see [10].

Theorem 1.2. If X_n , X are random measures then the following are equivalent:

(1) $X_n \to X$

(2) $X_n(f) \to X(f)$ (dist.) for all $f \in \mathfrak{F}_0$

(3) $L_{X_n}(f) \to L_X(f)$ for all $f \in \mathfrak{F}_0$

(4) $(X_n(B_1),\ldots,X_n(B_x)) \to (X(B_1),\ldots,X(B_k))$ (dist.)
for all $B_1,\ldots,B_k \in \mathcal{B}_0$ with $X(\partial B_i) = 0$ a.s.
for $i = 1,\ldots,k$, $k \geq 1$.

If X is a random measure then the *intensity* is defined by $(EX)(B) = E(X(B))$. EX is always a measure on $(S,\mathcal{B})$, but it is not necessarily Radon.

The most important counting random measure is the Poisson point random field X with intensity $\lambda \in M$. This means that X has independent increments (i.e. the measures of disjoint sets in $\mathcal{B}_0$ are independent random variables) and for each $B \in \mathcal{B}_0$, $X(B)$ is a Poisson random variable with parameter $\lambda(B)$. The Laplace function of X is

$$L_X(f) = \exp[\int(e^{-f} - 1)d\lambda] . \tag{1.3}$$

If $S = \{1,2,\ldots,n\}$, then random measures are just random vectors with nonnegative components. We are most interested in the case when $S = \mathbb{R}^d$. In this case a random measure is called *stationary* if for all $B_1,\ldots,B_k \in \mathcal{B}_0$ and $x \in \mathbb{R}^d$ we have
$(X(B_1 + x),\ldots,X(B_k + x)) \doteq (X(B_1),\ldots,X(B_k))$.

Random measures which arise as specific models in the applications are often described in terms of their transforms, moments, product densities, etc., see [5]. Rarely is it the case that the joint distribution of the masses assigned to the various subregions of S can be determined explicitly; the Poisson case being a notable exception. In the present paper we consider the problem of determining large scale approximations to the distribution for an important class of random measures in the form of the *central limit problem*. The approach used here is an adaptation of methods introduced by Newman in [13] based on notions of positive dependence, in particular, *association*. Special versions of this problem have already been solved in [2] and in [9]. The methods in this latter reference are based on mixing properties. The general scaling limit problem is treated in [1] for larger classes of random measures than those to be considered here and for non-central limit situations.

2. *Random Clusters and Infinitely Divisible Random Fields*

Let X and Y be random measures on $(S,\mathcal{B})$ with distributions P_X and P_Y, respectively. In the case that X and Y are independent, the distribution of the *independent superposition*, $(X + Y)(B) = X(B) + Y(B)$, for $B \in \mathcal{B}$, will be denoted as $P_X * P_Y$. This defines the "convolution" operation $*$ used below.

2.1 *Random Clusters*

Given locally compact Polish spaces S and S' with Borel σ-fields $\mathcal{B}$ and $\mathcal{B}'$, respectively, a *cluster field* is a measurable map $x' \to \rho_{x'}$ from S' into $\mathbb{P}$. The probability distribution $\rho_{x'}$ is called the member distribution at $x' \in S'$.

For a counting measure $\mu' \in N'$, define

$$\rho(\mu') = \mathop{*}_{x' \in S' \,:\, \mu'(\{x'\}) > 0} \rho_{x'}^{*\mu'(\{x'\})} \tag{2.1}$$

provided that the right hand side exists as a probability measure on $(M,\mathfrak{M})$. If X' is a point random field on S' such that $\rho(X')$ exists a.s., then we will say that X' supports the random cluster field $\rho = \{\rho_{x'} : x' \in S\}$. The probability measure $P_{X',\rho}$ on $(M,\mathfrak{M})$, defined by the property that $\rho(X')$ provides a version of the conditional distribution given X', is called the *cluster measure distribution* with *center field* X' and *member distribution* ρ. S' is called the *base space*. Conditions for the existence and uniqueness of such a distribution can be found in various places in the literature; see [15], [11]. A random measure X with distribution $P_{X',\rho}$ will be referred to as a *random cluster measure*.

2.2 *Infinitely Divisible Random Measures*

A random measure X is called infinitely divisible if for each integer n there is a random measure X_n such that X may be represented as an n-fold independent superposition of random measures distributed as X_n; i.e. $P_X = P_{X_n}^{*n}$. Many of the important properties of infinitely divisible random measures can be deduced from the following representation of the Laplace functional due to Kersten, Lee, and Matthes. Namely, X is an infinitely divisible random measure if and only if there is an $\alpha \in M$ and a σ-finite measure λ on $(M,\mathfrak{M})$ such that

$$\lambda(\{\mu \in M : \mu(S) = 0\}) = 0 \tag{2.2}$$

$$\int_M [1 - \exp\{-\mu(B)\}]\lambda(d\mu) < \infty \quad \text{for each} \quad B \in \mathcal{B} \tag{2.3}$$

and

$$-\log L_X(f) = \int_S f d\alpha + \int_M [1-\exp\{-\int_S f d\mu\}]\lambda(d\mu). \tag{2.4}$$

The pair (α,λ) is uniquely determined by the distribution of X. We write $X \sim I(\alpha,\lambda)$ to indicate the corresponding parameters. The representation (2.4) is referred to as the KLM-representation; an exhaustive treatment can be found in [11].

It is a simple matter to check using (2.4) that if X_1 and X_2 are independent random measures such that $X_1 \sim I(\alpha_1,\lambda_1)$ and $X_2 \sim I(\alpha_2,\lambda_2)$, then $X_1 + X_2$ is infinitely divisible and $X_1 + X_2 \sim I(\alpha_1 + \alpha_2, \lambda_1 + \lambda_2)$. Likewise, using (1.3) one sees immediately that Poisson random fields are infinitely divisible. Moreover, random cluster measures having an infinitely divisible center field X' are infinitely divisible. In particular, all random cluster measures with Poisson center field are infinitely divisible. That this latter class exhausts the class of infinitely divisible distributions is an important and well-known fact which can be recorded as follows; see [10].

Theorem 2.5. Suppose that X is an infinitely divisible random measure centered so that $X \sim I(0,\lambda)$.

(1) If $\rho_0 : M \to \mathbb{P}$ is the mapping $\mu \mapsto \delta_\mu$ and if X' is a Poisson random field on M with intensity measure λ than X has the cluster representation $X = \rho_0(X')$.

(2) X is a cluster random measure with a Poisson center field and base space $S' = N = \{1,2,\dots\}$.

(3) If $S = \mathbb{R}^d$ than X is a cluster random measure with Poisson center field and base space $S' = S = \mathbb{R}^d$.

An infinitely divisible random measure $(X \sim I(0,\lambda)$ is called *regular* if $\lambda\{\mu \mid \mu(S) = +\infty\} = 0$ and *singular* if $\lambda\{\mu \mid \mu(S) < +\infty\} = 0$. Every infinitely divisible random measure is the independent sum of a [deterministic] measure, a regular infinitely divisible random measure, and a singular infinitely divisible random measure; see [11]. In [15] these properties are described in the framework of a single example.

3. *Positive Dependence Properties*

There is a natural ordering of $\mathbb{R}^n$ by $\xi \leq \eta$ iff $\eta - \xi \in \mathbb{R}^{n+}$. A set $\mathbf{C}$ of random variables is called *associated* if for any $n \geq 1$ and $\xi_1 \xi_2, \dots, \xi_n \in \mathbf{C}$, $\mathrm{Cov}(f(\xi_1,\dots,\xi_n),\ g(\xi_1,\dots,\xi_n)) > 0$, for all $f,g : \mathbb{R}^n \to [0,1]$ measurable and non-decreasing with respect to the above partial order. This is a strong positive dependence property whose definition in this form originated in [6] and independently in [8]. We wish to extend this notion to random measures.

Put a partial ordering on M by $\mu \leq \nu$ if $\nu - \mu \in M$. This means there is no bounded Borel set whose μ measure is greater than its ν measure. We say a measurable subset $M_0 \subseteq M$ is increasing if for each $\mu \in M_0$, $\nu > \mu \Rightarrow \nu \in M_0$. If X is a random measure say that X is *associated* if whenever $F_1, F_2 : M \to \mathbb{R}$ are increasing and measurable then $\mathrm{Cov}(F_1(X), F_2(X)) > 0$ whenever the covariance is defined. Notice if the underlying space is $S = \{1,2,\ldots,n\}$ then this reduces to the usual definition of association for nonnegative random vectors. Also note that as usual these definitions may be applied to point random fields.

It is not difficult to check that a random measure X is associated in the sense of the above definition if and only if the family $\mathcal{C} = \{X(B) : B \in \mathcal{B}_0\}$ is a family of associated random variables. Moreover, the property of association can also be expressed as

$$P_X[M_1 \cap M_2] \geq P_X[M_1]P_X[M_2] \tag{3.1}$$

for any increasing $M_1, M_2 \in \mathfrak{M}$. The condition (3.1) is a consequence of an important formula usually attributed to Lehmann [12]. Namely,

$$\begin{aligned} \mathrm{Cov}\{F_1(X), F_2(X)\} &= \int_{-\infty}^{\infty} \int_{-\infty}^{\infty} \{P_X[M_1(r) \cap M_2(s)] \\ &= P_X[M_1(r)]P_X[M_2(s)]\}\, dr\, ds \end{aligned} \tag{3.2}$$

where $M_i(r) = \{\mu : F_i(\mu) > r\}$, $i = 1,2$.

In the case of point random fields X on $S = \mathbb{R}^d$ a useful tool for establishing association are the so-called *strong FGK inequalities* for the absolute product densities; see [8], [2], and [3]. The standard arguments for association usually involve combinations of the facts recorded in Theorem 3.3. The ideas for proofs can be found in [6] and in [4].

Theorem 3.3.

(1) If X has independent increments then X is associated; in particular the Poisson random fields are associated.

(2) If X, Y are associated then so is their independent superposition.

(3) If $F : M \to M$ is increasing and measurable and if X is associated then so is $F(X)$.

(4) If $X_n \to X$ and if each X_n is associated then X is associated.

The following is one of the main new results presented here. This generalizes a previous result in [2]. Complete proofs are to appear in [4] although a sketch is given in section 4.

Theorem 3.4. If $X = (X')$ is a random cluster measure and if X' is Poisson on S' then X is associated.

Since an infinitely divisible random measure is the super-position of a deterministic measure and a Poisson random cluster measure we have:

Corollary 3.5. If X is infinitely divisible then X is associated.

Corollary 3.6. If $(\xi_1,\ldots,\xi_n)$ is a nonnegative infinitely divisible random vector then $(\xi_1,\ldots,\xi_n)$ is associated.

In view of Theorem 3.4 it is natural to wonder whether in general a random cluster measure having an associated center field must be associated. The following example shows that in general this is not sufficient. Take $S = \{1,2\}$, $S' = \{1\}$, and

$$\rho(1) = \begin{cases} \delta_1 & \text{with probability } \frac{1}{2} \\ \delta_2 & \text{with probability } \frac{1}{2} \end{cases} .$$

Then $X' = \delta_1$ is associated but the random cluster measure X with center field X' and cluster member ρ is not associated.

4. *The Central Limit Problem*

In this section all random measures X will be

stationary on $S = \mathbb{R}^d$ with finite intensity. X satisfies the *Central Limit Theorem* with parameter σ^2 if for all disjoint $B_1,\ldots,B_n \in \mathcal{B}_0$ the normalized random vector

$$\left(\frac{X(tB_1) - E[X(tB_1)]}{t^{d/2}}, \ldots, \frac{X(tB_n) - E[X(tB_n)]}{t^{d/2}}\right) \qquad (4.1)$$

converges in distribution to a mean 0, independent Gaussian random vector with component variances $\sigma^2|B_1|,\ldots,\sigma^2|B_n|$; here $|B|$ is the Lebesgue measure of B. This is equivalent to convergence of X_t to Gaussian white noise with variance σ^2 as generalized random fields, where X_t is defined by

$$X_t(f) = \frac{X(f_t) - EX(f_t)}{t^{d/2}} \quad \text{and} \quad f_t(\underline{x}) = f(t\underline{x}) .$$

X is said to have *summable covariance* if

$$\sum_{\underline{k}\in Z^d} |\operatorname{cov}(X(I)\ ,\ X(I + \underline{k}))| < \infty . \qquad (4.2)$$

($I = [0,1)^d$ is the unit cube in $\mathbb{R}^d$.)

If X has summable covariance then X satisfies the Central Limit Theorem with parameter $\sigma^2 > 0$ if and only if $\sum_{\underline{k}} \operatorname{cov}(X(I)\ ,\ X(I + \underline{k})) = {}^2$ and we have convergence of (4.1) for all $B_1,\ldots,B_n$ bounded rectangles (i.e. products of finite intervals). The following theorem is a direct consequence of [14] and [2].

Theorem 4.3. Let X be an associated stationary random measure on $\mathbb{R}^d$ with $\sum_{\underline{k}\in Z^d} \operatorname{cov}[X(I),X(I + \underline{k})] = \sigma^2 < \infty$. Then X satisfies the central limit theorem with parameter σ^2.

It is Theorem 4.3 together with the result in Theorem 3.4 which solves the central limit problem for stationary infinitely divisible random measures. Although complete proofs are to appear in [4], the main steps of the solution will be given here. This includes a sketch of why Theorem 3.4 is true. First, the precise statement of the result is as follows.

Theorem 4.4. Suppose that X is a stationary infinitely divisible random measure. Then X satisfies the central limit theorem with parameter σ^2 if and only if X is a random cluster measure with Poisson centers X' having intensity parameter λ, say, and the total mass of the member distribution ρ has finite second moment ξ. In this case $\sigma^2 = \lambda \cdot \xi$.

If X is stationary and infinitely divisible with $X \sim I(\alpha,\lambda)$ then it is enough to consider the case $\alpha = 0$. Since by Corollary 3.5 X is associated, all covariances are nonnegative. If X satisfies the central limit theorem then we must have that $\mathrm{VAR}(\frac{X(nI)}{n^{d/2}}) \to \sigma^2$ as $n \to \infty$ through the integers. This implies that X has summable covariance. A simple calculation shows that X must then be regular, from which the cluster representation follows.

The converse in the case that $S = \mathbb{R}^d$ is proved in [2]. The general result stated here is proved by constructing a sequence of discrete approximations to X. Specifically, by a discrete cluster measure X we mean that the center field X' is Poisson on a finite set S' and the cluster member structure ρ is deterministic. Such cluster measures are infinitely divisible and easily seen to be associated. Association of infinitely divisible random measures follows from the following theorem from [4]. Once association is established the central limit problem is solved by an appeal to Theorem 4.3.

Theorem 4.5. $X \sim I(0,\lambda)$ if and only if there is a sequence $\{X_n\}$ of discrete random cluster measures such that $X_n \to X$ in distribution as $n \to \infty$.

If X is a cluster random measure with deterministic member distribution ρ (i.e. for each $x' \in X'$ $\rho_{x'}$ is a [nonrandom] measure) then X may be viewed as an increasing function of X'. Therefore, X is associated if X' is associated by Theorem 3.3(3). Thus discrete random cluster measures are associated and by Theorem 3.3(4) so are their limits.

To see why the converse is true first notice, for example by using Laplace functionals, that X is discrete if and only if there are measures $\mu_1,\dots,\mu_k \in M$ and independent Poisson random variables $\xi_1,\dots,\xi_k$ such that $X = \sum_{i=1}^{k} \xi_i\mu_i$. Let $X = \rho(X')$ where X' is Poisson on S' with intensity measure λ.

Now we want to give a finite approximation to each $\mu \in M$. For each $n = 1,2,\dots$ let $\mathcal{B}_n$ be a collection of disjoint bounded Borel sets with the following properties:
(i) for each $A \in \mathcal{B}_0$ there is an n with $A \subseteq \cup\mathcal{B}_n$,
(ii) every $B \in \mathcal{B}_n$ is a union of sets in $\mathcal{B}_{n+1}$, and
(iii) $\lim_{n\to\infty} \sup\{\mathrm{diam}(B) \mid B \in \mathcal{B}_n\} = 0$ where $\mathrm{diam}(B)$ is the diameter of B with respect to some fixed compatible metric. For each $B \in \mathcal{B}_n$ choose a point $x_B \in B$. If $\mu \in M$ set $\mu_n = \sum_{B\in\mathcal{B}_n} \min(n,[n\mu(B)]/n)\delta_{x_B}$ where [ ] is the greatest integer function. Note that μ_n must be one of a finite number of measures. If $f \in \mathfrak{F}_0$ then $\int f d\mu_n \to \int f d\mu$, i.e. $\mu_n \to \mu$ in the vague topology. If Y is a random measure define Y_n by $Y_n(\omega) = Y(\omega)_n$. $Y_n \to Y$ pointwise so that $Y_n \to Y$ in distribution.

Now let X be a random cluster measure with member distribution ρ and Poisson center field X' on S' with intensity measure λ. Let X_n be a random cluster measure with center field X' and member $\rho_{(n)}$ where $\rho_{(n)x'} = (\rho_{x'})_n$ is the nth approximation of the distribution $\rho_{x'}$. Then $X_n \to X$ (see also]7]). To finish the sketch we need only show X_n is discrete for each n.

Fix n and let (i) for $i \in J$ be a finite enumerator of the possible measures μ_n. Define $\nu_1(x') = \text{Prob}[\rho_{(n)x'} = \mu(i)]$ for each $i \in J$. The $\nu_i(x')$ are measureable functions and $\sum_i \nu_i(x') = 1$. A calculation with Laplace functionals shows $\tilde{X}_n \doteq \sum_{i \in J} \xi_i \mu(i)$ where the ξ_i are independent random variables with parameter $\int_S \nu_i(x')d\lambda(x')$.

5. *References*

[1] Burton, R.M. and E. Waymire, Scaling limits for point random fields. J. Mult. Anal., 15(2)(1984), 237-251.

[2] Burton, R.M. and E. Waymire, Scaling limits for associated random measures. Ann. of Prob., in press, (1985).

[3] Burton, R.M. and E. Waymire, A sufficient condition for association of a renewal process. Ann. of Prob., to appear (1985).

[4] Burton, R.M. and E. Waymire, Association of Infinitely Divisible Random Measures and the Central Limit Theorem. Pre-print, (1985).

[5] Daley, D.J. and D. Vere-Jones, A summary of the theory of point processes, in *Stochastic Point Processes: Statistical Analysis, Theory and Applications*, ed. P.A.W. Lewis, Wiley, New York, 1972, 299-383.

[6] Esary, J., F. Proschan, and D. Walkup, Association of random variables with applications. Ann. Mth. Stat., *38* (1967), 1466-1474.

[7] Fleischman, K., Continuity properties of clustered stochastic point processes and applications to spatially homogeneous branching processes. Collquia Mathematica Societeatis Janos Bolyai: Point processes and queing problems, *24* (1978), 49-67. Keszthely (Hungary)

[8] Fortuin, C., P. Kastelyn, and J. Ginibre, Correlation inequalities on some partially ordered sets, Comm. Math. Phys., *22* (1971), 89-103.

[9] Ivanoff, G., Central limit theorems for point processes. Stoch. Proc. Appl., *12* (1982), 171-186.

[10] Kallenberg, O., Random Measures. Academic Press, New York, 1976.

[11] Kersten, J., K. Matthes, and J. Mecke, Infinitely divisible point processes. Wiley, New York, 1978.

[12] Lehman, E.L., Some concepts of dependence. Ann. Math. Stat., *37* (1966), 1137-1153.

[13] Newman, C.M., Normal fluctuations and the FKG - Inequalities. Comm. Math. Phys., *74* (1980), 119-128.

[14] Newman, C.M. and A.L. Wright, An invariance principle for certain dependent sequences. Ann. of Prob., *9* (1981), 671-675.

[15] Waymire, E. and V.K. Gupta, An Analysis of the Polya Point Process. Adv. Appld. Probab., *15* (1983), 39-53.

[16] Westcott, M., On existence and mixing results for cluster point processes. J. R. Stat. Soc. B, *31* (1971), 290-300.

Robert M. Burton(1)
Department of Mathematics
Oregon State University
Corvallis, OR 97331

Ed Waymire(2)
Department of Mathematics
Oregon State University
Corvallis, OR 97331

(1) Supported in part by the Alexander von Humboldt Foundation and NSF grant MCS-8301702. The first author would like to thank the Institut für Mathematische Stochastik, Universität Gottingen for their hospitality during his sabbatical year.

(2) Supported in part by NSF grant CEE-8303864.

SECTION 5

MOVING AVERAGES IN STABLE RANDOM VARIABLES

WEAK CONVERGENCE OF MOVING AVERAGES

WITH INFINITE VARIANCE

by

Florin Avram

and

Murad S. Taqqu*

1. Introduction

A. We survey results concerning the asymptotic behavior of moving averages that belong to the stable domain of attraction. Some of the methods are highlighted and new results announced.

Let X_i, $i = \ldots, -1, 0, 1, \ldots$ be an i.i.d. sequence of random variables. We say that it belongs to the α-stable domain of attraction $\mathcal{D}(\alpha)$, $0 < \alpha < 2$, if the following conditions hold:

$$P\{|X_i| > x\} = x^{-\alpha}L(x) \tag{1.1a}$$

where $L(x)$ is a s.v.f. (slowly varying function) at infinity, and

$$\lim_{x\to\infty} \frac{P\{X_i > x\}}{P\{X_i < -x\}} = c, \quad 0 < c < \infty. \tag{1.1b}$$

We suppose also

$$\begin{cases} EX_i = 0 & \text{if } \alpha > 1 \\ EX_i = 0 \text{ or } X_i \text{ is symmetric} & \text{if } \alpha = 1. \end{cases} \tag{1.2}$$

Assuming $EX_i = 0$ when $\alpha = 1$ effectively restricts the choice of s.v.f's in (1.1a).

Under the above conditions which will be assumed throughout the paper, the classical central limit theorem holds. Letting

$$a_n = \inf\{x : P\{|X_i| > x\} \leq \frac{1}{n}\}, \tag{1.3}$$

* Research supported in part by NSF Grant ECS-84-08524

we have

$$(1.4) \qquad f.d.d. - \lim_{n\to\infty} a_n^{-1} \sum_{i=1}^{[nt]} X_i = X_\alpha(t)$$

where $X_\alpha(t)$ is the Levy α-stable process (which has independent and stationary increments), and where f.d.d. denotes the convergence of the finite dimensional distributions. In fact, $a_n = n^{1/\alpha}L'(n)$ where L' is s.v.f., and also (1.4) holds in the sense of weak-convergence in the J_1-Skorokhod topology on $D[0,1]$ (Theorem 2.7, Skorokhod (1957)).

B. Introduce now the moving average:

$$(1.5) \qquad Y_k = \mathrm{MA}_k(c_i, X_i) = \sum_{i=-\infty}^{+\infty} c_{k-i}X_i$$

It is well-defined if there exists ν, $\nu < \alpha$, such that

$$(1.6) \qquad \sum_{i=-\infty}^{+\infty} |c_i|^\nu < \infty,$$

since by the Bahr-Esseen inequality (see (3.1) below),

$$E|\mathrm{MA}_k(c_i, X_i)|^\nu \leq 2E|X_1|^\nu \sum_i |c_i|^\nu,$$

and thus, when (1.6) holds, (1.5) can be defined as an L_ν limit (and also as an a.s. limit (see Kawata (1972), Ths. 12.11.2, 12.10.4).) Through appropriate choices of the c_i's various dependence structures for the Y_k can be obtained.

We are interested in limit theorems for

$$(1.7) \qquad \begin{aligned} S_n(t) &= \sum_{j=1}^{[nt]} Y_j \\ &= \sum_{i=-\infty}^{+\infty} X_i\Big(\sum_{j=1-i}^{[nt]-i} c_j\Big). \end{aligned}$$

The limiting behavior of $S_n(t)$ depends on that of

$$(1.8) \qquad D_n^{(\alpha)}(c_\cdot) = \sum_{i=-\infty}^{+\infty} |\sum_{j=1-i}^{n-i} c_j|^\alpha.$$

(This is the variance of $S_n(1)$ if $EX_i^2 = 1$ when $\alpha = 2$ and, if $0 < \alpha \le 2$, one has $\ln E[e^{iuS_n(1)}] = |u|^\alpha D_n^{(\alpha)}(c_\cdot)$ when the X_i are symmetric α-stable. In fact, (see (3.3) below), $E|S_n(1)|^{\alpha-\eta} \le 2D_n^{(\alpha-\eta)}(c_\cdot)E|X_i|^{\alpha-\eta}$ for small enough $\eta > 0$.)

Alternatively, the limiting behavior of $S_n(t)$ depends on that of the kernel

$$k_{n,t}(u) = \sum_{j=1}^{[nt]} c_{j-[nu]} \tag{1.9}$$

because, if $b_n^{-1} k_{n,t}$ converges to k_t in an appropriate sense, then

$$f.d.d. - \lim_{n\to\infty} b_n^{-1} a_n^{-1} \sum_{j=1}^{[nt]} Y_j = f.d.d. - \lim_{n\to\infty} \sum_{i=-\infty}^{+\infty} b_n^{-1} k_{n,t}\left(\frac{i}{n}\right)\frac{X_i}{a_n} = \int_{-\infty}^{+\infty} k_t(u)\,dX_\alpha(u).$$

C. Convergence of adequately normalized sums of moving averages in $\mathcal{D}(\alpha)$ has been studied by Astrauskas (1983), Maejima (1983) and Davis and Resnick (1985). Three different cases have been found to occur.

Before describing each case in detail, we note that the limiting process must necessarily have stationary increments and be self-similar (Lamperti (1962)). Recall that $X(t)$ is self-similar with parameter H if $X(at)$ and $a^H X(t)$ have the same finite-dimensional distributions for all $a > 0$. Here, $0 < H < 1$ unless $H = 1/\alpha$ (see Taqqu and Wolpert (1983)).

We distinguish between *Case I:* $1/\alpha < H < 1$, where necessarily $1 < \alpha < 2$; *Case II:* $H = 1/\alpha$; *Case III:* $0 < H < 1/\alpha$. We now specify three different sets of conditions on the coefficients c_i of the moving average, one for each of the three cases.

Case I $(1/\alpha < H < 1)$. Suppose $1 < \alpha < 2$ and

$$c_n = \begin{cases} c_+ n^{\beta-1} L_1(n) & \text{if } n \ge 0 \\ c_- |n|^{\beta-1} L_1(n) & \text{if } n < 0 \end{cases}$$

where

$$0 < \beta < 1 - 1/\alpha \tag{1.10}$$

and $L_1(n)$ is a s.v.f. (Note that $\Sigma|c_i| = \infty$ and that the upper-bound on β ensures that the moving average is well-defined.)

Setting

(1.11a)
$$k_t(u) = \int_0^t q(v-u)dv,$$

(1.11b)
$$q(v) = c_+(v_+)^{\beta-1} + c_-(v_-)^{\beta-1},$$

where $v_+ = \max(v,0)$ and $v_- = -\min(v,0)$, one gets (see Theorem 5a below)

(1.12)
$$f.d.d. - \lim_{n\to\infty} \frac{1}{n^\beta L_1(n) a_n} \sum_{i=1}^{[nt]} Y_i = X_{\alpha,\beta}(t)$$

where

(1.13)
$$X_{\alpha,\beta}(t) = \int_{-\infty}^{+\infty} k_t(u) dX_\alpha(u).$$

This was shown by Astrauskas (1983) in the case $c_+ = 1, c_- = 0$ and by Maejima (1983) in the case $c_+ = -c_-$ and $L_1(n) = 1$. (It is easy to see that when $L_1(n) = 1$, one has

$$\begin{aligned} n^{-\beta} k_t^{(n)}(u) &= n^{-\beta} \sum_{j=0}^{[nt]} q(j - [nu]) \sim n^{-\beta} \sum_{j=0}^{[nt]} q(j - nu) \\ &= \frac{1}{n} \sum_{j=0}^{[nt]} q(\frac{j}{n} - u) \to k_t(u), \end{aligned}$$

but obviously, this is not enough to establish (1.12).)

The process $X_{\alpha,\beta}(t)$ is called *fractional stable process* (or fractional Levy motion). Its self-similarity parameter is

$$H = \frac{1}{\alpha} + \beta \in (\frac{1}{\alpha}, 1)$$

The process has continuous paths because $H > \frac{1}{\alpha}$ and weak convergence with respect to the sup-norm topology has been established by both Astrauskas (1983) and Maejima (1983).

Case II. $(H = 1/\alpha)$. Suppose $\sum_i |c_i| < \infty$ with $0 < \alpha < 2$. Here $D_n^{(\alpha)} = 0(n)$ (see (3.9) below) and

$$\lim_{n\to\infty} k_{n,t}(u) = \Big(\sum_{i=-\infty}^{+\infty} c_i\Big) 1_{[0,t]}(u).$$

The finite-dimensional distributions of $S_n(t)$ normalized by a_n converge to those of a Levy α-stable process (see Astrauskas (1983), Davis and Resnick (1985) and Theorem 1 below). The issue of tightness is more delicate than in the previous case (see Theorems 2 and 3 below).

Case III. $(0 < H < 1/\alpha)$. Let $0 < \alpha < 2$,

$$\max(0, \frac{1}{\alpha} - 1) < \beta < \frac{1}{\alpha}$$

and suppose either

$$(A) \qquad c_i = 0 \text{ for } i < 0, \ \sum_{i=0}^{n} c_i = n^{-\beta} L_1(n)$$

or

$$(B) \qquad c_o = 0, \ c_i = \text{sgn}(i)|i|^{-\beta-1} \text{ for } i > 1, \ L_1 \equiv 1.$$

Assumption (A) is used by Astrauskas (1983) and (B) by Maejima (1983). The analysis is similar to that of case I. We have $\lim_{n\to\infty} \frac{n^\beta}{L_1(n)} k_{n,t}(\frac{u}{n}) = k_t(u)$ where respectively,

$$\text{(A)} \qquad k_t(u) = (t-u)^{-\beta} 1_{\{u<t\}} - (-u)^{-\beta} 1_{\{u\le 0\}},$$

$$\text{(B)} \qquad k_t(u) = \frac{1}{\beta}(|t-u|^{-\beta} - |u|^{-\beta}).$$

Note that one can write $k_t(u) = \int_0^t q(v-u)dv$ in case (B) but not in case (A). The kernel $q(u)$ in case (B) is $q(\beta) = (sgn\ u)|u|^{-\beta-1}$.

The above authors show that

$$f.d.d. - \lim_{n\to\infty} \frac{n^\beta}{a_n L_1(n)} \sum_{j=0}^{[nt]} Y_i = \int_{-\infty}^{+\infty} k_t(u) dX_\alpha(u).$$

This limit is again called a fractional stable process, but this time the self-similarity parameter is

$$H = \frac{1}{\alpha} - \beta \in (0, \frac{1}{\alpha}).$$

Maejima (1983b) shows that in this case the fractional stable process has nowhere bounded paths and thus, no tightness is possible.

2. Results

We first state Theorem 1, which is a vector version of the results obtained by Astrauskas (1983) and Davis and Resnick (1985). The proof which is given in Section 3 illustrates one set of techniques. The scheme of discrete stochastic integrals described below illustrates a second approach.

Suppose throughout this section that $\{X_i\} \in \mathcal{D}(\alpha)$, $0 < \alpha < 2$, satisfies (1.1) and (1.2), and that a_n is as in (1.3). Introduce

$$Y_k^{(j)} = \sum_{i=-\infty}^{+\infty} c_{k-i}^{(j)} (X_i)^j \ , \ j = 1, \ldots, m, \tag{2.1}$$

where the $c_i^{(j)}$ satisfy $\sum_i |c_i^{(j)}|^{\nu_j} < \infty$ with $\nu_j < \alpha/j$. Let $X_\alpha^{(1)}(t) = X_\alpha(t)$ be the Levy α-stable process given in (1.4) and let $X_\alpha^{(j)}(t)$ be its j-variation, that is

$$X_\alpha^{(j)}(t) = f.d.d. - \lim_{n\to\infty} \sum_{k=0}^{[nt]} \left[X_\alpha^{(1)}(\frac{k+1}{n}) - X_\alpha^{(1)}(\frac{k}{n}) \right]^j. \tag{2.2}$$

Theorem 1. *If* $\nu_1 \le 1$, *then*

$$f.d.d. - \lim_{n\to\infty} \sum_{k=1}^{[nt]} \left(a_n^{-1} Y_k^{(1)}, a_n^{-2} Y_k^{(2)}, \ldots, a_n^{-m} Y_k^{(m)} \right) \tag{2.3}$$

$$= \Big((\sum_i c_i^{(1)}) X_\alpha^{(1)}(t),\ (\sum_i c_i^{(2)}) X_\alpha^{(2)}(t), \ldots, (\sum_i c_i^{(m)}) X_\alpha^{(m)}(t) \Big).$$

Is there convergence in function space? We conjecture that in the vector case $m > 1$, and when the moving averages are non-trivial none of the four Skorohod topologies J_1, J_2, M_1, M_2 is adequate. (See Skorohod (1957) for a definition of the topologies.) One has the implications

$$\begin{array}{ccccc} & & J_2 & & \\ & \nearrow & & \searrow & \\ J_1 & & & & M_2 \\ & \searrow & & \nearrow & \\ & & M_1 & & \end{array}$$

J_1, which is the commonly used Skorohod topology, is inadequate even in the unidimensional case $m = 1$. Indeed, let $\{X_i\}$, a_n and $\{Y_k^{(1)}\}$ be defined as Theorem 1, and let $c_i = c_i^{(1)}$ and $\nu = \nu_1$.

Theorem 2. *If there are integers $i_o \neq 0, K > 0$ such that $c_o \neq 0, c_{i_o} \neq 0, c_i = 0$ for $i < 0$ and $i > K$, then the sequence of partial sums $a_n^{-1} \sum_{k=1}^{[nt]} Y_k^{(1)}$ does not converge weakly in the J_1 sense.*

However, under some additional conditions, the unidimensional moving average converges in the M_1 sense and hence in the M_2 sense as well.

Theorem 3. *Let $c_i \geq 0$ with $\sum_i c_i^{\nu} < \infty$ and suppose either*

(i) $\alpha > 1$, $\nu \leq 1$ and there exists $\eta > 0$ with $\alpha - \eta > 1$ and

$$\lim_{n\to\infty} \left(\sum_{i=n}^{\infty} c_i^{\nu}\right)\left(\sum_{i=n}^{\infty} c_i\right)^{\alpha-\eta} (\ln n)^{1+\alpha} = 0,$$

or

(ii) $\nu < \alpha \leq 1$.

Then

$$a_n^{-1} \sum_{k=1}^{[nt]} Y_n^{(1)}(t) \xrightarrow{w} (\sum_i c_i) X_\alpha(t)$$

where the weak convergence is in the sense of the M_1 Skorohod topology on $D[0,1]$.

The proofs of Theorems 2 and 3 will appear elsewhere. Note that M_1 is weaker than J_1 but is strong enough to make continuous the widely used functionals $\max_{0<t<1}$ and $\min_{0<t<1}$. Note also that if, in Theorem 2, the moving average contains only a single non-zero coefficient, then the Y_k are proportional to the X_k and by Theorem 2.7 of Skorohod (1957), $a_n^{-1} \sum_{k=1}^{[nt]} Y_k^{(1)}$ converges weakly in the J_1 sense.

We formalize now an idea, due to Surgailis (1983), which, in principle, provides a method of treating all the three cases of moving averages considered in the last section. Consider the space of functions

$$L_{\alpha,\eta} = \{f : \|f\|_{\alpha,\eta} < \infty\}$$

where

$$\|f\|_{\alpha,\eta} = \max\left\{ \int_{-\infty}^{+\infty} |f(x)|^{\alpha-\eta} dx, \left(\int_{-\infty}^{+\infty} |f(x)|^{\alpha+\eta} dx \right)^{\frac{\alpha-\eta}{\alpha+\eta}} \right\}.$$

Note that when $\alpha > 1$ and $\alpha - \eta > 1$ then $\|f\|_{\alpha,\eta}^{1/(\alpha-\eta)}$ is a norm, and when $\alpha \leq 1$, $\|f\|_{\alpha,\eta}$ is a metric.

Theorem 4. (Scheme of discrete stochastic integrals). *Suppose that for each $j = 1, \ldots, m$, the sequence of functions $f_n^{(j)}$ satisfies*

(i) $f_n^{(j)}(u)$ is constant on each interval $[\frac{k}{n}, \frac{k+1}{n})$ for $k = \ldots, -1, 0, 1, \ldots$.

(ii) $f_n^{(j)} \to f^{(j)}$ in $L_{\frac{\alpha}{j},\eta}$ as $n \to \infty$.

Then, as $n \to \infty$

$$\left(\sum_{i=-\infty}^{+\infty} f_n^{(1)}\left(\frac{i}{n}\right)\frac{X_i}{a_n}, \sum_{i=-\infty}^{+\infty} f_n^{(2)}\left(\frac{i}{n}\right)\frac{X_i^2}{a_n^2}, \ldots, \sum_{i=-\infty}^{+\infty} f_n^{(m)}\left(\frac{i}{n}\right)\frac{X_i^m}{a_n^m} \right)$$

$$\xrightarrow{d} \left(\int_{-\infty}^{+\infty} f^{(1)}(u) dX_\alpha^{(1)}(u), \int_{-\infty}^{+\infty} f^{(2)}(u) dX_\alpha^{(2)}(u), \ldots, \int_{-\infty}^{+\infty} f^{(m)}(u) dX_\alpha^{(m)}(u) \right)$$

where the processes $X_\alpha^{(j)}$, $j = 1, \ldots, m$ are defined in (2.2) and d denotes convergence in distribution in $\mathbf{R}^m$.

Details about Theorem 4 and Theorem 5 below will appear elsewhere.

We now apply Theorem 4 to the joint convergence of normalized sums of

$$Z_k^{(j)} = \sum_{i=-\infty}^{+\infty} (c_{k-i})^j X_i^j,$$

$j = 1, \ldots, m$, where the c_i's satisfy the conditions of either Case I or of Case II described in the previous section. Let

$$k_{n,t}^{(j)} = \sum_{k=1}^{[nt]} (c_{k-[nu]})^j,$$

so that

$$a_n^{-j} \sum_{k=1}^{[nt]} Z_k^{(j)} = \sum_{i=-\infty}^{+\infty} k_{n,t}^{(j)}\left(\frac{i}{n}\right) \frac{X_i^j}{a_n^j}.$$

Set $f_{n,t}^{(1)}(u) = \frac{1}{n^\beta L_1(n)} k_{n,t}^{(1)}(u)$ if the c_i's are as in Case I, $f_{n,t}^{(1)}(u) = k_{n,t}^{(1)}(u)$ if the c_i's are as in Case II, and set $f_{n,t}^{(j)} = k_{t,n}^{(j)}(u)$ for $j = 2, \ldots, m$. One can show that the $f_{n,t}^{(j)}$ converge in $L_{\frac{\alpha}{j},\eta}$ for small enough η. Using Theorem 4 one gets

Theorem 5. (a) *If the c_i are defined as in Case I, then*

$$f.d.d. - \lim_{n\to\infty} \sum_{k=1}^{[nt]} \left(\frac{1}{n^\beta L_1(n) a_n} Z_k^{(1)}, \frac{1}{a_n^2} Z_k^{(2)}, \ldots, \frac{1}{a_n^m} Z_k^{(m)} \right)$$

$$= \left(X_{\alpha,\beta}(t), (\sum_i c_i^2) X_\alpha^{(2)}(t), \ldots, (\sum_i c_i^m) X_\alpha^{(m)}(t) \right)$$

(b) If the c_i are defined as in Case II, then

$$f.d.d. - \lim_{n\to\infty} \sum_{k=1}^{[nt]} \left(a_n^{-1} Z_k^{(1)}, a_n^{-2} Z_k^{(2)}, \ldots, a_n^{-m} Z_k^{(m)} \right)$$

$$= \left((\sum_i c_i) X_\alpha^{(1)}(t), (\sum_i c_i^2) X_\alpha^{(2)}(t), \ldots, (\sum_i c_i^m) X_\alpha^{(m)}(t) \right).$$

(c) The conclusions in (a),(b) still hold if $Z_k^{(j)}$ is replaced by $Z_k^j = \left(\sum_{i=-\infty}^{+\infty} c_{k-i} X_i \right)^j$

Remarks (1) Theorem 5b is a special case of Theorem 1. (2) Part (c) holds because sums of cross terms are asymptotically negligible.

3. Techniques

A. Some of the techniques of the proof will now be reviewed and Theorem 1 will be established. The following moment inequality is useful.

Lemma 1. *Suppose either (i) $0 < \nu \le 1$ and X_i is any sequence of random variables or (ii) $1 < \nu \le 2$ and X_i is a martingale difference sequence. Then*

$$E|\sum_{i=1}^{N} c_i X_i|^\nu \le 2 \sum_{i=1}^{N} |c_i|^\nu E|X_i|^\nu. \tag{3.1}$$

The inequality holds also when $N = \infty$ provided that $\sum_i |c_i|^\nu < \infty$.

The inequality (3.1) is elementary when $0 < \nu \le 1$ and is a consequence of the Bahr-Esseen inequality (see Chatterji (1969), Lemma 1) when $1 < \nu \le 2$.

We follow in the sequel the notations and assumptions introduced in Section 1: X_i satisfies (1.1),(1.2), a_n is as in (1.3), Y_k as in (1.5) with $\Sigma|c_i|^\nu < \infty$, $\nu < \alpha$, $S_n(t)$ as in (1.7), $D_n^{(\alpha)}(c_.)$ as in (1.8).

Using Lemma 1, we get

Lemma 2. *If there is an a such that*

$$\sum_{k=-\infty}^{+\infty} |\sum_{i=a}^{a+k} c_i|^\nu < \infty \tag{3.2}$$

then, as $n \to \infty$,

$$n^{-\delta} \sum_{k=1}^{n} Y_k \to 0$$

in probability for every $\delta > 0$.

Remarks (1) Interpret the inner sum in (3.2) as $\sum_{i=a+k}^{a}$ if $a + k < a$. (2) Let $c_{-1} = -1$, $c_1 = 1$ and $c_i = 0$ for $|i| \ne 1$. Then (3.2) holds if $a = -2$ but does not hold if $a = 0$. Thus, one must be careful in choosing the "right" a.

Proof of Lemma 2. Assume for convenience that $c_i = 0$ for $i < 0$ and $a = 0$ in

(3.1). By Lemma 1,

$$
\begin{aligned}
E|\sum_{i=1}^{n} Y_i|^\nu &= E|\sum_{k=-\infty}^{+\infty} X_k(\sum_{i=1-k}^{n-k} c_i)|^\nu \\
&\le 2E|X_k|^\nu \sum_{k=-\infty}^{+\infty} |\sum_{i=1-k}^{n-k} c_i|^\nu \\
&= 2E|X_k|^\nu \left\{\sum_{k=0}^{n} |\sum_{i=0}^{n-k} c_i|^\nu + \sum_{k=-\infty}^{-1} |\sum_{i=0}^{n-k} c_i - \sum_{i=0}^{-k} c_i|^\nu\right\} \\
&= 2E|X_k|^\nu \left\{\sum_{k=0}^{n} |\sum_{i=0}^{k} c_i|^\nu + \sum_{k=1}^{\infty} |\sum_{i=0}^{n+k} c_i - \sum_{i=0}^{k} c_i|^\nu\right\} \\
&\le 2E|X_k|^\nu \left\{\sum_{k=0}^{n} |\sum_{i=0}^{k} c_i|^\nu + 2^\nu \sum_{k=1}^{\infty} \left(|\sum_{i=0}^{n+k} c_i|^\nu + |\sum_{i=0}^{k} c_i|^\nu\right)\right\} \\
&\le 2E|X_k|^\nu \left\{2^\nu \sum_{k=0}^{n} |\sum_{i=0}^{k} c_i|^\nu + 2^\nu \left(\sum_{k=n+1}^{\infty} |\sum_{i=0}^{k} c_i|^\nu + \sum_{k=0}^{\infty} |\sum_{i=0}^{k} c_i|^\nu\right)\right\} \\
&= 2^{\nu+2} E|X_k|^\nu \sum_{k=0}^{\infty} |\sum_{i=0}^{k} c_i|^\nu .
\end{aligned}
$$

The result follows since this bound is independent of n. ∎

Also, as an immediate consequence of Lemma 1, we get

$$
E|S_n(1)|^{\alpha-\eta} \le 2D_n^{(\alpha-\eta)}(c_.)E|X_i|^{\alpha-\eta} < \infty \tag{3.3}
$$

for small enough $\eta > 0$. However, it will become apparent that such a bound is not sharp enough for our purposes. Assume in the sequel that $0 < \epsilon \le 1$.

Lemma 3. (Splitting lemma) *Let $X_{i,n} = X_i/a_n$. Then for η small enough, one can find a constant M, independent of n, b_i, ϵ, such that:*

(a)
$$
E|\sum_{i=-\infty}^{\infty} b_i X_{i,n}|^{\alpha-\eta} \le M \max\left\{\frac{1}{n}\sum_i |b_i|^{\alpha-\eta}, \left(\frac{1}{n}\sum_i |b_i|^{\alpha+\eta}\right)^{\frac{\alpha-\eta}{\alpha+\eta}}\right\}
$$

(b)
$$
P\left\{|\sum_{i=-\infty}^{\infty} b_i X_{i,n}| > \epsilon\right\} \le \frac{M}{n}\epsilon^{-(\alpha+\eta)} \max\left\{\sum_i |b_i|^{\alpha-\eta}, \sum_i |b_i|^{\alpha+\eta}\right\}
$$

Proof. The proof is done by splitting $X_{i,n}$ into a small part and a big part.

(i) If $\alpha < 1$, we let

$$X^{\leq}_{i,n} = X_{i,n} 1\{|X_{i,n}| \leq 1\}, \tag{3.4a}$$

$$X^{>}_{i,n} = X_{i,n} 1\{|X_{i,n}| > 1\}. \tag{3.4b}$$

Let $\eta > 0$ be such that $\alpha - \eta > 0$, $\alpha + \eta < 1$. Then, by (3.1),

$$E|\sum_i b_i X^{\leq}_{i,n}|^{\alpha+\eta} \leq 2E|X^{\leq}_{i,n}|^{\alpha+\eta} \sum_i |b_i|^{\alpha+\eta} \tag{3.5a}$$

and

$$E|\sum_i b_i X^{>}_{i,n}|^{\alpha-\eta} \leq 2E|X^{<}_{i,n}|^{\alpha-\eta} \sum_i |b_i|^{\alpha-\eta}. \tag{3.5b}$$

The tail behavior of the stable distribution ensures that

$$M' = \sup_n \left\{ nE|X^{\leq}_{i,n}|^{\alpha+\eta} \right\} \vee \sup_n \left\{ nE|X^{>}_{i,n}|^{\alpha-\eta} \right\} < \infty \tag{3.6}$$

[see Feller (1971), p. 579 or Astrauskas (1983), Lemma 1]. Part (a) follows then from

$$\begin{aligned} E|\sum_i b_i X_{i,n}|^{\alpha-\eta} &\leq E|\sum_i b_i X^{\leq}_{i,n}|^{\alpha-\eta} + E|\sum_i b_i X^{>}_{i,n}|^{\alpha-\eta} \\ &\leq \left(E|\sum_i b_i X^{\leq}_{i,n}|^{\alpha+\eta}\right)^{\frac{\alpha-\eta}{\alpha+\eta}} + E|\sum_i b_i X^{>}_{i,n}|^{\alpha-\eta} \\ &\leq 2\max\left\{ \left(nE|X^{\leq}_{i,n}|^{\alpha+\eta} \frac{1}{n} \sum_i |b_i|^{\alpha+\eta}\right)^{\frac{\alpha-\eta}{\alpha+\eta}}, nE|X^{>}_{i,n}|^{\alpha-\eta} \frac{1}{n} \sum_i |b_i|^{\alpha-\eta} \right\} \end{aligned}$$

and (3.6).

To prove part (b), start with Chevychev's inequality

$$\begin{aligned} P\left\{|\sum_i b_i X_{i,n}| > \epsilon\right\} &\leq P\left\{|\sum_i b_i X^{\leq}_{i,n}| > \frac{\epsilon}{2}\right\} + P\left\{|\sum_i b_i X^{>}_{i,n}| > \frac{\epsilon}{2}\right\} \\ &\leq \frac{2^{\alpha+\eta}}{\epsilon^{\alpha+\eta}} \left\{ E|\sum_i b_i X^{\leq}_{i,n}|^{\alpha+\eta} + E|\sum_i b_i X^{>}_{i,n}|^{\alpha-\eta} \right\}, \end{aligned}$$

and conclude by using (3.5), (3.6) and by setting $M = 2^{\alpha+\eta+1}M'$.

(ii) When $\alpha \geq 1$ and $EX_i = 0$, let η be such that $\alpha - \eta \geq 1$; and set

$$\overline{X}_{i,n}^{\leq} = X_{i,n}^{\leq} - EX_{i,n}^{\leq}, \tag{3.7a}$$

$$\overline{X}_{i,n}^{>} = X_{i,n}^{>} + EX_{i,n}^{\leq}. \tag{3.7b}$$

Thus, $E\overline{X}_{i,n}^{\leq} = 0$. Since $EX_{i,n} = 0$, we have also $E\overline{X}_{i,n}^{>} = 0$. Hence by (3.1), relations (3.5a) and (3.5b) still hold in this case but with $X_{i,n}^{\leq}$, $X_{i,n}^{>}$ replaced by $\overline{X}_{i,n}^{\leq}$, $\overline{X}_{i,n}^{>}$. Since

$$\begin{aligned} E|\overline{X}_{i,n}^{\leq}|^{\alpha+\eta} &\leq 2^{\alpha+\eta}\Big(E|X_{i,n}^{\leq}|^{\alpha+\eta} + |EX_{i,n}^{\leq}|^{\alpha+\eta}\Big) \\ &\leq 2^{\alpha+\eta+1}E|X_{i,n}^{\leq}|^{\alpha+\eta} \end{aligned}$$

and

$$\begin{aligned} E|\overline{X}_{i,n}^{>}|^{\alpha-\eta} &\leq 2^{\alpha-\eta}\Big(E|X_{i,n}^{>}|^{\alpha-\eta} + |EX_{i,n}^{\leq}|^{\alpha-\eta}\Big) \\ &= 2^{\alpha-\eta}\Big(E|X_{i,n}^{>}|^{\alpha-\eta} + |EX_{i,n}^{>}|^{\alpha-\eta}\Big) \\ &\leq 2^{\alpha+\eta+1}E|X_{i,n}^{>}|^{\alpha-\eta}, \end{aligned}$$

and since (3.6) holds, we get

$$M'' = \sup_n\Big\{nE|\overline{X}_{i,n}^{\leq}|^{\alpha+\eta}\Big\} \vee \sup_n\Big\{nE|\overline{X}_{i,n}^{>}|^{\alpha-\eta}\Big\} < \infty. \tag{3.8}$$

The proof now proceeds exactly as in the case $\alpha < 1$.

(iii) When $\alpha = 1$ and X_i are symmetric, we use the split (3.4) as in the case $\alpha < 1$; since $EX_{i,n}^{\leq} = 0$, then (3.5a) will hold by (3.1), and the rest of the proof is the same. ∎

Applying the Splitting Lemma 3 to the moving average $Y_k = \mathrm{MA}_k(c_i, X_i)$ yields

Corollary 1. *For $\eta > 0$ small enough,*

$$P\Big\{|a_n^{-1}\sum_{i=1}^{n} Y_i| \geq \epsilon\Big\} \leq \frac{M}{n}\epsilon^{-(\alpha+\eta)}D_n^{(\alpha\pm\eta)}(c_\cdot),$$

where

$$D^{(\alpha\pm\eta)} = \max\Big\{D_n^{(\alpha+\eta)}(c_\cdot),\ D_n^{(\alpha-\eta)}(c_\cdot)\Big\}.$$

B. We focus now on Case II where $\Sigma|c_i| < \infty$ and thus ν in (1.6) can be assumed to satisfy $\nu \le 1$. In this case $D_n^{(\alpha)}(c_.) = 0(n)$. In fact, one has

$$D_n^{(\alpha)}(c_.) \le ns(\alpha, \nu, c_.) \tag{3.9}$$

where

$$s(\alpha,\nu,c_.) = \begin{cases} \sum_i |c_i|^\alpha & \text{if } \alpha \le 1 \\ (\sum_i |c_i|)^{\alpha-\nu} \sum_i |c_i|^\nu & \text{if } \alpha > 1. \end{cases} \tag{3.10}$$

Indeed, if $\alpha \le 1$,

$$D_n^{(\alpha)}(c_.) = \sum_{i=-\infty}^{\infty} | \sum_{j=1-i}^{n-i} c_j|^\alpha \le \sum_{i=-\infty}^{\infty} \sum_{j=1-i}^{n-i} |c_j|^\alpha = n \sum_{j=-\infty}^{\infty} |c_j|^\alpha.$$

If $\alpha > 1$,

$$\begin{aligned} D_n^{(\alpha)}(c_.) &\le \Big(\sum_{j=-\infty}^{\infty} |c_j| \Big)^{\alpha-\nu} \Big(\sum_{i=-\infty}^{\infty} \Big| \sum_{j=1-i}^{n-i} c_j \Big|^\nu \Big) \\ &\le \Big(\sum_{j=-\infty}^{\infty} |c_j| \Big)^{\alpha-\nu} \Big(\sum_{i=-\infty}^{\infty} \sum_{j=1-i}^{n-i} |c_j|^\nu \Big) = ns(\alpha,\nu,c_.). \end{aligned}$$

Relation (3.9) and Corollary 1 yield

Corollary 2. *If $\nu \le 1$, then for $\eta > 0$ small enough,*

$$P\{a_n^{-1} | \sum_{i=[nt_1]}^{[nt_2]} Y_i| > \epsilon\} \le M\epsilon^{-(\alpha+\eta)}(t_2 - t_1)s(\alpha \pm \eta, \nu, c_.) \tag{3.11}$$

where

$$s(\alpha \pm \eta, \nu, c_.) = \max\{s(\alpha+\eta,\nu,c_.), s(\alpha-\eta,\nu,c_.)\}.$$

Remark. If we had applied the bound $D_n^{(\alpha)}(c_.) \le ns(\alpha,\nu,c_.)$ to the crude inequality (3.3), we would have obtained

$$\begin{aligned} P\{a_n^{-1}|\sum_{i=1}^{n} Y_i| > \epsilon\} &\le 2E|X_i|^{\alpha-\eta}(\epsilon a_n)^{-(\alpha-\eta)} D_n^{(\alpha-\eta)}(c_.) \\ &\le M'\epsilon^{-(\alpha-\eta)} s(\alpha-\eta,\nu,c_.) n (n^{1/\alpha} L'(n))^{-(\alpha-\eta)} \\ &\le M'\epsilon^{-(\alpha-\eta)} s(\alpha-\eta,\nu,c_.) n^{\eta'} \end{aligned} \tag{3.12}$$

where M' is a constant and $\eta' > \eta/\alpha$. Relation (3.11) however provides a much better bound than (3.12) because the bound in (3.11) does not depend on n. This is why it is better to use the splitting technique (which yields (3.11)) than the straightforward application of the Bahr-Esseen inequality which yields (2.12).

Proof of Theorem 1. By applying the truncation method of Billingsley (1968, Thm 4.2), it is enough to prove the result (2.3) for the case where $c_i^{(j)} = 0$ for $|i| \geq K$. Indeed, let

$$c_i^{(j),\leq K} = \begin{cases} c_i^{(j)} & \text{for } |i| < K \\ 0 & \text{otherwise,} \end{cases}$$

$c_i^{(j),>K} = c_i^{(j)} - c_i^{(j),\leq K}$, and let $Y_k^{(j),\leq K}$ be $Y_k^{(j)}$ with $c_i^{(j)}$ replaced by $c_i^{(j),\leq K}$. One must verify

$$\lim_{K\to\infty} \overline{\lim}_{n\to\infty} P\{a_n^{-j}\left|\sum_{k=1}^{[nt]}(Y_k^{(j)} - Y_k^{(j),\leq K})\right| > \epsilon\} = 0. \tag{3.13}$$

But (3.13) holds by Corollary 2, since

$$Y_k^{(j)} - Y_k^{(j),\leq K} = \mathrm{MA}_k((c_i^{(j),>K}),\ X_i^j),$$

and the constants $a_n^{(j)} = \inf\{y : P\{|X_i|^j > y\} \leq n^{-1}\}$ are asymptotically proportional to a_n^j, so that by applying Corollary 2, we get

$$\text{L.H.S. of (3.13)} \leq M\left(\epsilon^{-(\alpha/j+\eta)}t\right) \lim_{K\to\infty} s(\frac{\alpha}{j} \pm \eta, \nu_j,\ c^{(j),>K}) = 0.$$

(Note: it is important here that the bound in Corollary 2 does not depend on n.)

It remains now only to prove that (2.3) holds when there exists K such that $c_i = 0$ for $|i| \geq K$. Let $s^{(j)} = \sum_{i=-\infty}^{\infty} c_i^{(j)}$ and let

$$\overline{Y}_k^{(j)} = s^{(j)}X_k^j = \sum_{i=-\infty}^{\infty} s^{(j)}\delta_o(k-i)X_i^j,$$

for $j = 1,\ldots m$. Since by Avram and Taqqu (1984)

$$f.d.d. - \lim_{n\to\infty} \sum_{k=1}^{[nt]}\left(a_n^{-1}\overline{Y}_k^{(1)},\ldots a_n^{-m}\overline{Y}_k^{(m)}\right) = \left(s^{(1)}X_\alpha^{(1)}(t),\ldots s^{(m)}X_\alpha^{(m)}(t)\right),$$

it is sufficient to prove that for $j = 1, \dots, m$,

$$a_n^{-j} \sum_{k=1}^{[nt]} \left(Y_k^{(j)} - \overline{Y}_k^{(j)} \right) \to 0$$

in probability. To show this, note that

$$Y_k^{(j)} - \overline{Y}_k^{(j)} = \mathrm{MA}_k\left(c_i^{(j)} - s^{(j)} \delta_o(i), X_i^j \right).$$

$Y_k^{(j)} - \overline{Y}_k^{(j)}$ are moving averages with two special properties: they have finitely many coefficients and the sum of their coefficients is 0. Since (3.2) holds with $a = -K$, Lemma 2 shows that sums of such sequences go to 0 in probability when they are normalized by any sequence that goes to infinity. This concludes the proof of Theorem 1. ∎

References

[1] Astrauskas, A. (1983). Limit theorems for sums of linearly generated random variables. *Lithuanian Mat. Journal* 23(2), 127-134.

[2] Avram, F. and Taqqu, M.S. (1984). Symmetric polynomials of random variables attracted to an infinitely divisible law. School of Operations Research and Industrial Engineering Technical Report No. 622, Cornell University, Ithaca. To appear in *Z. Wahrscheinlichkeitstheorie verw. Geb.*

[3] Billingsley, P. (1968). *Convergence of Probability Measures*. Wiley: New York.

[4] Chatterji, S.D. (1969). An L^p-convergence theorem. *Ann. Math. Stat.* 40, 1068-1070.

[5] Davis, R. and Resnick, S. (1985). Limit theory for moving averages with regularly varying tail probabilities. *Ann. Prob.* 13, 179-195.

[6] Feller, W. (1971). *An Introduction to Probability Theory and Its Applications.* Vol. II, 2nd Ed. Wiley: New York.

[7] Kawata, T. (1972). *Fourier Analysis in Probability Theory*. Academic Press: New York.

[8] Lamperti, J.W. (1962). Semi-stable stochastic processes. *Trans. Amer. Math. Soc.* 104, 62-78.

[9] Maejima, M. (1983). On a class of self-similar processes. *Z. Wahrscheinlichkeitstheorie verw. Geb.* 62, 235-245.

[10] Maejima, M. (1983b). A self-similar process with nowhere bounded sample paths. *Z. Wahrscheinlichkeitstheorie verw. Geb.* 65, 115-119.

[11] Skorohod,A.V. (1957) Limit theorems for stochastic processes with independent increments. *Theor. Probability Appl.* 2, 138-171.

[12] Surgailis, D. (1983). Zones of attraction of self-similar multiple integrals. *Lithuanian Mat. Journal* **22**, 327-430.

[13] Taqqu, M.S. and Wolpert, R. (1983). Infinite variance self-similar processes subordinate to a Poisson measure. *Z. Wahrscheinlichkeitstheorie verw. Geb.* 62, 53-72.

Florin Avram
Department of Mathematics
Purdue University
West Lafayette, IN 47907

Murad S. Taqqu
Department of Mathematics
Boston University
Boston, MA 02215

Limit Theory for the Sample Correlation Function of Moving Averages*

by

Richard Davis and Sidney Resnick

§1. Introduction

The sample autocorrelation function (acf) is an important statistic in time series analysis. It is frequently relied upon for assessing the dependence structure of a time series and may also be used for model identification and parameter estimation in the class of ARMA models. In this paper, we review some of the main asymptotic results for sample acfs of infinite order moving averages. While the classical theory concerning sample acfs requires the process to have at least finite second moments, our main interest in this paper will be the case when the process has an infinite variance. It turns out that in the infinite variance case, the sample acf can have desirable large sample properties and these can be helpful in estimating various parameters associated with the model.

Consider the infinite order moving average process $\{X_t\}$,

$$X_t = \sum_{j=-\infty}^{\infty} c_j Z_{t-j}, \quad t = 1, 2, \ldots$$

where $\{Z_t, -\infty < t < \infty\}$ is an iid sequence of random variables with common distribution F which belongs to the domain of attraction of a stable law with index α, $0 < \alpha \leq 2$. We abbreviate this as $Z_1 \in \mathscr{D}(\alpha)$. The conditions on the coefficients $\{c_j\}$ will be specified in Section 2. Defining

$$C(h) = \sum_{t=1}^{n} X_t X_{t+h}, \quad h \geq 0 \tag{1.1}$$

the sample acf is given by

$$\hat{\rho}(h) = C(h)/C(0), \quad h \geq 1.$$

If we follow standard conventions the sum in (1.1) should terminate at n-h, but it is notationally more convenient if we use n instead and it is readily seen that our results apply to the sums terminated at n-h as well. Define

$$\rho(h) = (\Sigma\, c_j c_{j+h})/(\Sigma\, c_j^2),\ h \geq 1$$

which corresponds to $\mathrm{Corr}(X_t, X_{t+h})$ if $\mathrm{Var}(Z_t) < \infty$. However $\rho(h)$ is still a meaningful quantity even in the $\mathrm{Var}(Z_t) = \infty$ situation provided the series converge.

If $0 < \alpha < 2$ it was shown in Davis and Resnick (1985) that $\hat{\rho}(h) \xrightarrow{P} \rho(h)$. Note that in this case, $\mathrm{Var}(Z_t) = \infty$. A more refined result for AR(p) processes was given by Kanter and Stieger (1974) and Hannan and Kanter (1977). They proved that for any $\delta > \alpha$,

$$n^{1/\delta}(\hat{\rho}(h) - \rho(h)) \xrightarrow{P} 0. \tag{1.2}$$

Yohai and Maronna (1977) showed that $n^{1/2}(\hat{\rho}(h)-\rho(h))$ is bounded in probability for AR(p) processes provided the Z_t's are symmetrically distributed and $E \log^+ |Z_t| < \infty$. On the other hand if $\mathrm{Var}(Z_t) < \infty$, then $n^{1/2}(\hat{\rho}(h) - \rho(h))$ is asymptotically normal (cf. Anderson, 1971, p. 489).

Basic to the understanding of the weak limit behavior of $\hat{\rho}(h)$ is understanding weak limits of vectors of the form $\sum_{t=1}^{n} (Z_t^2, Z_t Z_{t+h}, 1 \leq h \leq \ell)$. This is discussed in Section 2. In Section 3 these results are then used to show that under appropriate assumptions, there exists a slowly varying function $L_1(n)$ such that $n^{1/\alpha}L_1(n)(\hat{\rho}(h) - \rho(h))$ converges in distribution. In particular if the tail distribution of $|Z_1|$ is asymptotically equivalent to a Pareto (as is the case when the Z_i's have a stable distribution) then we may take $L_1(n) = (\log n)^{-1/\alpha}$ and the limit is equal in distribution to the ratio of two independent stable random variables with indices α and $\alpha/2$. The limit random variable can easily be expressed in terms of the limit random vector of $\sum_{t=1}^{n} (Z_t^2, Z_t Z_{t+h}, 1 \leq h \leq \ell)$. Some applications of these results to time series are given in Section 4.

§2. Sample covariance function of $\{Z_t\}$

In order to describe the limit laws of the partial sums $\sum_{t=1}^{n} Z_t Z_{t+h}$ and subsequently $\hat{\rho}(h)$, we consider three separate cases. As before we assume $Z_1 \in \mathcal{D}(\alpha)$, $0 < \alpha \leq 2$ and in addition suppose $E Z_1 = 0$ for $1 < \alpha \leq 2$ and for $\alpha = 1$, we assume for brevity that Z_1 has a symmetric distribution. Details of the following results including the nonsymmetric case with $\alpha = 1$ may be found in Davis and Resnick (1984, 1985).

Case 1. $\alpha = 2$. $Z_1 \in \mathcal{D}(2)$ if and only if $L(t) = E Z_1^2 1_{[|Z_1| \leq t]}$ is slowly varying (Feller, 1971). If the a_n's are a sequence of constants chosen so that

$$n a_n^{-2} L(a_n) \to 1$$

then $a_n^{-1} \sum_{t=1}^{n} Z_t \Rightarrow N(0, 1)$. A result by Maller (1981) gives $Z_1 Z_2 \in \mathcal{D}(2)$ which is obvious only if $\mathrm{Var}(Z_1) < \infty$. Now let $\tilde{a}_n$ be a sequence of constants satisfying

$$n \tilde{a}_n^{-2} E|Z_1 Z_2|^2 1_{[|Z_1 Z_2| \leq \tilde{a}_n]} \to 1.$$

Then using standard arguments for m-dependent sequences and a result characterizing domains of attraction of a multivariate normal distribution (cf. de Haan, Omey, and Resnick, 1984), we have in $\mathbb{R}^{\ell}$

$$\tilde{a}_n^{-1}\left(\sum_{t=1}^{n} Z_t Z_{t+h},\ 1 \leq h \leq \ell\right) \Rightarrow (S_h,\ 1 \leq h \leq \ell) \tag{2.1}$$

where $S_1, S_2, \ldots S_{\ell}$ are iid N(0, 1) rv's. Observe that if $\mathrm{Var}(Z_1) < \infty$, then all of the components in the vector (2.1) are uncorrelated in which case (2.1) can be established quite easily. Also, in this case we may take $a_n = n^{1/2} \sigma$, and $\tilde{a}_n = n^{1/2} \sigma^2$, $\sigma^2 = \mathrm{Var}(Z_1)$. Combining (2.1) with the weak law of large numbers, (cf. Feller, 1971, Theorem 2, p. 236)

$$a_n^{-2} \sum_{t=1}^{n} Z_t^2 \xrightarrow{P} 1,$$

we obtain

$$\left(a_n^{-2} \sum_{t=1}^{n} Z_t^2,\ \tilde{a}_n^{-1} \sum_{t=1}^{n} Z_t Z_{t+h},\ 1 \leq h \leq \ell\right)$$
$$\Rightarrow (1,\ S_h,\ 1 \leq h \leq \ell).$$

Case 2. $0 < \alpha < 2$ and $E|Z_1|^\alpha = \infty$. $Z_1 \in \mathcal{D}(\alpha)$ is equivalent to

$$P(|Z_1| > x) = x^{-\alpha}L(x)$$

where $L(x)$ is a slowly varying function at ∞ and the tail balancing condition

$$\frac{P(Z_1 > x)}{P(|Z_1|>x)} \to p, \quad \frac{P(Z_1 < -x)}{P(|Z_1| > x)} \to 1 - p$$

as $x \to \infty$, $0 \le p \le 1$. Moreover from a theorem of Embrechts and Goldie (1980) as extended by Cline (1983), we have $Z_1Z_2 \in \mathcal{D}(\alpha)$. If the sequences a_n and $\tilde{a}_n$ are defined by

$$a_n = \inf\{x: \ P(|Z_1| > x\} \le n^{-1}\} \tag{2.2}$$
$$\tilde{a}_n = \inf\{x: P(|Z_1Z_2| > x) \le n^{-1}\},$$

then it was shown using point process methods in Davis and Resnick (1984) that

$$(a_n^{-2} \sum_{t=1}^{n} Z_t^2, \ \tilde{a}_n^{-1} \sum_{t=1}^{n} Z_tZ_{t+h}, \ 1 \le h \le \ell) \Rightarrow (S_0, S_h, \ 1 \le h \le \ell)$$

where $S_0, S_1, \ldots, S_\ell$ are mutually independent rv's, S_0 is stable with index $\alpha/2$ and $S_1, \ldots, S_\ell$ are identically distributed with a stable distribution of index α. While both sequences of constants $\{a_n\}$ and $\{\tilde{a}_n\}$ are regularly varying with index α^{-1}, we have $\tilde{a}_n/a_n \to \infty$. For example if $|Z_1|$ has a Pareto distribution then we may take $a_n = n^{1/\alpha}$ and $\tilde{a}_n = (n \log n)^{1/\alpha}$.

Case 3. $0 < \alpha < 2$ and $E|Z_1|^\alpha < \infty$. In addition, we need to further assume

$$\frac{P(|Z_1Z_2| > t)}{P(|Z_1| > t)} \to 2E|Z_1|^\alpha \quad \text{as } t \to \infty. \tag{2.3}$$

It was shown in Chover, Ney and Wainger (1973) that if the limit in (2.3) exists then it must be $2E|Z_1|^\alpha$. On the other hand a limit distribution of $\sum_{t=1}^{n} Z_tZ_{t+h}$ does not exist without having convergence of $P(|Z_1Z_2| > t)/P(|Z_1| > t)$ (cf. Davis and Resnick, 1985). If the sequence a_n is as defined in (2.2), then

$$(a_n^{-2} \sum_{t=1}^{n} Z_t^2, \ a_n^{-1} \sum_{t=1}^{n} Z_t Z_{t+h}, \ 1 \le h \le \ell) \Rightarrow (S_0, S_h, \ 1 \le h \le \ell)$$

where S_0 is stable with index $\alpha/2$ and $S_1, \ldots, S_\ell$ are identically stable distributed with index α. However, unlike Case 2, the S_h's are not independent. The dependence structure between the S_h's can be exhibited via a representation of S_h as a sum of points of a Poisson process. For example suppose $0 < \alpha < 1$ and let $\{j_k, k \ge 1\}$ denote the points of a Poisson process on $\mathbb{R}$ with intensity measure given by $\nu(dx)=(p\alpha(x)^{-\alpha-1} 1_{(0,\infty)}(x)+(1-p)\alpha(-x)^{-\alpha-1} 1_{(-\infty,0)}(x))dx$. Note that this intensity is the Levy measure of the stable law which attracts $\sum_{t=1}^{n} Z_t$. We may then represent S_0, S_h, $h \ge 1$ by

$$S_0 = \sum_{k=1}^{\infty} j_k^2 \text{ and } S_h = \sum_{k=1}^{\infty} j_k(Z_k^{(h)} + Z_{-k}^{(h)}), \ h \ge 1$$

where $\{Z_k^{(h)}, -\infty < k < \infty\}$, $h = 1, 2, \ldots$ are independent copies of the sequence $\{Z_t\}$. The S_h's are not independent due to the presence of the j_k's in each of their representations. Moreover, it is easy to see by checking characteristic functions that $S_0, S_1, S_2, \ldots, S_\ell$ are jointly stable; S_0 has corresponding Levy measure $\nu_0(dx) = (\alpha/2)x^{-\alpha/2-1} 1_{(0,\infty)}(x)(dx)$ and S_h has Levy measure $\nu_1(dx) = E|Z_1 + Z_2|^\alpha \nu(dx)$.

§3. Sample correlation function of the moving average process

In all three cases of Section 2, we have in $\mathbb{R}^{\ell+1}$

$$(3.1) \qquad (a_n^{-2} \sum_{t=1}^{n} Z_t^2, \ \tilde{a}_n^{-1} \sum_{t=1}^{n} Z_t Z_{t+h}, \ 1 \le h \le \ell)$$
$$\Rightarrow (S_0, \ S_h, \ 1 \le h \le \ell)$$

for some suitable choice of normalizing constants $a_n > 0$ and $\tilde{a}_n > 0$ and where $S_0 \equiv 1$ if $\alpha = 2$. This result enables us to derive the limiting distribution of the sample acf, $\hat{\rho}(h)$, of the moving average process $X_t = \sum_{j=-\infty}^{\infty} c_j Z_{t-j}$. The assumptions on the coefficients $\{c_j\}$ are as follows:

If $\alpha = 2$, assume

$$\sum_j |c_j| < \infty \text{ and } \sum_j |c_j|^{\delta}|j| < \infty \text{ with } \begin{cases} \delta = 2 \text{ if } \mathrm{Var}(Z_1) < \infty \\ \delta < 2 \text{ if } \mathrm{Var}(Z_1) = \infty. \end{cases}$$

If $0 < \alpha < 2$ assume

$$\sum_j |c_j|^{\delta}|j| < \infty \text{ with } \begin{cases} 0 < \delta < \alpha \text{ if } \alpha \le 1 \\ \delta = 1 \quad\quad \text{ if } \alpha > 1. \end{cases}$$

Under these assumptions, we then have in all three cases of Section 2,

$$(3.2) \qquad (\tilde{a}_n^{-1}\, a_n^2\, (\hat{\rho}(h) - \rho(h)),\ 1 \le h \le \ell) \Rightarrow (Y_h,\ 1 \le h \le \ell)$$

where $Y_h = \sum_{j=1}^{\infty} (\rho(h+j) + \rho(h-j) - 2\rho(j)\rho(h))S_j/S_0$. Since a_n and $\tilde{a}_n$ are both regularly varying with index α^{-1}, $\tilde{a}_n^{-1}\, a_n^2 = n^{1/\alpha}\, L_1(n)$ for some slowly varying function $L_1(\cdot)$.

There are two main steps in proving (3.2) which works for all three cases. The first is to show

$$(3.3) \qquad a_n^{-2}\, (C(0) - \sum_j c_j^2 \sum_{t=1}^{n} Z_t^2) \xrightarrow{P} 0$$

and the second is establishing

$$(3.4) \qquad \tilde{a}_n^{-1}\, a_n^2\, (\hat{\rho}(h) - \rho(h)$$

$$- (C(0))^{-1} \sum_{t=1}^{n} \sum_{i \ne j} c_i(c_{j+h} - c_j\rho(h))Z_{t-i}Z_{t-j}) \xrightarrow{P} 0.$$

Then from (3.1) we have $a_n^{-2}\, C(0) \Rightarrow \sum c_j^2\, S_0$ and by using a continuous mapping like argument and (3.1) once again, we obtain

$$\tilde{a}_n^{-1}\, a_n^2\, (C(0))^{-1} \sum_{t=1}^{n} \sum_{i \ne j} c_i(c_{j+h} - c_j\rho(h))Z_{t-i}Z_{t-j}$$

$$\Rightarrow \sum_{i \ne j} c_i(c_{j+h} - c_j\rho(h))S_{|i-j|}/(\sum c_j^2\, S_0)$$

$$= Y_h$$

which combined with (3.4) establishes (3.2).

In the classical case ($\alpha = 2$ and $\mathrm{Var}(Z_1) < \infty$), (3.2) becomes

$$(n^{1/2}(\hat{\rho}(h) - \rho(h)),\ 1 \le h \le \ell) \Rightarrow \underset{\sim}{Y} = (Y_1, \ldots, Y_\ell)$$

where $\underset{\sim}{Y}$ has a multivariate normal distribution with mean zero and covariance matrix given by

$$r_{gk} = \sum_{j=1}^{\infty} (\rho(g+j) + \rho(g-j) - 2\rho(j)\rho(g))(\rho(k+j) + \rho(k-j) - 2\rho(j)\rho(k))$$

which agrees with Bartlett's celebrated formula (cf. Anderson, 1971).

The limit laws for the mean corrected version of the sample acf remain the same for $1 < \alpha \le 2$. That is if

$$\tilde{\rho}(h) = \sum_{t=1}^{n} (X_t - \bar{X})(X_{t+h} - \bar{X}) / \sum_{t=1}^{n} (X_t - \bar{X})^2$$

where $\bar{X} = \sum_{t=1}^{n} X_t/n$, then for $1 < \alpha \le 2$,

$$(\tilde{a}_n^{-1} a_n^2 (\tilde{\rho}(h) - \rho(h)), 1 \le h \le \ell) \Rightarrow (Y_h, 1 \le h \le \ell).$$

On the other hand if $0 < \alpha < 1$, then

$$(n(\tilde{\rho}(h) - \rho(h)), 1 \le h \le \ell) \Rightarrow$$
$$((\rho(h) - 1), 1 \le h \le \ell)(\sum_j c_i \; S)^2/(\sum c_i^2 \; S_0)$$

where $S = \sum_{k=1}^{\infty} j_k$ is the sum of the points of a Poisson process with intensity ν given above.

§4. Applications.

Throughout this section, assume the conditions of either Case 1 or Case 2 of Section 2 are met, and for simplicity that Z_1 has a symmetric distribution. We do not consider Case 3 here because the dependence structure between the S_h's makes the limit distribution in (3.2) harder to work with. In Cases 1 and 2, we have for the limit random variables Y_h in (3.2)

$$Y_h \overset{d}{=} \left(\sum_{j=1}^{\infty} |\rho(h+j) + \rho(h-j) - 2\rho(j)\rho(h)|^{\alpha}\right)^{1/\alpha} S_1/S_0.$$

In the $\alpha = 2$ case, the coefficient of S_1 is just the standard deviation of Y_h.

1) <u>MA(q)</u>. Suppose $\{X_t\}$ is the finite moving average

$$X_t = Z_t + \theta_1 Z_{t-1} + \dots + \theta_q Z_{t-q}.$$

Then, since $\rho(h) = 0$ for $|h| > q$, we have for $h > q$

$$n^{1/\alpha} L_1(n)(\hat{\rho}(h) - \rho(h)) \Rightarrow (1 + 2 \sum_{j=1}^{q} |\rho(j)|^\alpha)^{1/\alpha} S_1/S_0.$$

2) <u>Estimation of θ in a MA(1)</u>. For the MA(1) process $X_t = Z_t + \theta Z_{t-1}$, $\rho(1) = \theta/(1 + \theta^2)$. A method of moments type estimator for θ is found by solving the latter equation for θ. Choosing the solution with the constraint $|\theta| \le 1$ (cf. Fuller, 1976), gives (set $\hat{\rho} = \hat{\rho}(1)$)

$$\theta = \begin{cases} (1 - (1 - 4\hat{\rho}^2)^{1/2})/(2\hat{\rho}) & \text{if } 0 \le |\hat{\rho}| \le .5 \\ -1 & \hat{\rho} < -.5 \\ 1 & \hat{\rho} > .5 . \end{cases}$$

Letting $g(\rho)$ denote the inverse of the function $\theta/(1 + \theta^2)$ with $|\theta| \le 1$, we have by the mean value theorem

$$\hat{\theta} - \theta = g(\hat{\rho}) - g(\rho) = g'(\rho)(\hat{\rho} - \rho) + o_p(\hat{\rho} - \rho).$$

Hence

$$\begin{aligned} n^{1/\alpha} L_1(n)(\hat{\theta} - \theta) &\Rightarrow \\ &(1-\theta^2)^{-1}(1+\theta^2)^2((1-2\rho^2(1))^\alpha + |\rho(1)|^\alpha)^{1/\alpha} S_1/S_0 \\ &= (((1 + \theta^4)^\alpha + |\theta|^\alpha(1 + \theta^2)^\alpha)^{1/\alpha}/(1 - \theta^2))S_1/S_0. \end{aligned}$$

By setting $\alpha = 2$, the coefficient of S_1 is the standard deviation of $\hat{\theta}$ (cf. Fuller, 1976, p. 343). Note that while this estimate of θ is inefficient in the finite variance case, its performance in the $0 < \alpha < 2$ case seems to be good. For example, in a simulation experiment, 100 replications of the process $Z_t - .4Z_{t-1}$, $t = 1, 2, \dots 100$ were generated where Z_t is Cauchy distributed. The mean of the $\hat{\theta}$'s was -.40074 with a standard deviation of .0790. This compares favorably with the asymptotic standard deviation of $((1-\theta^2)/n)^{1/2} = .0917$ for the maximum likelihood estimator of θ in a MA(1) model assuming the noise sequence is normally distributed. While comparing variances in this situation may be a bit misleaading, it nevertheless gives an indication of the reasonable good performance of $\hat{\theta}$ in the $0 < \alpha < 2$ case.

3) AR(1). Let $\{X_t\}$ be the AR(1) process $X_t = \varphi X_{t-1} + Z_t$ where $|\varphi|<1$. In this case, $\rho(h) = \varphi^{|h|}$ and estimating φ by $\hat{\varphi} = \hat{\rho}(1)$, we have

$$n^{1/\alpha}L_1(n)(\hat{\varphi} - \varphi) \Rightarrow \left(\sum_{j=1}^{\infty} (\varphi^{1+j} + \varphi^{j-1} + 2\varphi^j\varphi)^{\alpha}\right)^{1/\alpha} S_1/S_0$$

$$= \frac{1 - \varphi^2}{(1 - \varphi^{\alpha})^{1/\alpha}} S_1/S_0.$$

4) Yule-Walker estimates. The Yule-Walker matrix equation for the AR(p) model $X_t = \varphi_1 X_{t-1} \quad .. + \varphi_p X_{t-p} + Z_t$, assuming $1 - \varphi_1 z - \varphi_2 z^2 - \ldots - \varphi_p z^p \neq 0$, $|z| \leq 1$, is

(4.1) $$R \underset{\sim}{\varphi} = \underset{\sim}{\rho}$$

where R is the p × p matrix $[\rho(i-j)]_{i,j=1}^{p}$, $\underset{\sim}{\varphi} = (\varphi_1, \ldots, \varphi_p)'$ and $\underset{\sim}{\rho} = (\rho(1), \ldots,, \rho(p))'$. The Yule-Walker estimate of $\underset{\sim}{\varphi}$ is then defined as the solution of (4.1) with R and $\underset{\sim}{\rho}$ replaced by $\hat{R} = [\hat{\rho}(i - j)]_{i,j=1}^{p}$ and $\hat{\underset{\sim}{\rho}} = (\hat{\rho}(1), \ldots \hat{\rho}(p))'$, respectively. As in Yohai and Maronna (1977), for $\underset{\sim}{z} \in \mathbb{R}^p$ define $\varphi(\underset{\sim}{z}) = R(\underset{\sim}{z})^{-1}z$ where $R(\underset{\sim}{z}) = [z_{|i-j|}]_{i,j=1}^{p}$ and $z_o \equiv 1$. Since $\hat{R} \xrightarrow{P} R$ and R is non-singular, this implies $\varphi(\hat{\underset{\sim}{\rho}})$ is well defined for large n. The mean value theorem then gives

$$\hat{\underset{\sim}{\varphi}} - \underset{\sim}{\varphi} = D(\hat{\underset{\sim}{\rho}} - \underset{\sim}{\rho}) + o_p(\hat{\underset{\sim}{\rho}} - \underset{\sim}{\rho})$$

where D is the p × p matrix of partial derivatives of φ evaluated at ρ. Consequently,

$$n^{1/\alpha}L_1(n)(\hat{\underset{\sim}{\varphi}} - \underset{\sim}{\varphi}) \Rightarrow D \underset{\sim}{Y}$$

where $\underset{\sim}{Y} = (Y_1, Y_2, \ldots, Y_p)'$ with $Y_h = \sum_{j=1}^{\infty} (\rho(h+j) + \rho(h-j) - 2\rho(j)\rho(h))S_j/S_o$, $h = 1, \ldots, p$.

*This research was partially supported by NSF grant MCS 8501763.

References

[1] Anderson, T.W. (1971). Statistical Analysis of Time Series. Wiley, New York.

[2] Chover, J., Ney, P. and Wainger, S. (1973). Functions of probability measures. J. Analyse Math. 26, 255-302.

[3] Cline, D. (1983). Estimation and linear prediction for regression, autoregression and ARMA with infinite variance data. Ph.D. Thesis, Department of Statistics, Colorado State University, Fort Collins, CO 80523.

[4] Davis, R.A. and Resnick, S. (1984). Limit theory for the sample covariance and correlation functions of moving averages. (To appear in Annals of Statistics).

[5] Davis, R.A. and Resnick, S. (1985). More limit theory for the sample correlation function of moving averages. (To appear in Stochastic Processes and Their Applications.)

[6] Embrechts, P. and Goldie, C. (1980). On closure and factorization properties of subexponential and related distributions. J. Austral. Math. Soc. (Series A), 29, 243-256.

[7] Feller, W. (1971). An Introduction to Probability Theory and its Applications. Vol. II, 2nd edition, Wiley, New York.

[8] Fuller, W. (1976). Introduction to Statistical Time Series. Wiley, New York.

[9] Haan, L. de, Omey, E. and Resnick, S. (1984). Domains of attraction and regular variation in $\mathbb{R}^d$. J. Mult. Analysis, 14, 17-33.

[10] Hannan, E.J. and Kanter, M. (1977). Autoregressive processes with infinite variance. J. Appl. Probability, 14, 411-415.

[11] Kanter, M. and Steiger, W.L. (1974). Regression and autoregression with infinite variance. Advances in Appl. Probability, 6, 768-783.

[12] Maller, R. (1981). A theorem on products of random variables with applications to regression. Austral. J. Statist., 23, 177-185.

[13] Yohai, V. and Maronna, R. (1977). Asymptotic behavior of least-squares estimates for autoregressive processes with infinite variances. Ann. Statistics, 5, 554-560.

SECTION 6

EXTREME VALUES

Extreme values for stationary sequences

George L. O'Brien

1. Introduction.

Let $(X_n)_{n=1}^{\infty}$ be a strictly stationary real-valued sequence and let $F(x) = P[X_1 \le x]$. Let $M_{i,j} = \max(X_{i+1},\ldots,X_j)$ and $M_j = \max(X_1,\ldots,X_j) = M_{0,j}$. Let $(c_n)_{n=1}^{\infty}$ be a sequence of real numbers. The purpose of this article is to review the theory of the asymptotic behaviour of $P[M_n \le c_n]$ as $n \to \infty$. We mainly consider aspects which are not treated in Leadbetter, Lindgren and Rootzén [4]. To avoid trivialities we assume that $F(c_n) < 1$ for all n, $F(c_n) \to 1$, and $P[X_1 = \sup\{x : F(x) < 1\}] = 0$. All limits are "as $n \to \infty$."

If (X_n) is an i.i.d. sequence then obviously

$$P[M_n \le c_n] = (F(c_n))^n . \qquad (1)$$

We are interested in examining the effect of the dependence structure of (X_n) on $P[M_n \le c_n]$. This is very difficult without some kind of mixing or asymptotic independence assumption, such as the following condition. We say (X_n) has <u>asymptotic independence of maxima</u> relative to (c_n) (notation: (X_n) has AIM(c_n)) if there exists a sequence (q_n) of positive integers with (here and elsewhere we write "q" for "q_n", "p" for "p_n", etc.) $q = o(n)$ and

$$\alpha_n := \max|P[M_i \le c_n, M_{i+q,i+q+j} \le c_n] - P[M_i \le c_n]\, P[M_j \le c_n]| \to 0 ,$$

where the maximum is over all i and j with $i + q + j \leq n$.

This condition is weaker than strong mixing (cf. Bradley [1]) in that it only requires $|P(AB) - P(A)P(B)|$ to be small for events A and B that arise in the consideration of maxima.

The following argument of Loynes [5] is at the heart of all theorems dealing with $P[M_n \leq c_n]$. Let (p_n) be a sequence of positive integers satisfying

$$p = o(n) \ , \quad q = o(p) \quad \text{and} \quad n\alpha_n = o(p) \quad . \tag{2}$$

Separate the random variables $X_1, X_2, \ldots, X_n$ into $2r = 2n(p+q)^{-1}$ blocks of successive variables, alternately of length p and of length q . The second condition in (2) assures us that we may ignore (asymptotically) the possibility that the largest observation among $X_1, \ldots, X_n$ occurs in one of the blocks of length q ; the third condition in (2) then allows us to treat the distinct blocks of length p as independent. Thus, we obtain the preliminary approximation

$$P[M_n \leq c_n] - (P[M_p \leq c_n])^r \to 0 \quad . \tag{3}$$

From the preceding paragraph, we see that a consequence of the assumption that (X_n) has $AIM(c_n)$ is to preclude any long-range influences; we can therefore concentrate on the effect of short-range dependence, where "short-range" means "within a distance of p."

The short-range dependence affects $P[M_n \leq c_n]$ in that it allows for the clustering together of large values to a greater extent than in the independent case (where, asymptotically, the occurrence of large values has a Poisson character). If large values tend to come in pairs, say, then the second observation of the pair is wasted in that it does not affect M_n . The presence of such wasted values causes $P[M_n \leq c_n]$ to be larger than it would be in the independent case (with the same F) . In other words, clustering of values greater than c_n goes hand in hand with longer stretches with no such values, thus enlarging $P[M_n \leq c_n]$.

Several authors have proved theorems of the following form (with some other assumptions): if (X_n) satisfies some mixing condition, if

$$\lim\inf(F(c_n))^n > 0 \ , \tag{4}$$

and if

$$P[M_{1,p} \le c_n | X_1 > c_n] \to 1 \ , \tag{5}$$

(where $(p) = (p_n)$ satisfies (2)), then

$$P[M_n \le c_n] - (F(c_n))^n \to 0 \ . \tag{6}$$

Watson [10] proved such a result for m-dependent sequences (where p may be replaced by m in (5)), Loynes [5] extended it to strongly mixing sequences, and Leadbetter [2] proved it for sequences satisfying a condition slightly stronger than the $AIM(c_n)$ condition. Leadbetter's version has the nice feature that it incorporates earlier results for stationary Gaussian sequences.

Formula (5) essentially serves to preclude the possibility of more than one observation in excess of c_n among $X_1,\ldots,X_p$. The theorems cited above thus support the idea that clustering of large values is the main effect of the dependence. We will see in the next section that the conditional probability in (5) plays a more precise role than has been indicated so far.

2. The Main Theorem.

__Theorem 1__. Let (X_n) satisfy $AIM(c_n)$ and let (p_n) be as in (2). If either (4) holds or

$$\lim\inf P[M_{1,p} \le c_n | X_1 > c_n] > 0 \ , \tag{7}$$

then

$$P[M_n \le c_n] - (F(c_n))^{nP[M_{1,p} \le c_n | X_1 > c_n]} \to 0 \ . \tag{8}$$

This theorem is proved in [8]; Rootzén [9] has independently obtained a similar result. We see from (8) that the asymptotic behaviour of $P[M_n \le c_n]$ is determined by that of $(F(c_n))^n$, as in (6), and that of $P[M_{1,p} \le c_n | X_1 > c_n]$. The connection between this last quantity and the clustering of large values will be discussed further in the examples at the end of this article.

The assumption that either (4) or (7) holds serves to exclude the delicate situation where the second term converges to 0^0. The more careful analysis needed to deal with this case has not yet been done. We see from Theorem 1 that the assumption (4) was unnecessary in the previously cited results of Watson, Loynes and Leadbetter, since (7) is implied by (5).

When (4) or (7) holds, the conclusion (8) is equivalent to

$$P[M_n \le c_n] - \exp\{-nP[M_{1,p} \le c_n, X_1 > c_n]\} \to 0 . \tag{9}$$

This formula can also be proved under slightly different hypotheses; for example, it holds if (X_1) satisfies $\mathrm{AIM}(c_n)$ and $\liminf P[M_n \le c_n] > 0$.

Newell [6] was the first author to obtain results for $P[M_n \le c_n]$ which allow for clustering of large values; essentially, he obtained Theorem 1 in the m-dependent case. O'Brien [7] and Leadbetter [3] have obtained a variety of other special cases of Theorem 1. We refrain from listing these results but we do offer one new corollary of Theorem 1 as an example.

<u>Corollary</u>. Let (X_n) be φ-mixing and suppose that for some fixed m

$$P[X_i > c_n | X_1 > c_n] \to 0 \quad \text{for} \quad i > m . \tag{10}$$

Then (8) holds with p replaced by m.

If (X_n) is m-dependent, (10) reduces to $F(c_n) \to 1$. Thus, we get Newell's result. Actually, Newell also assumed (4), but in fact, the other hypotheses imply that either (4) or (7) holds.

3. Markov Sequences.

Let $X_n = f(J_n)$ where (J_n) is a stationary positive Harris Markov sequence on a separable measurable space S and f is a real-valued function on S. Then (X_n) has $AIM(c_n)$ for all sequences (c_n) for which $P(X_1 > c_n) \to 0$. If (Y_n) is aperiodic, (X_1) is even strongly mixing. Thus Theorem 1 and other results in this article apply to (X_1). Many examples considered in the extreme value literature are of this type. We give two such examples in Section 5.

The theory of extreme values carries over in large part to the case where the sequence (J_n) is replaced by a Markov sequence (K_n) which has the same transition function as (J_n) but a different initial distribution (and hence is not stationary). In fact, if we let $Y_n = f(K_n)$ and $R_n = \max(Y_1, Y_2, \ldots, Y_n)$, then

$$P[M_n \le c_n] - P[R_n \le c_n] \to 0 \tag{11}$$

for all sequences (c_n) for which

$$P[Y_k > c_n] \to 0 \ , \quad k = 1,2,\ldots \ .$$

The last restriction is essential since without it one might for example have $P[X_1 > c_n] = 0$ and $P[Y_1 > c_n] = 1$ for all n.

Rootzén [9] has studied the same class of Markov sequences by means of the regeneration theory of such sequences. His approach is more easily applied in some cases, notably for autoregressive sequences.

4. Phantom Distribution Functions.

Up to now, we have considered the long-run behaviour of $P[M_n \le c_n]$ for a single sequence (c_n). We now want to examine how information about $P[M_n \le c_n]$ for a suitable chosen sequence

(c_n) can provide considerable information about the asymptotic nature of $P[M_n \le e_n]$ for other sequences (e_n).

A distribution function (d.f.) G is called a phantom d.f. for (X_n) if

$$P[M_n \le e_n] - G^n(e_n) \to 0 \tag{12}$$

for every sequence (e_n). If such a G exists, then the distribution of M_n is rather like that of $\max(Y_1, Y_2, \ldots, Y_n)$ where (Y_n) is an i.i.d. sequence and $P[Y_1 \le y] = G(y)$. The following theorem from [8] shows that a phantom d.f. exists under quite general circumstances.

Theorem 2. Let (X_n) be a stationary sequence. Let $c_n = \inf\{x : P[M_n \le x] \ge e^{-1}\}$. Assume (X_n) has $AIM(c_{[tn]})$ for all $t > 0$ both as defined in Section 1 and with "$\le$" replaced thrice by "$<$". Then any d.f. G satisfying $G(x) = \exp\{n^{-1}\}$ if $c_n \le x < c_{n+1}$ is a phantom d.f. for (X_n).

The proof of this result depends mainly on formula (3) and on bounds obtained from the monotonicity of $P[M_n \le x]$ in both n and x. The theorem appears to give a means of constructing a phantom d.f. G, but it is not in general practicable to carry this out.

Suppose that for each $a \in (0,1)$ there is a sequence $(c_n(a))$ such that $(F(c_n(a)))^n \to a$ and $P[M_n \le c_n(a)] \to a^\tau$, for some fixed $\tau \ge 0$. Leadbetter, Lindgren and Rootzén [13] call the parameter τ the extremal index of (X_n). If $\tau > 0$, then

$$P[M_n \le e_n] - (G(e_n))^n \to 0$$

for all sequences (e_n), where $G(x) = (F(x))^\tau$. This is obvious if $e_n = c_n(a)$ for some a and easily extends to other sequences (e_n). Thus g is a phantom d.f. for (X_n). The existence of an extremal index τ for (X_n) and its value can often be established using

Theorem 1.

A phantom d.f. can also be constructed using Rootzén's aforementioned approach to functions of Markov sequences. In fact, Rootzén points out that there is a phantom d.f. of the form $G(x) = (P[M \le x])^{\tau}$, where M is a random variable representing the maximum value of X_n between the first two regeneration epochs and τ^{-1} is the expected time between these epochs.

We note that Theorem 2 gives a phantom d.f. which increases by jumps, whereas, in the two special cases just considered, G may well turn out not to be of this type. It is clear that phantom d.f.'s are not unique since only their right tails matter. More precisely, if G is a phantom d.f. for (X_n) and $x_0 = \sup\{x : G(x) < 1\}$ then a d.f. H is a phantom d.f. for (X_n) iff

$$(1 - H(x))(1 - G(x))^{-1} \to 1 \quad \text{as} \quad x \to x_0 - .$$

5. Examples.

Our first example, taken from [8], displays in a simple way that the long term behaviour of $P[M_n \le c_n]$ can be affected by the presence of clusters of large values and by the relative sizes of the values in the clusters. Let (Y_n) be an i.i.d. sequence with Y_1 uniform on $(-1,0)$. Let $X_n = \max(Y_n, b^{-1}Y_{n+1})$, where $0 < b \le 1$. Then (X_n) is stationary and 1-dependent. Incidentally, (X_n) is a function of the Markov chain $(J_n) = ((Y_n, Y_{n+1}))$. The d.f. of X_1 is $F(x) = (1+x)(1+bx)$ for $-1 \le x \le 0$. If Y_n is large, which is to say close to zero, then X_{n-1} and X_n are both large, so large values tend to come in pairs, although the two members of a pair may not be equal in value. If (c_n) is a sequence such that $(F(c_n))^n \to a \in (0,1)$ and (p_n) satisfies (2), then

$$|P[M_{1,p} \le c_n | X_1 > c_n] - P[X_2 \le c_n | X_1 > c_n]|$$
$$\le P[M_{2,p} > c_n | X_1 > c_n]$$

$$= P[M_{2,p} > c_n] < p\, P[X_1 > c_n] \to 0 .$$

Thus, (8) becomes

$$P[M_n \le c_n] - (F(c_n))^{nP[X_2 \le c_n | X_1 > c_n]} \to 0 .$$

It is easily calculated that

$$\lim P[X_2 \le c_n | X_1 > c_n] = \lim_{x \to 0-} P[X_2 \le x | X_1 > x] = (1+b)^{-1} .$$

Thus, (X_n) has extremal index $(1+b)^{-1}$ and phantom d.f. $F^{(1+b)^{-1}}$. The special case $b = 1$ of this example was given by Newell [6]. It is important to realize that the relative sizes of the values in the clusters of large observations are as important as the sizes of the clusters themselves. Also, the clusters need not consist of adjacent observations; to see this, consider defining $X_n = \max(Y_n, b^{-1} Y_{n+2})$. More complicated examples are also possible.

There are many examples for which one cannot use the extremal index to find a phantom d.f., since the extremal index may not exist or may equal zero. One example in [8] has the property that $(F(c_n))^n \to a \in (0,1)$ does not occur for any sequence (c_n) but still a uniform phantom d.f. exists. The following example is also taken from [8].

Let (X_n) be the stationary Markov chain with state space $\{1,2,...\}$ and transition matrix T given by $T_{i,i+1} = t_i$ and $T_{i,1} = 1 - t_i$, where $0 < t_i \le 1$ for each i and $\mu := 1 + t_1 + t_1 t_2 + ... < \infty$. Let H be the d.f. of the first return epoch R to 1, starting at 1. Then $E(R) = \mu$ and, as discussed previously, G is a phantom d.f. for (X_n), where $(G(x))^{\mu} = H(x)$. If $t_i = t \in (0,1)$ for all i, then $\mu = (1-t)^{-1}$ and $G(x) = (F(x))^{1-t}$. Now suppose $t_i \to 1$ as $i \to \infty$. If (c_n) is any sequence such that $(F(c_n))^n \to a \in (0,1)$ (such sequences exist for all $a \in (0,1)$), then

$$P[M_{1,p} \le c_n | X_1 > c_n] \le P[X_2 \le c_n | X_1 > c_n] \to 0$$

so (X_1) has extremal index 0. If the function $1 - H(x)$ is regularly varying with exponent $\gamma < -1$ at ∞, then G and F are related by $x(1 - (G(x))^{\mu})(1 - F(x))^{-1} \to -(\gamma+1)\mu$. This can be shown to hold if, for example, $t_i = (i+1)(i+3)^{-1}$, in which case $\gamma = -2$ and $\mu = 3$.

References.

[1] Bradley, R. (1985). This volume.

[2] Leadbetter, M.R. (1974). On extreme values in stationary sequences. Z. Wahrscheinlichkeitstheorie und verw. Gebiete 28, 289-303.

[3] Leadbetter, M.R. (1983). Extremes and local dependence in stationary sequences. Z. Wahrscheinlichkeitstheorie und verw. Gebiete 65, 291-306.

[4] Leadbetter, M.R., Lindgren, G., and Rootzén, H. (1983). Extremes and related properties of random sequences and processes. Springer Statistics Series. Berlin.

[5] Loynes, R.M. (1965). Extreme values in uniformly mixing stationary stochastic processes. Ann. Math. Statist. 36, 993-999.

[6] Newell, G.F. (1964). Asymptotic extremes for m-dependent random variables,. Ann. Math. Statist. 35, 1322-1325.

[7] O'Brien, G.L. (1974). The maximum term of uniformly mixing stationary processes. Z. Wahrscheinlichkeitstheorie und verw. Gebiete 30, 57-63.

[8] O'Brien, G.L. (1984). Extreme values for stationary and Markov sequences. To appear.

[9] Rootzén, H. (1985). Maxima and exceedances of stationary Markov chains. To appear.

[10] Watson, G.S. (1954). Extreme values in samples from m-dependent stationary stochastic processes. Ann. Math., Statist. 25, 798-800.

EXTREME VALUES AND RARE EVENTS OF NON-STATIONARY RANDOM SEQUENCES

Jürg Hüsler

1. Introduction

The aim of this paper is to summarize some of the recent results of extreme values in non-stationary sequences. The classical extreme value theory considers mainly the limit distributions of the maxima or minima for iid. sequences. This theory was extended consecutively by a number of papers (e.g. Watson [24], Loynes [18], Leadbetter [15]) to apply for a wide class of stationary sequences. The state of theory is well described in Leadbetter, Lindgren and Rootzen [17].

In many applications of the extreme value theory the random sequence cannot be considered to be stationary (e.g. Horowitz [10]). Therefore the classical extreme value theory has to be extended also to non-stationary random sequences. In the general case of random sequences it is reasonable to introduce some structural conditions; e.g. we use the requirement that makes the influence of each single random variable on the limit distribution of the extreme value arbitrarly small (cf. formula (2.1)). With such conditions, the analogous classical case of limit distributions of the maxima of independent, non-identically distributed random variables was discussed by Juncosa [13] and Meijzler [19] (see also §3.). Without such conditions, any distribution function can be the limit distribution of the maxima of a suitably chosen independent, non-identically distributed random sequence. A few papers dealt with the general case of nonstationary sequences, e.g. Galambos [8] and Turkman and Walker [23] giving results for rather general random sequences, Tiago de Oliveira [22] considering periodic disturbances, Deo [7] dealing with particular Gaussian sequences as also Horowitz [10], and Resnick and Neuts [20] treating a particular chain-dependent sequence.

Denote by $\{X_i, i\geq 1\}$ a general random sequence with marginal distribution functions $F_i(x) = P\{X_i \leq x\}$, $i\geq 1$. Let $M_n = \max(X_i, 1\leq i\leq n)$. Extreme value theory is mainly concerned with the question whether $(M_n - b_n)/a_n$ converges in distribution with suitable normalizing sequences $\{a_n, n\geq 1\}, \{b_n, n\geq 1\}$, $a_n>0$. Instead of the linear normalization we can consider any norming sequence $\{u_n, n\geq 1\}$ where $u_n = u_n(x)$. Since $P\{M_n \leq u_n\} = P\{X_i \leq u_n,\ i\leq n\}$ we may use a yet more general norming sequence $\{u_{ni},\ i\leq n,\ n\geq 1\}$ which is called a "boundary" for the extreme values. Hence we deal with the behaviour of $P\{X_i \leq u_{ni},\ i\leq n\}$ as n tends to ∞. One of the main questions is concerned with the relation to the independent case i.e. whether

$$P\{X_i \leq u_{ni},\ i\leq n\} = \prod_{i=1}^{n} F_i(u_{ni}) + o(1) \quad \text{as } n\to\infty \tag{1.1}$$

Related to the extreme values are the number of exceedances $N_n = \{X_i > u_{ni},\ i\leq n\}$ and the point process Y_n generated by the exceedances. We may consider the convergence of these sequences N_n and Y_n and obtain further distributional results on the maxima as well as on the r-th largest values as consequences.

Section 2 contains the general asymptotic results on the exceedances under certain mixing type conditions which extend the results of Leadbetter [15], [16] to nonstationary random sequences. In Section 3 we deal with the consequences of the general results for the limit distributions of the maxima extending the mentioned work of Juncosa and Meijzler. Section 4 deals with a general approach of rare events which is appropriate to discuss the joint limit behaviour of minima and maxima or of the exceedances above a certain number of different boundaries. Finally, we discuss the important case of nonstationary Gaussian sequences in the last section.

2. Number of exceedances

In this section we deal with the number N_n of exceedances of the boundary $\{u_{ni}\}$ by the random sequence $\{X_i\}$. In general, let

$$N_n(I) = \sum_{i=1}^{n} 1\{X_i > u_{ni},\ i \in I\}$$

be the number of exceedances in the subset I of $\{1,...,n\}$. As mentioned in the introduction we use an uniformity assumption for the boundary. Define $x_{0,i} = \sup\{x: F_i(x)<1\}$ and let $F_i(x_{0,i}-) = 1$ for all i. We suppose always that

$$\bar{F}_{max} = \max\{\bar{F}_i(u_{ni}), i \leq n\} \to 0 \quad \text{as } n\to\infty \tag{2.1}$$

where $\bar{F}_i(x) = 1 - F_i(x)$.

One of the most general results in the stationary case was provided by Leadbetter [15] by weakening the strong mixing condition used in Loynes [18] to a distributional mixing condition $D = D(u_n)$. This condition together with a local condition, called $D' = D'(u_n)$, was sufficient to prove

$$P\{M_n \leq u_n\} - (F(u_n))^n \to 0 \quad \text{as } n\to\infty$$

where u_n is a suitable normalization (e.g. $n\bar{F}(u_n) \to \tau$ as $n\to\infty$). A similar approach was used by Hüsler [11] to solve the convergence problem (1.1) in the nonstationary case. The considered random sequences were slightly more restricted than in this paper, since we assumed identical marginal distributions. The sufficient conditions for the limit result in the general case are the following.

<u>Condition A</u>: Let $F_n^* = F_n^*(\{u_{ni}\}) = \sum_{i=1}^{n} \bar{F}_i(u_{ni})$. Then assume (2.1) and

$$\limsup_{n\to\infty} F_n^* < \infty \quad \text{and} \quad \liminf_{n\to\infty} F_n^* > 0 \ . \tag{2.2}$$

Condition A restricts the possible boundaries such that the mean number of exceedances F_n^* remains finite. It excludes also the non-interesting case $F_n^* \to 0$. For, if $F_n^* \to 0$, we find easily without further conditions that $P\{X_i \leq u_{ni},\ i\leq n\} \to 1$ and $\Pi_{i\leq n} F_i(u_{ni}) \to 1$; hence (1.1) follows immediately. We state now the distributional mixing condition D and the local condition D'.

<u>Condition $D(\{u_{ni}\})$:</u> For any integer $1\leq i_1<i_2<\ldots<i_p<j_1<\ldots<$ $<j_q\leq n$ for which $j_1-i_p\geq m$, let $I = \{i_\ell,\ \ell=1,\ldots,p\}$, $J = \{j_\ell,\ \ell=1,\ldots,q\}$, $B(I) = \{X_i \leq u_{ni},\ i\in I\}$ and similarly $B(J)$ and $B(I\cup J)$. Then we assume that

$$\sup_{I,J} |P(B(I\cup J)) - P(B(I))P(B(J))| \leq \alpha_{n,m}$$

where $\alpha_{n,m_n^*} \to 0$ as $n\to\infty$ for some sequence m_n^* such that

$$m_n^*\, \bar{F}_{max} \to 0 \quad \text{as } n\to\infty. \tag{2.3}$$

<u>Condition $D'(\{u_{ni}\})$:</u> Let n,r be integers and $I=\{i:\ i_1\leq i\leq i_2\}$ a subinterval of $\{1,\ldots,n\}$ such that $\sum_{i\in I} \bar{F}_i(u_{ni}) \leq F_n^*/r$. Then assume that $\alpha_{n,r}^*$ and $g(r)$ exist such that

$$\max_{I}\ \min_{I^*\subset I} \sum\sum_{i<j\in I^*} P\{X_i>u_{ni},\ X_j>u_{nj}\} \leq \alpha_{n,r}^* \tag{2.4}$$

such that

$$\lim_{r\to\infty} \limsup_{n\to\infty}\ r\,\alpha_{n,r}^* = 0 \tag{2.5}$$

where the min in (2.4) is considered on subsets I^* of I with

$$\sum_{i\in I-I^*} \bar{F}_i(u_{ni}) \leq g(r)/r \qquad \text{for all } n\geq n_0(r)$$

and $g(r) \to 0$ as $r\to\infty$.

The advantage of introducing the associated subsets I^* with respect to I in (2.4) is that the inequality for the double-sum might not hold for I itself because of some "unimportant" (w.r.t. $\bar{F}_i$) points. This fact is very helpful in the Gaussian case. In the general case we get the following result (Hüsler [11],[12]).

Theorem 2.1. Let $\{X_i,\ i\geq 1\}$ be a random sequence and $\{u_{ni}\}$ a boundary. Assume that the conditions A, D and D' hold with respect to $\{u_{ni}\}$. Then

$$F_n^* \to \tau \quad \text{as } n\to\infty \tag{2.6}$$

is equivalent to the fact that N_n has asymptotically a Poisson distribution with parameter τ, $\tau > 0$.

Since $\{N_n=0\} = \{X_i \leq u_{ni},\ i \leq n\}$ we get immediately

Corollary 2.2. Assume the conditions of Theorem 2.1. Then i) for $\tau>0$

$$P\{X_i \leq u_{ni},\ i\leq n\} \to \exp(-\tau) \quad \text{as } n\to\infty \tag{2.7}$$

is equivalent to (2.6), and

ii) if $\sum_{i\in I_n} \bar{F}_i(u_{ni}) \to \tau'$ as $n\to\infty$ for a sequence $\{I_n\}$ of subintervals of $\{1,\dots,n\}$, then

$$P\{X_i \leq u_{ni},\ i\in I_n\} \to \exp(-\tau') \quad \text{as } n\to\infty$$

and the number $N_n(I_n)$ of exceedances in I_n has asymptotically a Poisson distribution with parameter τ'.

Remark: Obviously, (2.7) shows that under the stated conditions $P\{X_i \leq u_{ni},\ i\leq n\} - \prod_{i\leq n} F_i(u_{ni}) \to 0$ as $n\to\infty$. To prove only this statement, we do not need the convergence condition (2.6) (cf. Hüsler [11],[12]).

The second part of the corollary suggets that we can consider even the point process Y_n of exceedances on $(0,1]$. For each n Y_n consists of the points i/n whenever $X_i > u_{ni}$, $i\leq n$. By using a general convergence result of Kallenberg [14] we get for the simple point process

Theorem 2.3. Let $\{X_i, i\geq 1\}$ be a random sequence and $\{u_{ni}\}$ be a boundary. Assume that the conditions A, D and D' hold with respect to $\{u_{ni}\}$. If

$$\sum_{i\leq nt} \bar{F}_i(u_{ni}) \to \lambda(t) \quad \text{for all } 0<t\leq 1, \tag{2.8}$$

then Y_n converges in distribution to a Poisson process on $(0,1]$ with intensity measure defined by λ.

Dropping the assumption D' in the above results, makes possible a variety of clustering behaviour of exceedances. In the stationary case, the mean number of exceedances in a cluster of a certain length is asymptotically equal to the reciprocal of the extremal index, defined in Leadbetter et al. [17] and dealt with in detail by Leadbetter [16]. Since we drop also the stationarity of the random sequence, the clustering and the generalized extremal index

$$\theta = -(\log \lim_{n\to\infty} P\{X_i \leq u_{ni}, i\leq n\})/\tau$$

where $F_n^* \to \tau$, depends on the type of boundary, on τ and on the dependence structure of the random sequence. A detailed discussion in Hüsler [12] shows that general results are not possible without reasonable further requirements. We mention the following interesting result for stationary sequences, where we assume that the extremal index exists with respect to the boundary $u_{ni} = u_n(\tau_0)$ with $n\bar{F}_1(u_n(\tau_0))\to\tau_0>0$. Define the class of boundaries related to $u_n(\tau_0)$:

$$\mathcal{B}(u_n(\tau_0)) = \{\{u_{ni}\} \text{ such that (2.6) with } \tau,\ 0<\tau\leq\tau_0, \text{ and}$$
$$\text{either } \sum_{i\leq n} \bar{F}_1(u_{ni})1(u_{ni}^*\leq u_{ni}) \to 0 \quad \text{or}$$
$$\sum_{i\leq n} \bar{F}_1(u_{ni})1(u_{ni}^*> u_{ni}) \to 0 \quad \text{hold}$$
$$\text{where } u_{ni}^* = u_n(\tau) = u_{[n\tau/\tau_0]}(\tau_0)\} \quad .$$

Theorem 2.4. Let $\{X_i, i\geq 1\}$ be a stationary sequence such that the extremal index $\theta = \theta(u_n(\tau_0))$ exists and (2.6) holds for $u_n(\tau_0)$, $\tau_0 > 0$. Then $\theta(u_{ni}) = \theta$ and therefore $P\{X_i \leq u_{ni}, i\leq n\} \to \exp(-\theta\tau)$ as $n\to\infty$, for all $\{u_{ni}\} \in \mathcal{B}(u_n(\tau_0))$.

Thus the extremal index describes not only the behavior of the exceedances above a high level but also for a certain class of related not constant boundaries.

3. Limit distributions for the maxima

In this section we consider the particular case of the maxima M_n of a nonstationary sequence $\{X_i\}$. If a fixed level normalization $\{u_n, i\leq n\}$ exists such that (2.6) holds, then we get immediately from Corollary 2.2

Corollary 3.1. If the random sequence $\{X_i, i\geq 1\}$ satisfies the condition D, D' and (2.1) with respect to the boundary $\{u_n\}$, then (2.6) is equivalent to

$$P\{M_n \leq u_n\} \to \exp(-\tau) \quad \text{as } n\to\infty.$$

Instead of an arbitrary normalization u_n, we consider now the usual linear normalization $u_n(x) = a_n x + b_n$ with sequences $\{a_n\}$, $\{b_n\}$ $(a_n>0)$. Also in this case the limit results are immediate consequences of Corollary 2.2. But in this particular case, the class of possible limit distributions is characterized by the following condition (3.2), given in a paper of Meijzler [19] (see also Galambos [8]).

Assume that for all t, $0<t\leq 1$, and x

$$\lim_{n\to\infty} \sum_{i\leq nt} \bar{F}_i(a_n x + b_n) = w(t,x) \quad \text{and} \quad \lim_{n\to\infty} \bar{F}_{max} = 0 \qquad (3.1)$$

Then for an independent random sequence $\{X_i^*\}$ with the same marginal distributions F_i,

$$(M_n - b_n)/a_n \overset{d}{\to} H(\cdot) \quad \text{as } n\to\infty,$$

with H a non-degenerate distribution where either

i) $\log H(x)$ is concave or (3.2)

ii) $x_0(H) < \infty$ and $\log H(x_0(H)-e^{-x})$ is concave or

iii) $x_1(H)$ is finite and $\log H(x_1(H)+e^{x})$ is concave

with $x_1(H)=\inf\{x:H(x)>0\}$, $x_0(H)=\sup\{x:H(x)<1\}$.

Obviously $H(x) = \exp(-w(1,x))$ and

Corollary 3.2. If the random sequence $\{X_i,\ i\geq 1\}$ satisfies the conditions D, D' and (3.1) with respect to a constant boundary $u_n(x) = a_n x + b_n$, for all x, then $(M_n - b_n)/a_n \xrightarrow{d} H(\cdot)$ as $n\to\infty$, where $H(\cdot)$ satisfies (3.2).

Hence under these conditions, M_n converges to the same limit law as the maxima of an associated independent random sequence with the same marginal distributions. Again, this relation is violated if we drop the condition D'. The class of chain-dependent sequences defined in a paper of Denzel and O'Brien [5] suggests that the properties of the extremal index of stationary sequences can be verified also in the nonstationary case. These chain-dependent sequences are in general not stationary, depending on the initial distribution of the associated ergodic Markov chain. But they satisfy the following condition, which defines a class of almost stationary random sequences with respect to the extreme values.

We assume that for all t, $0<t\leq 1$, $\Delta>0$, $t+\Delta\leq 1$,

$$P\{M_{(nt,n(t+\Delta)]} \leq u_n\} - P\{M_{n\Delta} \leq u_n\} \to 0 \quad \text{as } n\to\infty \tag{3.3}$$

where $M_{(k,m]} = \max\{X_i,\ k<i\leq m\}$ and $u_n = u_n(x) = a_n x + b_n$.

Theorem 3.3. Let $\{X_i,\ i\geq 1\}$ be a random sequence satisfying the condition D, (3.1) and (3.3) for all $u_n(x) = a_n x + b_n$, $x\in R$. If for all x and t

$$w(t,x) = tw(1,x) \tag{3.4}$$

and $w(1,x)$ is continous in x, then there exist θ and θ' with $0<\theta\leq\theta'\leq 1$ such that for all x

$$\limsup_{n\to\infty} P\{M_n \leq u_n(x)\} = \exp(-\theta w(1,x)) = H^{\theta}(x)$$

$$\liminf_{n\to\infty} P\{M_n \leq u_n(x)\} = \exp(-\theta' w(1,x)) = H^{\theta'}(x).$$

This result extends Theorem 3.7.1 in Leadbetter et al. [17] to the class of almost stationary sequences.

4. Rare events

The point process Y_n related to the extreme values considers the instants at which the random sequence takes on values from the set (u_n,∞). In dealing jointly also with the smallest values of the random sequences, we have to introduce another set $(-\infty,-v_n)$. Both sets are shrinking and define rare events of the random sequence. In the other mentioned situation of exceedances of multiple levels, we have to consider K sets $(u_n^{(j)},\infty)$, $j=1,\ldots,K$. Both cases may be discussed by the following general approach for K rare events. For stationary Gaussian sequences, Zähle [25] proved a general point process result for one rare event. For general random sequences the same approach is used also (e.g. Berman [3]); for a discussion related to extreme values and further references see Galambos [9]. The following is related to a result of Adler [1], where we restrict to a finite number of rare events but include also a possible dependence among the rare events.

We deal jointly with the K Borel sets $A_{i,n,k} \subset \bar{R}$, $i\le n$, $k\le K$ which may depend also on i as in Section 2. We assume

$$P_{max} = \max\{P\{X_i \in A_{i,n,k}\},\ i\le n,\ k\le K\} \to 0 \quad \text{as } n\to\infty \qquad (4.1)$$

and that the K sets are disjoint for each fixed i and n. Similar to the condition in Section 2 we assume that

$$\sum_{i\le nt} P\{X_i \in A_{i,n,k}\} \to \lambda_k(t) \qquad \text{as } n\to\infty \qquad (4.2)$$

for all t, $0<t\le 1$, and $k\le K$. Let $\lambda_k(s,t) = \lambda_k(t) - \lambda_k(s)$.

We use again a distributional mixing condition, called $D^* = D^*(\{A_{i,n,k}\})$, depending on the sequence of rare events. Thus we define the sequence $\{\xi_{i,n},\ i\le n,\ n\ge 1\}$ where $\xi_{i,n} = \sum_{k\le K} k1\{X_i \in A_{i,n,k}\}$. Since the sequences $\{X_i\}$ and $\{A_{i,n,k}\}$ are used only for the definition of the triangular array $\{\xi_{i,n}\}$, the following result can be interpreted also as a result for general triangular arrays $\{\xi_{i,n}\}$. We say that the sequence $\{X_i\}$ satisfies <u>Condition D*</u>, if for each n, m and each choice of integers $1 \leq i_1 < i_2 < \ldots < i_p < j_1 < \ldots < j_q \leq n$, $j_1 - i_p \geq m$, and each choice of non-empty subsets B_ℓ, C_ℓ of $\{1,\ldots,K\}$

$$|P\{\xi_{i_\ell} \notin B_\ell, \ell \leq p, \xi_{j_\nu} \notin C_\nu, \nu \leq q\} - P\{\xi_{i_\ell} \notin B_\ell, \ell \leq p\} P\{\xi_{j_\ell} \notin C_\ell, \ell \leq q\}|$$
$$\leq \alpha_{n,m}$$

where $\alpha_{n,m_n^*} \to 0$ as $n \to \infty$ for some sequence m_n^* for which $m_n^* P_{n\,max} \to 0$ and $\xi_i = \xi_{i,n}$. Assume $\alpha_{n,m}$ to be non-increasing in m.

Obviously, D* is a distributional mixing type condition since $\{\xi_i \notin B\} = \{X_i \notin \cup_{k \in B} A_{i,n,k}\}$. Davis [4] has shown that asymptotically the minima and maxima of a stationary random sequence are in general not independent. In the following general result we use a local dependence condition which admits such a dependence. But we restrict the dependence of the rare events that they are at most pairwise dependent (cf. (4.8)) and we exclude the asymptotic clustering of any rare event (cf. (4.9)). Obviously, we might use the following method also for more complicated dependency, by extending the Bonferroni inequalities.

In the following we denote $\mathcal{K} = \{1,\ldots,K\}$. Define for given n, $I \subset \{1,\ldots,n\}$ and $B \subset \mathcal{K}$

$$S_1(I,B) = \sum_{i \in I} P\{\xi_{i,n} \in B\}$$

$$S_2(I,B) = \sum\sum_{i<j \in I} P\{\xi_{i,n} \in B, \xi_{j,n} \in B\} \quad \text{and}$$

$$S_3(I,B) = \sum\sum\sum_{i<j<\ell \in I} P\{\xi_{i,n} \in B, \xi_{j,n} \in B, \xi_{\ell,n} \in B\}.$$

In addition we denote $\lambda(s,t,B) = \sum_{k \in B} \lambda_k(s,t)$ for any $B \subset \mathcal{K}$.

In Section 2 we introduced the subsets I* in the local condition D' to deal with certain "unimportant" points. As consequences, all timepoints $i (\leq n)$ were deleted if $i \in I - I^*$. In the more general situation of several rare events we cannot delete a timepoint i, because i does not need to be "unimportant" for all rare events. Hence we introduce a new triangular array $\{\xi^*_{i,n}, i \leq n, n \geq 1\}$ of random variables with values in $\{0,1,\ldots,K\}$, which are related to the $\xi_{i,n}$'s. The relation is motivated by the fact that if i is an "unimportant" point for the k-th rare event, we might define $\xi^*_{i,n}=0$ if $\xi_{i,n}=k$, and conversely, if i is an "important" point for the k-th rare event, then $\xi^*_{i,n} = \xi_{i,n} = k$ if this occurs. Hence we assume that for all $i \leq n$, $n \geq 1$, $k>0$

$$\{\xi^*_{i,n} = k\} \subset \{\xi_{i,n} = k\} . \tag{4.3}$$

If we replace in the above defined sums $S_j(I,B)$, $j=1,2,3$, the $\xi_{i,n}$'s by the $\xi^*_{i,n}$'s, we indicate it by a "*", e.g.

$$S_1^*(I,B) = \sum_{i\in I} P\{\xi^*_{i,n} \in B\}.$$

<u>Condition $D'(\{A_{i,n,k}\})$:</u> Let $s,t \in (0,1]$, $s<t$ be arbitrary. For fixed n and r let $I_\ell = I_{\ell,n,r}$, $\ell=1,\ldots,r$, be disjoint subintervals of $(ns,nt]$ such that

$$S_1(I_\ell,K) < S_1((ns,nt],K)/r \tag{4.4}$$

and

$$\limsup_{n\to\infty} \left| \sum_{\ell=1}^{r} S_1(I_\ell,K) - \lambda(s,t,K) \right| \to 0 \text{ as } r\to\infty. \tag{4.5}$$

Then assume that a sequence $\{\xi^*_{i,n},\ i\leq n,\ n\geq 1\}$ exists such that i) (4.3) holds and

$$\limsup_{n\to\infty} \left| \sum_{\ell=1}^{r} S_1^*(I_\ell,K) - \lambda(s,t,K) \right| \to 0 \text{ as } r\to\infty; \tag{4.6}$$

ii) if $K'=\{k: \lambda_k(s,t)>0\}\neq\emptyset$, assume that for each $\{k,k'\}\subset K'$ $h(s,t,\{k,k'\})$ exists such that

$$\begin{aligned} &\limsup_{n\to\infty} \sum_{\ell=1}^{r} S_2^*(I_\ell,\{k,k'\}) \to h(s,t,\{k,k'\}) \quad \text{as } r\to\infty, \\ &\liminf_{n\to\infty} \sum_{\ell=1}^{r} S_2^*(I_\ell,\{k,k'\}) \to h(s,t,\{k,k'\}) \quad \text{as } r\to\infty; \end{aligned} \tag{4.7}$$

iii) if $K'' = \{k\in K': h(s,t,\{k,k'\})\neq 0 \text{ for some } k'\in K'\}\neq\emptyset$, then

$$\sum_{\ell=1}^{r} \limsup_{n\to\infty} S_3^*(I_\ell,K'') \to 0 \quad \text{as } r\to\infty. \tag{4.8}$$

In addition, we assume that for all s,t and k

$$h(s,t,\{k\}) = 0. \tag{4.9}$$

Remarks: 1) If $\lambda(s,t,\{k\}) = 0$, define $h(s,t,\{k,k'\}) = 0$ for all $k' \in K$.

2) It is easily shown that the same $\xi_{i,n}$ and $\xi^*_{i,n}$ satisfy the conditions (4.5), (4.6) and (4.8) also for any $B \subset K$ respectively $B \subset K''$.

3) The function $h(s,t,\{k,k'\})$ does not depend on the selected $\xi^*_{i,n}$'s, which can be shown by straightforward application of the Bonferroni inequalities.

4) (4.9) excludes the possibility of clustering of a single rare event as in the condition D' of Section 2. The function h plays the role of mutual dependence of two rare events. It defines the intensity measure for the asymptotic joint occurence of the k-th rare event together with the k'-th rare event. Furthermore, we define for any $B \subset K$

$$h(s,t,B) = \sum_{k<k' \in B} h(s,t,\{k,k'\}).$$

By (4.9), $h(s,t,B)$ is the limit value of $\sum_\ell S_2^*(I_\ell,B)$.

5) By the assumptions (4.6) - (4.8) it is easily shown that the functions

$$\lambda^*(s,t,k) = \lambda_k(s,t) - \sum_{k' \neq k} h(s,t,\{k,k'\})$$

are nonnegative. These functions define the intensity measures for the asymptotic single occurrence of the k-th rare event.

6) It is obvious that the condition $D'(\{A_{i,n,k}\})$ with $K=1$ is equivalent in this particular case to the condition D' of Section 2, by defining $\xi^*_{i,n} = \xi_{i,n}$ if $i \in I \cap I^{*c}$, and $= 0$ if $i \in I - I^*$ and $A_{i,n,1} = (u_{ni},\infty]$.

7) It may be proved that the functions $h(\cdot,\cdot,\cdot)$ do not depend on the selected sequence of intervals I_ℓ, if Condition D* is satisfied.

Under these conditions we get the following limit result for the point process Y_n on $(0,1] \times R$ consisting of the points $(i/n,k)$ whenever $\xi_{i,n} = k$, $k>0$. The limiting point process Y on $(0,1] \times R$ can be constructed as follows. Let $\tilde{Y}$ be a Poisson point process on $(0,1] \times R^K$. The points of $\tilde{Y}$ occur only on

$(0,1]\times\varepsilon_k$ with intensity measure defined by $\lambda^*(s,t,k)$ and on $(0,1]\times(\varepsilon_k+\varepsilon_{k'})$ with intensity measure defined by $h(s,t,\{k,k'\})$ where the ε_k's are the unit vectors in R^K. Then Y has a point at (s,k) if $\tilde{Y}$ has a point at (s,ε_k) or $(s,\varepsilon_k+\varepsilon_{k'})$ for some k'.

Theorem 4.1. Let $\{X_i,\ i\geq 1\}$ be a random sequence and $\{A_{i,n,k},\ i\leq n,\ n\geq 1,\ k\leq K\}$ be sequence of Borel sets, disjoint for fixed i and n, $K<\infty$. Assume that (4.1), (4.2), Condition D* and D' hold. Then Y_n on $(0,1]\times R$ converges in distribution to the limit point process Y, defined above.

The proof follows the usual lines by using the block-method and a convergence result for simple point processes by Kallenberg [14]. Therefore we sketch it only.

Because of Theorem 4.7 of Kallenberg [14] and the assumption (4.2) it is sufficient to prove that for any disjoint intervals $(s_j,t_j]$ of $(0,1]$ and subsets D_j of R, $j=1,\ldots,J$, $J<\infty$

$$P\{Y_n(\cup_{j\leq J}(s_j,t_j]\times D_j)=0\}\rightarrow\exp\{-\sum_{j=1}^{J}(\lambda(s_j,t_j,B_j)-h(s_j,t_j,B_j))\}$$

where $B_j = D_j\cap\mathcal{K}$. We treat only the case $J=1$, since for $J>1$ the statement follows by the following separation method and the condition D*. Hence, we consider with $B=D\cap\mathcal{K}$

$$P\{Y_n((s,t]\times D)=0\}\rightarrow\exp(-\lambda(s,t,B)+h(s,t,B)). \qquad (4.10)$$

If $\lambda(s,t,B)=0$, then also $h(s,t,B)=0$ and (4.10) follows easily, since $P\{Y_n((s,t]\times D)=0\}\geq 1-S_1((ns,nt],B)$. Therefore we assume without loss of generality that $\lambda(s,t,B)>0$ and such that $\lambda(s,t,\{k\})>0$ for each $k\in B$. Define $B'=B\cap\mathcal{K}''$ and $B''=B\cap(\mathcal{K}'-\mathcal{K}'')$.

We construct r intervals I_ℓ, $\ell=1,\ldots,r$ for any fixed r. If n is sufficiently large, then $S_1((ns,nt],\mathcal{K})>0$. Hence for such n, let $I_\ell = (n_0+i_{\ell-1},\ n_0+i_\ell]$ such that $S_1(I_\ell,\mathcal{K})\leq S_1((ns,nt],\mathcal{K})/r$, where $n_0=[ns]$, $i_0=0$ and the i_ℓ's are consecutively selected such that $S_1(I_\ell\cup\{n_0+i_\ell+1\},\mathcal{K})>S_1((ns,nt],\mathcal{K})/r$.

Furthermore, split each I_ℓ into $I_{\ell,1}$ and $I_{\ell,2}=(n_0+i_\ell-m_\ell,\ n_0+i_\ell]$ such that $S_1(I_{\ell,2},K) \leq \varepsilon S_1((ns,nt],K)/r$ and $S_1(I_{\ell,2}\cup\{n_0+i_\ell-m_\ell\},K) > \varepsilon S_1((ns,nt],K)/r$. Because of this construction we get $(m_\ell+1)K\,P_{max} > \varepsilon S_1((ns,nt],K)/r$ for all $\ell\leq r$. Hence by setting $\varepsilon = \varepsilon(n,r) = r(m_n^*+1)K\,P_{max}/S_1((ns,nt],K)$, we have $m_\ell \geq m_n^*$ and $\varepsilon\to 0$ as $n\to\infty$ for any r, by Condition D* (cf. [11]).

Similar to Hüsler [12] the following statements are proved: i) $0 \leq P\{\xi_{i,n} \notin B,\ i\in \cup I_{\ell,1}\} - P\{Y_n((s,t]\times D)=0\} \to 0$ as $n\to\infty$ for any r.

ii) Because of D*, we have for any r

$$|P\{\xi_{i,n}\notin B,\ i\in\cup I_{\ell,1}\} - \Pi_\ell P\{\xi_{i,n}\notin B,\ i\in I_{\ell,1}\}| \leq r\alpha_{n,m_n^*} \xrightarrow[n\to\infty]{} 0.$$

iii) $0 \leq \Pi_\ell P\{\xi_{i,n}\notin B,\ i\in I_{\ell,1}\} - \Pi_\ell P\{\xi_{i,n}\notin B,\ i\in I_\ell\}\to 0$, as $n\to\infty$, for all r.

The final step is to approximate $P\{\xi_{i,n}\notin B,\ i\in I_\ell\}$. Observe that by the construction $1-P\{\xi_{i,n}\notin B,\ i\in I_\ell\}\leq S_1((ns,nt],K)/r \to 0$ as $r\to\infty$, uniformly in $\ell\leq r$. Define $E_1 = \{\xi^*_{i,n}\notin B',\ i\in I_\ell\}$, $E_2 = \{\xi^*_{i,n}\notin B'',\ i\in I_\ell\}$. Hence by the property of the $\xi^*_{i,n}$'s, we have $E_1\cap E_2 \supset \{\xi_{i,n}\notin B,\ i\in I_\ell\}$ and thus $0\leq P(E_1\cap E_2) - P\{\xi_{i,n}\notin B,\ i\in I_\ell\} \leq P\{\exists i\in I_\ell:\ \xi_{i,n}\in B,\ \xi^*_{i,n}\notin B\} \leq$
$\leq S_1(I_\ell,B) - S_1^*(I_\ell,B)$.

Now $1 - P(E_1\cap E_2) = P(E_1^c) + P(E_2^c) - P(E_1^c\cap E_2^c)$. For each event we use the following Bonferroni inequalities:

$$P(E_1^c) \leq S_1^*(I_\ell,B') - S_2^*(I_\ell,B') + S_3^*(I_\ell,B')$$
$$P(E_1^c) \geq S_1^*(I_\ell,B') - S_2^*(I_\ell,B')$$
$$P(E_2^c) \leq S_1^*(I_\ell,B'')$$
$$P(E_2^c) \geq S_1^*(I_\ell,B'') - S_2^*(I_\ell,B'')$$
$$P(E_1^c\cap E_2^c) \leq \sum_{k'\in B'}\ \sum_{k''\in B''} S_2^*(I_\ell,\{k',k''\}).$$

Combining these approximations in the usual way, we find with the assumptions of Condition D' that

$$\limsup_{n\to\infty}\ (\liminf_{n\to\infty})\ \sum_{\ell=1}^{r}(1-P\{\xi_{i,n}\notin B, i\in I_\ell\})\to\lambda(s,t,B)-h(s,t,B)$$

as $r\to\infty$. Hence the statement (4.10) is proved.

5. Gaussian sequences

In this section we assume that $\{X_i, i\geq 1\}$ is a Gaussian sequence where $EX_i = \mu_i$, $VarX_i = \sigma_i^2$ and $r(i,j)$ denotes the correlation function. Without loss of generality we can assume that $\mu_i = 0$ and $\sigma_i = 1$ for all i, because the usual normalization transforms a boundary value u_{ni} into $u^*_{ni} = (u_{ni} - \mu_i)/\sigma_i$. Furthermore we assume in the following that the considered boundaries satisfy the conditions (2.1) and (2.2). Denote by $\delta(n) = \sup\{|r(i,j)|, |i-j|\geq n\}$ and let $\delta(1) < 1$. Then we proved in Hüsler [11] that the conditions D and D' of Section 2 hold with respect to any boundary $\{u_{ni}\}$ if Berman's condition

$$\delta(n)\log n \to 0 \quad \text{as } n\to\infty \tag{5.1}$$

is satisfied. Following Section 2, the number of exceedances of a Gaussian sequence has asymptotically a Poisson distribution for any boundary which satisfies (2.6). If in addition also (2.8) holds, i.e. if

$$\sum_{i\leq nt} \bar{\Phi}(u_{ni}) \to \lambda(t) \qquad \text{for any } t\in(0,1] \tag{5.2}$$

where Φ the unit normal law, then the point process Y_n of exceedances converges in distribution to a Poisson point process.

Concerning the rare events let us consider exceedances of the p levels $\{u_{ni}^{(j)}\}$, $j=1,\ldots,p$, by $\{X_i\}$ and of the q levels $\{v_{ni}^{(j)}\}$, $j=1,\ldots,q$, by $\{-X_i\}$. Assume that for fixed n and i

$$-v_{ni}^{(q)} < -v_{ni}^{(q-1)} < \ldots < -v_{ni}^{(1)} < u_{ni}^{(1)} < \ldots < u_{ni}^{(p)}$$

and that each boundary satisfies again (2.1) and (2.2). Then we associate K rare sets with these boundary values by defining

$$A_{i,n,k} = [-v_{ni}^{(q-k+2)}, -v_{ni}^{(q-k+1)}) \text{ for } 1\leq k\leq q \text{ with } v_{ni}^{(q+1)}=\infty$$

$A_{i,n,k} = (u_{ni}^{(k-q)}, u_{ni}^{(k-q+1)}]$ for $q<k\leq q+p=K$ with $u_{ni}^{(p+1)} = \infty$. Obviously, these sets are disjoint and we assume that (4.2) holds. Furthermore, (4.2) is equivalent to the assumption that (5.2) holds for each boundary $\{u_{ni}^{(j)}\}$ and $\{v_{ni}^{(j)}\}$, where the limit λ depends also on each boundary. Again we can show that Condition D* and D' with respect to $\{A_{i,n,k}\}$ are satisfied, if Berman's condition (5.1) holds.

Theorem 5.1. Let $\{X_i, i\geq 1\}$ be a Gaussian sequence where the correlation function satisfies (5.1). If the boundaries $\{u_{ni}^{(j)}\}$ and $\{v_{ni}^{(j)}\}$ satisfy (5.2), then the point process Y_n associated with the rare sets $\{A_{i,n,k}\}$ defined above, converges in distribution to the limit point process Y consisting of K independent Poisson processes.

The particular case p=q=1 but with general random sequences is treated in the thesis of Schüpbach [21]. A consequence of this result is that the number of exceedances above $\{u_{ni}^{(1)}\}$ and low level crossings related to $\{v_{ni}^{(1)}\}$ are even in nonstationary Gaussian sequences asymptotically independent, if (5.1) holds. This extends the results of Berman [2] for stationary Gaussian sequences, and of Deo [7] for particular nonstationary Gaussian sequences. If in addition $u_{ni}^{(1)} = v_{ni}^{(1)}$ for all $i\leq n$, then we can deduce the asymptotic distribution of the maxima of the absolute values of $\{X_i\}$. Hence our result extends also another statement of Deo [6] for general Gaussian sequences, even under the weaker condition (5.1). Finally we mention that Theorem 5.1 extends also a result in Leadbetter et al. [17] for the exceedances of multiple levels (e.g. q=0, p>1) for nonstationary Gaussian sequences, satisfying (5.1).

References

[1] Adler, R.J. (1978) Weak convergence results for extremal processes generated by dependent random variables. Ann. Probab. 6,660-667.

[2] Berman, S.M. (1971) Asymptotic independence of the number of high and low level crossings of stationary Gaussian sequences. Ann. Math. Statist. 42, 927-946.

[3] Berman, S.M. (1983) Sojourns of stationary processes in rare sets. Ann. Probab. 11, 847-866.

[4] Davis, R.A. (1983) Limit laws for upper and lower extremes from stationary mixing sequences. J. Multivariate Anal. 13, 273-286.

[5] Denzel, G.E. and O'Brien, G.L. (1975) Limit theorems for extreme values of chain-dependent processes. Ann. Probab. 3, 773-779.

[6] Deo, C.M. (1972) Some limit theorems for maxima of absolute values of Gaussian sequences. Sankhya Ser. A 34, 289-292.

[7] Deo, C.M. (1972) Some limit theorems for maxima of nonstationary Gaussian sequences. Ann. Statist. 1, 981-984.

[8] Galambos, J. (1972) On the distribution of the maximum of random variables. Ann. Math. Statist. 43, 516-521.

[9] Galambos, J. (1978) The asymptotic theory of extreme order statistics. Wiley, New York.

[10] Horowitz, J. (1980) Extreme values from a nonstationary stochastic process: An application to air quality analysis. Technometrics 22, 469-478.

[11] Hüsler, J. (1983) Asymptotic approximation of crossing probabilities of random sequences. Z. Wahrsch. verw. Gebiete 63, 257-270.

[12] Hüsler, J. (1986) Extreme values of non-stationary sequences. To be published in J. Appl. Probab.

[13] Juncosa, M.L. (1949) The asymptotic behaviour of the minimum in a sequence of random variables. Duke Math. J. 16, 609-618.

[14] Kallenberg, O. (1976) Random measures. Akademie-Verlag, Berlin, Academic Press, London-New York.

[15] Leadbetter, M.R. (1974) On extreme values in stationary sequences. Z. Wahrsch. verw. Gebiete 28, 289-303.

[16] Leadbetter, M.R. (1983) Extremes and local dependence in stationary sequences. Z. Wahrsch. verw. Gebiete 65, 291-306.

[17] Leadbetter, M.R., Lindgren, G. and Rootzen, H. (1983) Extremes and related properties of random sequences and processes. Springer, New York, Heidelberg, Berlin.

[18] Loynes, R.M. (1965) Extreme values in uniformly mixing stationary stochastic processes. Ann. Math. Statist. 36, 993-999.

[19] Meijzler, D.G. (1956) On the problem of limit distribution for the maximal term of a variational series. L'vov Politechn. Inst. Naucn. Zp. (Fiz.-Mat.) 38, 90-109 (in Russian).

[20] Resnick, S.I. and Neuts, M.F. (1970) Limit laws for maxima of a sequence of random variables defined on a Markov chain. Adv. Appl. Probab. 2, 323-343.

[21] Schüpbach, M. (1986) Limit laws for minima and maxima of nonstationary random sequences. Ph.D.thesis, University of Bern.

[22] Tiago de Oliveira, J. (1976) Asymptotic behaviour of maxima with periodic disturbances. Ann. Inst. Statist. Math. 28, 19-23.

[23] Turkman, K.F. and Walker, A.M. (1983) Limit laws for the maxima of a class of quasi-stationary sequences. J. Appl. Probab. 20, 814-821.

[24] Watson, G.S. (1954) Extreme values in samples from m-dependent stationary stochastic processes. Ann. Math. Statist. 25, 798-900.

[25] Zähle, U. (1980) A Poisson limit theorem for rare events of a Gaussian sequence. Theory Probab. Appl. 25, 90-104.

Jürg Hüsler, Department of Statistics, University of Bern
Sidlerstr. 5, CH-3012 Bern, Switzerland

STATIONARY SELF-SIMILAR EXTREMAL PROCESSES AND RANDOM SEMICONTINUOUS FUNCTIONS

WIM VERVAAT

The main topic of this survey is the characterization of all limiting processes of

$$M_n(t) \ := \ a_n \bigvee_{k \leq [nt]} \xi_k + b_n \quad (a_n, b_n \in \mathbb{R},\, a_n > 0),$$

where $(\xi_k)_{k=1}^{\infty}$ may be any stationary sequence of random variables. The limits are identified as the stationary self-similar extremal processes, and some of their properties are investigated.

For this it is necessary to find an intrinsic definition of "extremal process". The right way to do this involves random upper semicontinuous functions and opens unexpected connections to the theory of random closed sets.

1. Introduction

Let $(\xi_k)_{k=1}^{\infty}$ be a sequence of random variables, and set

$$M_n \ := \ \bigvee_{k=1}^{n} \xi_k$$

for the partial maxima. Research on the asymptotics of M_n, or rather $a_n M_n + b_n$ for suitably chosen constants $a_n > 0$, b_n, has a long tradition in probability theory. The resulting theory developed along the same lines as its counterpart for partial sums $\sum_{k=1}^{n} \xi_k$. As for the latter, most is known with the ξ_k independent and identically distributed. All limit laws for $a_n M_n + b_n$ were already obtained by Fisher & Tippett (1928). The domains of attraction (= the collection of distributions of ξ_k producing a specified limit law for $a_n M_n + b_n$) were characterized by Gnedenko (1943) and De Haan (1970). Functional limit theorems for

$$a_n M_{[nt]} \ + \ b_n$$

as random functions of t were obtained first by Dwass (1964) and Lamperti (1964). The limit processes were called "the" extremal processes and became a separate topic of research. Pickands (1971) was the first who viewed them as functionals of planar Poisson processes. This approach was perfected in many later papers as Mori & Oodaira (1976), Resnick (1975,1985), Adler (1978), Goldie (1983) and Deheuvels (1981, 1983), and also made convergence considerations shift from random functions to planar point processes.

As could be expected, a second line of research has developed in the last decades on asymptotics for M_n with the assumption of independent identically distributed ξ_k relaxed to specific forms of dependence, mostly including stationarity. Research has been very productive. Without hope of being complete we mention work by Berman (1964), Loynes (1965), Newell (1964), Adler (1978), the monograph Leadbetter, Lindgren & Rootzén (1983), and O'Brien (1984).

However, if one looks to the results with special interest in finding new limiting extremal processes, one gets a bit disappointed. With very few exceptions all limiting processes are the same as in the iid case (so the dependence structure is "weak" in the sense that it vanishes in the limit), or marginal limit results are obtained for M_n, whose functional analogues for $M_{[nt]}$ have random constant functions as limits (so there is too strong dependence in the limit), or the limits are random mixtures of all these processes. An interesting early exception is Mori (1977), who obtains essentially new limit processes for nth largest values ($n \geq 2$). Without stationarity it is not so hard to find other limiting extremal processes, cf. Weissman (1975b) for the case of independent ξ_k.

In O'Brien, Torfs & Vervaat (1985+) *all* limiting processes of $a_n M_{[nt]} + b_n$ for stationary $(\xi_k)_{k=1}^{\infty}$ are characterized. It is the aim of the present paper to review these and some related results. For this it is necessary to answer the question "What is an extremal process?", where we want an intrinsic characterization, not a list of limit processes popping out of special cases.

We start this review with presenting Lamperti's (1962) fundamental theorem on self-similarity, which applies to all stochastic processes, not just extremal processes. Then we give an intrinsic but rather naive definition of extremal process, and identify all limiting processes for partial maxima of stationary sequences as the stationary self-similar extremal processes.

Later on we return to the definition of extremal process and find a surprising connection with random semicontinuous functions and random closed sets. Then we are able to answer the question whether all stationary self-similar extremal processes arise from planar point processes, as do the classical limits. The answer is negative. Consequently, the point process approach, which was so successful and clarifying in the last decade, is not general enough for handling all limiting extremal processes arising from stationary sequences. A related surprising phenomenon is that these extremal processes may have continuous sample paths.

Finally we discuss some topological aspects of the theory of random upper semicontinuous functions and random closed sets. The main papers in this direction are Salinetti & Wets (1985+), Norberg (1984), Vervaat (1982) and Gerritse (1985).

2. Lamperti's self-similarity theorem

Basic to our work is a result of much wider applicability, due to Lamperti (1962) and streamlined by Weissman (1975a) and Vervaat (1981). It applies to *all* processes, not only extremal processes, and explains why self-similar processes abound in nature.

First some notations and definitions. Let Aff denote the collection of affine transformations of R:

$$a\cdot+b:\quad x\mapsto ax+b \quad \text{with } a,b\in\mathbb{R},\ a>0.$$

Then Aff is a (noncommutative) group under composition of functions, with identity id $:= 1\cdot+0$ and inverse $(\cdot-b)/a$. It becomes a topological group by declaring

$$\text{Aff}\ni a\cdot+b \ \mapsto\ (\log a,\, b)\in\mathbb{R}^2$$

a homeomorphism. We say that $\gamma_n = a_n\cdot+b_n$ diverges to infinity in Aff if $|\log a_n|+|b_n|\to\infty$. Important for us are the subgroups of Aff which are homeomorphic and isomorphic to R with addition. They all have the form $\{\gamma^t: t\in\mathbb{R}\}$, where with $\gamma =: a\cdot+b$, $c := b/(1-a)$ if $a\neq 1$,

$$\gamma^t x := \begin{cases} x+tb & \text{if } a=1,\\ a^t(x-c)+c & \text{if } a\neq 1.\end{cases}$$

Let us call further a stochastic process $(X(t))_{t\in T}$, $T\subset\mathbb{R}$, *measurable in law* if the mapping

$$\lambda:\quad T\ni t\mapsto \text{law } X(t),$$

which assigns to each $t\in T$ the *marginal* distribution of $X(t)$ (an element of the Polish space of probability measures on Bor R) is Lebesgue measurable. This is a very weak regularity assumption, which does not guarantee at all that X is a well-behaved process. For instance, if $(X(t))_{t\in\mathbb{R}}$ is a collection of iid random variables, then λ is constant, hence measurable, whereas no version of X exists with measurable sample paths. In most applications (including ours), λ is trivially measurable, since $X(t)$ is constant on intervals $(k-1,\,k]$, $k\in\mathbb{Z}$.

Finally, let us agree that $=_d$ and $\to_d$ without further specification mean equality and convergence of finite-dimensional distributions. Later on, we will use these symbols also in connection with the attribute "in S", where S is a metric space of functions to which the random functions under consideration belong wp1. Then $=_d$ and $\to_d$ apply to these processes as S-valued random variables.

Theorem 1 (Lamperti). *Let $(X(t))_{t\in T}$ with $T = (0,\infty)$ (not $[0,\infty)$) be a stochastic process, and let $\gamma_a\in$ Aff for $a\in T$. If*

(1) *X is measurable in law,*

(2) *γ_a diverges to infinity in* Aff *as $a\to\infty$,*

(3) *$\gamma_a X(a\cdot)\to_d Y$ for some stochastic process $Y = (Y(t))_{t\in T}$,*

(4) *$Y(1)$ is nondegenerate,*

then there is a $\beta\in$ Aff, $\beta\neq$ id, such that

(5) $\gamma_n \gamma_{na}^{-1} \to \beta^{\log a}$ *in* Aff $(a \in T)$,

(6) $Y(a\cdot) =_d \beta^{\log a} Y(\cdot) \quad (a \in T)$.

Definition 1. *A process Y that satisfies* (6) *is called* β*-self-similar* (β*-ss*). *If, more particularly,* $\beta x = e^H x$ $(H \in \mathsf{R})$, *in which case* (6) *becomes*

(7) $Y(a\cdot) =_d a^H Y(\cdot)$,

then Y is called H-ss.

Remark. It is very well possible that $H<0$, as we will see in the next sections. However, for processes Y with *stationary increments,* the most favorite ss process in the literature, this is impossible unless $Y \equiv 0$ (cf. Vervaat (1985), Th. 1.3). In Lamperti's original (1962) version of Theorem 1 $H<0$ did not occur, because he considered $T = [0,\infty)$ rather than $T = (0,\infty)$ and, moreover, required Y to be continuous in probability.

3. A first definition of extremal process

With the point process approach of the last decade it has become standard to regard extremal processes as random functions of time *intervals* rather than time itself. So we are given a collection of random variables $(M(I))_{I \in \mathfrak{I}}$, where $\mathfrak{I}$ denotes the family of all intervals in R. We may think of $M(I)$ as the supremum of values of a certain phenomenon observed during time interval I. In this interpretation the following is obvious. Let $\bigcup_{n=1}^{\infty} I_n$ also be an interval (we do not assume the I_n to be disjoint). Then

(8) $$M(\bigcup_{n=1}^{\infty} I_n) = \bigvee_{n=1}^{\infty} M(I_n).$$

For the time being we *define* an extremal process to be a collection of R-valued (occasionally $[-\infty,\infty]$-valued) random variables $(M(I))_{I \in \mathfrak{I}}$ such that (8) holds in all cases where $\bigcup_{n=1}^{\infty} I_n$ is an interval. It follows that $M(I)$ is an increasing function of I (wp1 for any countable collection of I's; for all I simultaneously wp1 after taking a monotone version of the process M). In fact we may regard M as a function $M(t,u]$ of two reals $t<u$, that is increasing in u, decreasing in t and right-continuous in both variables.

Now a simple definition suggests itself for convergence in distribution of extremal processes. It is a straightforward generalization of a similar notion for monotone processes on R (cf. Lamperti (1964)). We say that extremal processes M_n converge in distribution to M as $n \to \infty$, if for each finite sequence of bounded intervals $(I_k)_{k=1}^{m}$ such that

(9) $$M(\operatorname{int} I_k) = M(\operatorname{cl} I_k) \;\text{ wp1}$$

we have that

$$(M_n(I_k))_{k=1}^m \to_d (M(I_k))_{k=1}^m \text{ in } \mathbb{R}^m.$$

So convergence in distribution is in fact convergence of finite-dimensional distributions for bounded I, except that we must avoid considering I such that $M(\mathrm{int} I) < M(\mathrm{cl} I)$ with positive probability. The set of endpoints of such I is countable (= finite or countably infinite). As we will be interested mainly in *stationary* extremal processes, i.e., $M(\cdot+b) =_d M(\cdot)$, the set of endpoints of such I must be countable and translation invariant, hence empty. So we need not worry about restriction (9), when we are dealing with stationary processes.

4. Stationary self-similar extremal processes

The next sections through Section 9 review results of O'Brien, Torfs & Vervaat (1985+).

We now turn to the problem we promised to solve in the introduction. Let $(\xi_k)_{k=-\infty}^{\infty}$ be a stationary sequence of random variables in $\mathbb{R}$, i.e., $(\xi_{k+1})_{k=-\infty}^{\infty} =_d (\xi_k)_{k=-\infty}^{\infty}$. Set for intervals $I \subset \mathbb{R}$

$$(10) \qquad M_n(I) := a_n \bigvee\nolimits_{k \in nI} \xi_k + b_n.$$

Then M_n is an extremal process which is "concentrated" at the grid $n^{-1}\mathbb{Z}$, in the sense that $M_n(I) = M_n(J)$ if $I \cap n^{-1}\mathbb{Z} = J \cap n^{-1}\mathbb{Z}$. We want to characterize all extremal processes M that are limits in distribution of M_n as above for some underlying stationary sequence (ξ_k) and some choice of constants $a_n > 0$, b_n such that $|\log a_n| + |b_n| \to \infty$.

The first observation we can make is that M_n is stationary for shifts over multiples of $1/n$, so M is stationary for shifts over $1/n$ for all n. Consequently M is stationary, i.e., stationary for all real shifts. For the second observation, we consider (10) first for $I = (0,t]$, $t>0$. Setting

$$X(t) := \bigvee\nolimits_{k=1}^{\lfloor t \rfloor} \xi_k,$$

we find

$$M_n(0,t] = a_n X(nt) + b_n.$$

By Theorem 1 each limit M of (10) must be such that $M(0,\cdot]$ is self-similar. By stationarity $M(b, b+\cdot]$ is ss as well, so $M(\cdot)$, parametrized by the real intervals, is self-similar (which means that M satisfies (6) for some β). We have already found half of the following characterization theorem.

Theorem 2. *An extremal process M is a limit in distribution of* (10) *for a stationary sequence of random variables $(\xi_k)_{k=-\infty}^{\infty}$ with $a_n > 0$, $|\log a_n| + |b_n| \to \infty$ iff M is stationary and β-ss for some $\beta \in \mathrm{Aff}\backslash\{\mathrm{id}\}$, i.e.,*

$$M(a\cdot+b) =_d \beta^{\log a} M(\cdot) \quad \text{for } a,b \in \mathrm{R},\ a>0.$$

In this case we have, with $\gamma_n := a_n\cdot + b_n$,

$$\gamma_n \gamma_{[na]}^{-1} \to \beta^{\log a} \quad \textit{for } a>0.$$

It remains to show that each stationary ss extremal process indeed occurs as a limit in (10). This is easy: take $\xi_k := M(k-1,k]$, $a_n\cdot+b_n := \beta^{-\log n}$.

5. Reduction to the 1-ss case

Let M be an extremal process with values in $J \subset [-\infty,\infty]$. Let ϕ be an increasing left-continuous function on J. Then also $\phi(M)$ satisfies (8), so is an extremal process. This enables us to reduce all stationary ss extremal processes to one standard case by simple transformations. We need the following lemma, which is not hard to prove.

Lemma 1. *Let M be an* R*-valued stationary* β*-ss extremal process. If*

$$\beta x = e^H(x-c)+c \quad \text{with } H\neq 0,$$

then M takes values wp1 in $[c,\infty)$ *if* $H>0$, *in* $(-\infty,c]$ *if* $H<0$.

Now the following is obvious. Let M be an R-valued stationary β-ss extremal process. If

$$\beta x = x+b,$$

then

$$e^{M/b}$$

is a $(0,\infty)$-valued stationary 1-ss extremal process. If

$$\beta x = e^H(x-c)+c,$$

then

$$|M-c|^{1/H} \operatorname{sgn} H$$

is a $[0,\infty)$-valued (if $H>0$) or $(0,\infty]$-valued (if $H<0$) stationary 1-ss extremal process.

So by two simple transformations we can reduce all stationary ss R-valued extremal processes to stationary 1-ss $[0,\infty]$-valued extremal processes. Stationarity and 1-self-similarity can be condensed into the condition

(11) $\quad M(a\cdot+b) =_d aM(\cdot) \quad (a,b \in \mathsf{R},\ a>0).$

We now concentrate on $[0,\infty]$-valued extremal processes satisfying (11).

6. Poincaré processes

The classical extremal process (reduced to the 1-ss case) arising from stationary *independent* ξ_k, can be represented in the following way. Let Π be a Poisson process in the halfplane $\mathsf{R}\times(0,\infty)$ with intensity

(12) $\quad \mathsf{E}\Pi(dt,dx) = c\,dt\,x^{-2}dx \quad (c>0).$

By abuse of notation we write Π both for the random integer-valued measure on $\mathrm{Bor}(\mathsf{R}\times(0,\infty))$ and the random point set on which this measure is concentrated. So

$$\Pi(B) = \#(\Pi\cap B) \quad \text{for } B \in \mathrm{Bor}(\mathsf{R}\times(0,\infty)).$$

Set for intervals $I\subset\mathsf{R}$

(13) $\quad M(I) := \sup\{x:\ (t,x)\in\Pi,\ t\in I\} = \sup((I\times(0,\infty)) \cap \Pi).$

Note that $M(I)<\infty$ wp1 for bounded I, since

$$\mathsf{E}\Pi(I\times(1,\infty)) = c\,\mathrm{Leb}\,I < \infty.$$

We now want to generalize this example by considering M in (13) based on other point processes Π in $\mathsf{R}\times(0,\infty)$. Of course, M must satisfy (11), which can be seen to be equivalent to

(14) $\quad \Pi =_d \{(at+b,ax):\ (t,x)\in\Pi\} \quad (a,b \in \mathsf{R},\ a>0),$

i.e., the random set Π is invariant in distribution for horizontal shifts and multiplication by scalars with the origin as center. Such point processes are called *Poincaré* by O'Brien & Vervaat (1985). The intensity $\mathsf{E}\Pi$ of Poincaré processes Π must have the same invariance as in (14), which entails that $\mathsf{E}\Pi$ is given by (12), possibly with $c=\infty$. We refer to O'Brien & Vervaat (1985) for many examples of Poincaré processes (in the context of ss processes with stationary increments). The most curious one is the so-called g-adic lattice process, in which the knowledge of one point determines wp1 the location of all other points at the same or lower level.

However, for later use we now present one method of transforming Poincaré processes into new ones.

7. Subordinated Poincaré processes

Let Π_0 be any Poincaré process, for instance the Poisson process in the beginning of the previous section. Let $\varkappa$ be some fixed subset of $\mathbf{R}\times(0,\infty)$. If $\varkappa$ is finite, we can construct a new point process Π by regarding the points of Π_0 as centers of "clouds" $\varkappa$, with the clouds expanded proportionally to the heights of their centers. In formula:

$$(15)\qquad \Pi := \bigcup_{(t,x)\in\Pi_0}((t,0)+x\varkappa).$$

The resulting process is Poincaré, i.e., satisfies (14) as is easily checked. We call Π the Poincaré process subordinated to Π_0 by $\varkappa$. A variant with more randomness is subordination by iid random clouds $(\varkappa_n)_{n=1}^{\infty}$. The particular case $\mathbf{E}\#\varkappa_n = \infty$ opens the standard way of constructing Poincaré processes with infinite intensity.

So far, all examples of stationary 1-ss extremal processes were based on point processes, more particularly Poincaré processes. Is this the only way to produce them? In order to answer this question we must know better what kind of mathematical object extremal processes are.

8. Back to the definition of extremal process

In Section 3 we defined extremal processes as random functions on $\mathfrak{J}$, the intervals in $\mathbf{R}$, such that (8) holds. We now consider *deterministic* functions that satisfy the same or similar requirements. It is convenient to replace $\mathfrak{J}$ by a larger family of subsets that is closed under arbitrary unions, and for our case the obvious choice is the open sets in $\mathbf{R}$. At this stage it does not make any difference to replace $\mathbf{R}$ by some general topological space E, with collection of open sets $\mathcal{G}$. Furthermore, it is convenient to have a compact range of values for extremal processes, so we replace the range $\mathbf{R}$ by $\overline{\mathbf{R}} := [-\infty,\infty]$.

Definition 2. *Let E be a topological space and $\mathcal{G}$ its topology. A function $m: \mathcal{G}\to\overline{\mathbf{R}}$ is called a **sup measure** if*

$$(16)\qquad m(\bigcup_{j\in J}G_j) = \bigvee_{j\in J}m(G_j)\ ,\quad m(\varnothing)=-\infty$$

for all collections $(G_j)_{j\in J}$ in $\mathcal{G}$.

For functions $f: E\to\overline{\mathbf{R}}$ we can construct sup measures $f^\vee$ by

$$f^\vee(G) := \bigvee_{t\in G}f(t)\quad \text{for } G\in\mathcal{G}.$$

Naturally the question comes up whether all sup measures can be obtained this way. If so, then there is a need for selecting canonical f, as many f may have the same $f^\vee$ in common (for instance, for all D dense in E we have $1_D^\vee\equiv 1$ on $\mathcal{G}\setminus\{\varnothing\}$). A bit of reflection makes one guess that the following is a good candidate.

Definition 3. *The **sup derivative** of a set function $m: \mathcal{G} \to \bar{\mathsf{R}}$ is the function $d^\vee m: E \to \bar{\mathsf{R}}$ defined by*

$$d^\vee m(t) := \inf\{m(G): G \in \mathcal{G},\ G \ni t\}.$$

Before presenting the main result we recall some facts about semicontinuous functions.

Definition 4. *A function $f: E \to \bar{\mathsf{R}}$ is **upper semicontinuous** (usc) if $f^{\leftarrow}[-\infty, x)$ is open for all $x \in (-\infty, \infty]$. A function f is **lower semicontinuous** (lsc) if $-f$ is usc.*

Lemma 2. *$f^* := d^\vee(f^\vee)$ is the smallest usc function $\geqslant f: E \to \bar{\mathsf{R}}$; f is usc iff $f = f^*$.*

We now state analogous results for set functions $m: \mathcal{G} \to \bar{\mathsf{R}}$.

Lemma 3.
(a) A set function $m: \mathcal{G} \to \bar{\mathsf{R}}$ is a sup measure, i.e., (16) holds, iff $m = (d^\vee m)^\vee$.
(b) Let m be a sup measure and set $D^\vee m := \{f: E \to \bar{\mathsf{R}}: f^\vee = m\}$. Then
(i) $d^\vee m \in D^\vee m$,
(ii) $d^\vee m = \sup D^\vee m$,
(iii) $f \in D^\vee m$ and f is usc iff $f = d^\vee m$.

Corollary. *(a) $(d^\vee m)^\vee$ is the largest sup measure $\leqslant m: \mathcal{G} \to \bar{\mathsf{R}}$.*
(b) The mappings $m \mapsto d^\vee m$ with inverse $f \mapsto f^\vee$ are bijections between SM, the sup measures on $\mathcal{G}$, and US, the usc functions on E. These bijections preserve infima and suprema.

We now want to define formally extremal processes.

Definition 5. *An extremal process is a random sup measure, i.e., an SM-valued random variable.*

However, for a valid definition something is missing. First we must make SM a measurable space, preferably in relation to a topology on SM. We will do this only in Section 10. Before this we continue to investigate extremal processes in a "semi-naive" way, exploiting the just found relationship with random usc functions, but neglecting all measurability problems.

9. Random sup derivatives

Let M be an extremal process and $X := d^{\vee}M$ its sup derivative. Then X is a random usc function on $\mathbf{R}$. Clearly M is $[0,\infty]$-valued, stationary and 1-ss, i.e., (11) holds, iff X is, i.e.,

(17) $\quad X(a\cdot+b) =_d a\,X(\cdot) \quad \text{for } a,b \in \mathbf{R},\ a>0$

with values in $[0,\infty]$.

It is instructive to see what X is in the examples of Section 6. If Π is a Poincaré process, for instance the Poisson process in the beginning of Section 6 that generates the classical limit extremal process, then

$$X(t) = 0 \vee \sup\{x\colon (t,x)\in\Pi\}.$$

So, if no vertical contains more than one point of Π, which is the case for Π Poisson, then Π is the part of the graph of X outside the horizontal axis.

We want to solve the problem whether all stationary 1-ss random usc functions X arise this way from Poincaré point processes Π. If one studies related general properties of random usc X, one finds the following result.

Theorem 3. *Let X be a $[0,\infty]$-valued stationary 1-ss random usc function. Then wp1 one of the following holds:*
(a) $X \equiv \infty$;
(b) $X \equiv 0$;
(c) $\{t\colon X(t)>0\}$ is a dense F_σ-set in $\mathbf{R}$ with Lebesgue measure 0.

So in all interesting cases the set $\{t\colon X(t)>0\}$ is thin and dense, but not necessarily countable.

We now construct an example in which this set is uncountable. Consequently, the resulting X is not generated by a point process. We use the subordination recipe of Section 7, starting with a primary Poincaré point process Π_0 of finite intensity, but this time with uncountable clouds as, for instance, the Cantor set

$$C := \left\{\textstyle\sum_{n=1}^{\infty}\varepsilon_n 3^{-n}\colon\ \varepsilon_n = 0 \text{ or } 2\right\}$$

at level 1: $\varkappa = C\times\{1\}$. Then (15) defines a random uncountable Poincaré subset Π of $\mathbf{R}\times(0,\infty)$, which still generates a stationary 1-ss extremal process M by (13). However, there may be a complication. The clouds may become so dense in $\mathbf{R}\times(0,\infty)$ that $M\equiv\infty$, thus producing a trivial extremal process in a nontrivial way. The complication is real. We obtain $M\equiv\infty$ for instance, if C is replaced by a full line segment. The following theorem gives a full characterization of $M\not\equiv\infty$ for linear horizontal clouds.

Theorem 4. *Let Π_0 be a Poincaré point process of finite intensity and let Π be defined by* (15) *with $\varkappa = C\times\{1\}$ for some $C\subset\mathbb{R}$. Then Π does not change if C is replaced by its closure. Moreover, M is finite on finite intervals wp1, if (with C closed) C is bounded and the lengths L_n of the disjoint open intervals whose union is $[\inf C, \sup C]\backslash C$ satisfy*

$$\sum_{n=1}^{\infty} L_n \,|\log L_n| < \infty.$$

Conversely, if Π_0 is such that $\Pi_0(B) = \infty$ wp1 if $\mathsf{E}\Pi_0(B) = \infty$ for $B \in \mathrm{Bor}(\mathbb{R}\times(0,\infty))$, then the conditions on C are necessary for M being finite wp1 on finite intervals.

Obviously, the Cantor set satisfies the conditions on C in the theorem. The Poisson process satisfies the condition on Π_0 in the second part of the theorem, but not all Poincaré processes do (cf. O'Brien & Vervaat (1985), Ex. 3.5.(f)). A related theorem for ss processes with stationary increments has been obtained by Vervaat (1985, Th. 4.5).

So, with Poisson Π_0 and Cantor C in Theorem 4 we have found our first example of a stationary 1-ss locally bounded extremal process M with uncountable $\{t: d^{\vee}M(t)>0\}$. By iterated subordination also other interesting examples can be constructed. Let

$$\varkappa := \{(0,1)\} \cup \{(-2^{-n}, 1-2^{-n}): n=1,2,\ldots\}.$$

Subordinate Π_0 by $\varkappa$ to Π_1, Π_1 by $\varkappa$ to Π_2, etc. Let $\Pi_\infty := \bigcup_{n=0}^{\infty}\Pi_n$. Then Π_∞ is Poincaré and generates a locally bounded stationary 1-ss extremal process M such that $t \mapsto M(0,t]$ is continuous wp1.

10. The sup vague topology

The remainder of this survey is taken from Vervaat (1982). See also Norberg (1984) and Salinetti & Wets (1985+).

As observed in the corollary of Lemma 3 in Section 8, there is a natural bijection between SM, the space of sup measures on $\mathcal{G}$ (the topology of E), and US, the usc functions on E. So, with topologizing one space, we topologize the other as well.

We do not assume E to be Hausdorff. Reasons for this generality are given with the example at the end of Section 11 and in the third last paragraph of Section 12. Quasicompactness means the finite open subcover property. Compactness means quasicompactness together with the Hausdorff property. Let $\mathcal{K}$ denote the quasicompact sets in E. We want to extend the domain $\mathcal{G}$ of sup measures so that it contains $\mathcal{K}$. In fact we can define $m(A)$ for all $A \subset E$ by

$$m(A) := \bigwedge_{G\supset A} m(G) = \bigvee_{t\in A} d^{\vee}m(t).$$

Definition 6. *The **sup vague topology** on SM is the smallest topology for which the evaluations*

$$SM \ni m \mapsto m(A) \in \overline{\mathsf{R}}$$

are usc for $A \in \mathcal{K}$, lsc for $A \in \mathcal{G}$.

Corollary. *A sequence or net (m_n) converges sup vaguely to m in SM iff*

$$\limsup m_n(K) \leqslant m(K) \quad \text{for } K \in \mathcal{K},$$
$$\liminf m_n(G) \geqslant m(G) \quad \text{for } G \in \mathcal{G}.$$

In the latter characterization we recognize the porte-manteau theorem for vague convergence of (σ-additive) Radon measures on a locally compact space. This explains the term "sup vague".

By identifying sup measures with their sup derivatives we carry over the sup vague topology from *SM* to *US*. So a sequence or net (f_n) of usc functions converges sup vaguely to an usc function f if

$$\limsup f_n^\vee(K) \leqslant f^\vee(K) \quad \text{for } K \in \mathcal{K},$$
$$\liminf f_n^\vee(G) \geqslant f^\vee(G) \quad \text{for } G \in \mathcal{G}.$$

The classes $\mathcal{K}$ and $\mathcal{G}$ in the above criterion can be thinned out to much smaller subclasses. For instance, if E is locally compact, metric and separable, then $\mathcal{K}$ and $\mathcal{G}$ may be replaced by the compact and open balls, more particularly the compact and open intervals for $E = \mathsf{R}$.

The spaces *SM* and *US* with the sup vague topology have very nice properties.

Theorem 5.
(a) SM and US are quasicompact.
(b) If E is locally quasicompact, then SM and US are Hausdorff, hence compact.
(c) If E is locally quasicompact with countable base, then SM and US are compact and metric.

Caution. E is locally quasicompact if each point in E has a neighborhood base of quasicompact sets. For non-Hausdorff E this is *not* implied by the existence of at least one quasicompact neighborhood for each point. In particular quasicompact E need not be locally quasicompact.

11. Convergence in distribution

We now restrict our attention to locally quasicompact E with countable base, in which case Theorem 5 tells us that *SM* and *US* are compact and metric. A mapping

$$M\colon\ (\Omega, \mathcal{A}) \to (SM, \mathrm{Bor}\, SM)$$

turns out to be measurable iff $\omega \mapsto M(G, \omega)$ is measurable for each $G \in \mathcal{G}$, or weaker, for each $G \in \mathcal{G}_0$, $\mathcal{G}_0$ a base of $\mathcal{G}$. Consequently, M has the right measurability for an extremal process as

SM-valued random variable iff $M(G)$ is an $\bar{\mathbb{R}}$-valued random variable for each $G \in \mathcal{G}_0$, $\mathcal{G}_0$ a base of $\mathcal{G}$.

Since SM is compact, convergence in distribution of extremal processes can be characterized rather easily, without complicated tightness conditions. We state the result for Hausdorff E (so E is metrizable, being locally compact with countable base).

Theorem 6. *Let M be an extremal process and $\mathcal{B}(M)$ the collection of compact balls B in E such that $M(\mathrm{int}\,B) = M(B)$ wp1. Then $M_n \to_d M$ in SM iff*

$$(M_n(B))_{B \in \mathcal{B}(M)} \to_d (M(B))_{B \in \mathcal{B}(M)},$$

where $\to_d$ means convergence of finite-dimensional distributions.

For $E = \mathbb{R}$, $\mathcal{B}(M)$ becomes the collection of compact intervals I such that $M(\mathrm{int}I) = M(I)$ wp1 (cf. our naive approach at the end of Section 3).

Everything translates immediately to random usc functions by identifying extremal processes with their sup derivatives: the random usc functions. Although the latter are random functions on E, their measurability and probability distributions are connected strongly with the corresponding random set functions on $\mathcal{G}$. For an instructive example, consider the random usc functions $X \equiv 0$ and $Y(t) = 1_{\{U\}}(t)$, with U uniformly distributed in $(0,1)$. Then X and Y as classical random functions of t have the same finite-dimensional distributions. On the other hand, $X^{\vee}(0,1) = 0$ wp1, $Y^{\vee}(0,1) = 1$ wp1, so X and Y have different distributions as random usc functions.

We did not assume E to be Hausdorff. The reader who finds this annoyingly esoteric should consider the following example, which shows that the generality is necessary for our formulation to cover weak convergence in the classical setting.

Example. $E = \mathbb{R}\downarrow$, the real line with the *lower* topology, whose nontrivial open sets are $(-\infty,x)$ for $x \in \mathbb{R}$. Then E is not Hausdorff. Further, $A \subset E$ is quasicompact iff $A = \varnothing$ or $\sup A \in A$. So E is locally quasicompact with countable base $\{(-\infty,x): x \in \mathbb{Q}\}$. Sup measures are functions of $(-\infty,x)$ for $x \in \mathbb{R}$, and can be regarded as functions of x. As such the sup measures coincide with the left-continuous nondecreasing functions on $\mathbb{R}$. The sup derivative is the corresponding right-continuous function. Sup vague convergence coincides with pointwise convergence at the continuity points of the limit, so with weak convergence when restricted to distribution functions on $\mathbb{R}$. Indeed US is compact (Helly's selection theorem) and metrizable (by a Lévy type metric), as Theorem 5 tells us.

12. More about the sup vague topology

In this section we give more background information about the sup vague topology on SM and US, especially its connections with other parts of mathematics. Let us denote by $\mathcal{F} = \mathcal{F}(E)$ the family of closed subsets of E. Furthermore, for functions $f: E \to \overline{\mathbb{R}}$ we define the *hypograph* of f by

$$\text{hypo } f := \{(t,x) \in E \times (-\infty,\infty]: x \leq f(t)\}.$$

Note that we consider $E \times (-\infty,\infty]$ rather than $E \times \overline{\mathbb{R}}$, so that hypo $(-\infty \cdot 1_E) = \emptyset$.

The following are well-known properties of usc functions.

Lemma 4.
(a) $f: E \to \overline{\mathbb{R}}$ *is usc iff* hypo f *is closed in* $E \times (-\infty,\infty]$.
(b) 1_A *with* $A \subset E$ *is usc iff* $A \in \mathcal{F}(E)$.

So we have the following table of injections.

$$\begin{array}{ccccc} SM & \underset{f^{\vee} \leftarrow f}{\overset{d^{\vee}}{\rightleftarrows}} & US & \xrightarrow{\text{hypo}} & \mathcal{F}(E \times (-\infty,\infty]) \\ & & \Big\uparrow {\scriptstyle F \mapsto 1_F} & & \\ & & \mathcal{F}(E) & & \end{array}$$

One may wonder whether they are homeomorphisms, or can be made so. First of all, SM and US are homeomorphic, as we identified them. The identification $F \mapsto 1_F$ makes $\mathcal{F}$ a subspace of US. The sup vague topology restricted from US to $\mathcal{F}$ turns out to be the topology considered by Matheron (1975) in connection with random closed sets: $F_n \to F$ in $\mathcal{F}$ iff for $K \in \mathcal{K}$, $G \in \mathcal{G}$:

$$\left.\begin{array}{l} F \cap K = \emptyset \Rightarrow F_n \cap K = \emptyset \\ F \cap G \neq \emptyset \Rightarrow F_n \cap G \neq \emptyset \end{array}\right\} \text{ for all but finitely many } n.$$

We may endow $\mathcal{F}(E \times (-\infty,\infty])$ with a similar topology. Will hypo become a homeomorphism? The answer is yes, if E is locally quasicompact. Because of this connection the sup vague topology is known as the topology of "hypo convergence" in optimization and decision theory (cf. Salinetti & Wets (1985+)).

We can say more about hypo. It maps US onto a subset of $\mathcal{F}(E \times (-\infty,\infty])$. Surprisingly, the image of US under hypo is itself the family of closed sets of a cruder topology on $E \times (-\infty,\infty]$: the product topology of $E \times (-\infty,\infty]\uparrow$, where $(-\infty,\infty]\uparrow$ has the *upper* topology with nontrivial open sets $(x,\infty]$. In fact, US is homeomorphic to $\mathcal{F}(E \times (-\infty,\infty]\uparrow)$, provided that E is locally quasicompact. It follows that it does not matter where one starts defining a topology, in US as we did, or in $\mathcal{F}$ like Matheron (1975), provided that one starts in a generality that includes non-Hausdorff E.

One may consider US and $\mathcal{F}$ as lattices, abstract sets in which one can take suprema and infima. For locally quasicompact E the lattice structures of US and $\mathcal{F}$ have nice properties that make them, with the order reversed, *continuous lattices* in the sense of Gierz & al. (1980). This monograph gives the state of the art (of 1980) of the theory of continuous lattices, a subject whose investigation started in the early seventies from quite different parts of mathematics.

We have seen that allowing E to be non-Hausdorff is essential for a general treatment of the subject. The natural regularity condition turned out to be local quasicompactness. Locally quasicompact spaces are studied in Hofmann & Mislove (1981) and Gierz & al. (1980). Older literature is virtually non-existent as far as the non-Hausdorff case is concerned.

13. Lattice-valued extremal processes

Several authors (Balkema & Resnick (1977), De Haan & Resnick (1977), Deheuvels (1978), Gerritse (1983)) have started considering extremal processes with values in $\mathbb{R}^d$ with componentwise order. In other words, there is a tendency to study extremal processes with values in L rather than $\bar{\mathbb{R}}$, where L is a lattice in which the order need no longer be linear. In this formulation we can also regard processes of random closed sets as lattice-valued extremal processes: take $L = \mathcal{F}(E')$ for some space E', with intersections as infimum and closure of unions as supremum.

This is a topic of research in Nijmegen (G. Gerritse (1985)) and Gothenburg (T. Norberg). For older work, see Penot & Thera (1982). A complication with the generalization of the relations between sup measures and usc functions is that the three equivalent characterizations of upper semicontinuity (cf. Definition 4 and Lemmas 2 and 4) diverge into different concepts for general L. Here again the notion of continuous lattice plays a central role.

References

[1] Adler, R.J. (1978): Weak convergence results for extremal processes generated by dependent random variables. *Ann. Probab.* **6** 660-667.

[2] Balkema, A.A. & S.I. Resnick (1977): Max-infinite divisibility. *Z. Wahrsch. verw. Gebiete* **14** 309-319.

[3] Berman, S.M. (1964): Limit theorems for the maximum term in stationary sequences. *Ann. Math. Statist.* **35** 502-516.

[4] de Haan, L. (1970): On regular variation and its application to the weak convergence of sample extremes. *Math. Centre Tracts* **32**, Mathematisch Centrum, Amsterdam.

[5] de Haan, L. & S.I. Resnick (1977): Limit theory for multivariate sample extremes. *Z. Wahrsch. verw. Gebiete* **40** 317-337.

[6] Deheuvels, P. (1978): Caractérisation complète des lois extrêmes multivariées et de la convergence des types extrêmes. *Pub. Inst. Stat. Univ. Paris* **23.3-4** 1-36.

[7] Deheuvels, P. (1981): The strong approximation of extremal processes. *Z. Wahrsch. verw. Gebiete* **58** 1-6.

[8] Deheuvels, P. (1983): The strong approximation of extremal processes (II). *Z. Wahrsch. verw. Gebiete* **62** 7-15.

[9] Dwass, M. (1964): Extremal processes. *Ann. Math. Statist.* **35** 1718-1725.

[10] Fisher, R.A. & L.H.C. Tippett (1928): Limiting forms of the frequency distribution of the largest or smallest member of a sample. *Proc. Cambridge Philos. Soc.* **24.2** 180-190.

[11] Gerritse, G. (1983): Supremum self-decomposable random vectors. Report 8341, Dept of Mathematics, Catholic Univ., Nijmegen. To appear in *Z. Wahrsch. verw. Gebiete.*

[12] Gerritse, G. (1985): Lattice-valued semicontinuous functions. Report 8532, Dept of Mathematics, Catholic Univ., Nijmegen.

[13] Gierz, G., K.H. Hofmann, K. Keimel, J.D. Lawson, M. Mislove & D.S. Scott (1980): *A compendium of continuous lattices.* Springer, Berlin.

[14] Gnedenko, B.V. (1943): Sur la distribution limite du terme maximum d'une série aléatoire. *Ann. of Math.* **44** 423-453.

[15] Goldie, C. (1983): On records and related topics in probability theory. Thesis, University of Sussex.

[16] Hofmann, K.H. & M.W. Mislove (1981): Local compactness and continuous lattices. In: *Continuous lattices (Proceedings, Bremen 1979)*, eds. B. Banaschewski & R.-E. Hofmann. *Lecture Notes in Math.* **871**, Springer, Berlin.

[17] Lamperti, J. (1962): Semi-stable stochastic processes. *Trans. Amer. Math. Soc.* **104** 62-78.

[18] Lamperti, J. (1964): On extreme order statistics. *Ann. Math. Statist.* **35** 1726-1737.

[19] Leadbetter, M.R., G. Lindgren & H. Rootzén (1983): *Extremes and related properties of random sequences and processes.* Springer, New York, Heidelberg, Berlin.

[20] Loynes, M.R. (1965): Extreme values in uniformly mixing stationary stochastic processes. *Ann. Math. Statist.* **36** 993-999.

[21] Matheron, G. (1975): *Random sets and integral geometry.* Wiley, New York.

[22] Mori, T. (1977): Limit distributions of two-dimensional point processes generated by strong mixing sequences. *Yokohama Math. J.* **25** 155-168.

[23] Mori, T. & H. Oodaira (1976): A functional law of the iterated logarithm for sample sequences. *Yokohama Math. J.* **24** 35-49.

[24] NEWELL, G.F. (1964): Asymptotic extremes for m-dependent random variables. *Ann. Math. Statist.* **35** 1322-1325.

[25] NORBERG, T. (1984): Random capacities and their distributions. Preprint 1984-03, Chalmers University of Technology, Göteborg. To appear in *Z. Wahrsch. verw. Gebiete.*

[26] O'BRIEN, G.L. (1984): Extreme values for stationary and Markov sequences. Preprint, York University.

[27] O'BRIEN, G.L., P.J.J.F. TORFS & W. VERVAAT (1985+): Stationary self-similar extremal processes. Forthcoming.

[28] O'BRIEN, G.L. & W. VERVAAT (1985): Self-similar processes with stationary increments generated by point processes. *Ann. Probab.* **13** 28-52.

[29] PENOT, J.P. & M. THERA (1982): Semi-continuous mappings in general topology. *Arch. Math.* **38** 158-166.

[30] PICKANDS III, J. (1971): The two-dimensional Poisson process and extremal processes. *J. Appl. Probab.* **8** 745-756.

[31] RESNICK, S.I. (1975): Weak convergence to extremal processes. *Ann. Probab.* **3** 951-960.

[32] RESNICK, S.I. (1985): Point processes, regular variation and weak convergence. *Adv. Appl. Probab.* (to appear).

[33] SALINETTI, G. & R.J.-B. WETS (1985+): On the convergence in distribution of measurable multifunctions, normal integrands, stochastic processes and stochastic infima. *Math. Oper. Res.* (to appear).

[34] SALINETTI, G., W. VERVAAT & R.J.-B. WETS (1985+): On the convergence in probability of random sets (measurable multifunctions). *Math. Oper. Res.* (to appear).

[35] VERVAAT, W. (1981): The structure of limit theorems in probability. Lecture notes, Catholic University, Leuven & Catholic University, Nijmegen.

[36] VERVAAT, W. (1982): Random semicontinuous functions and extremal processes. Handwritten manuscript. Being revised.

[37] VERVAAT, W. (1985): Sample path properties of self-similar processes with stationary increments. *Ann. Probab.* **13** 1-27.

[38] WEISSMAN, I. (1975a): On location and scale functions of a class of limiting processes with application to extreme value theory. *Ann. Probab.* **3** 178-181.

[39] WEISSMAN, I. (1975b): Multivariate extremal processes generated by independent non-identically distributed random variables. *J. Appl. Probab.* **12** 477-487.

DEPARTMENT OF MATHEMATICS
CATHOLIC UNIVERSITY
TOERNOOIVELD 5
6525 ED NIJMEGEN
THE NETHERLANDS

Progress in Probability and Statistics

1 CINLAR/CHUNG/GETOOR. Seminar on Stochastic Processes, 1981
ISBN 3-7643-3072-4
2 KESTEN. Percolation Theory for Mathematicians
ISBN 3-7643-3107-0
3 ASMUSSEN/HERING. Branching Processes
ISBN 3-7643-3122-4
4 CHUNG/WILLIAMS. Introduction to Stochastic Integration
ISBN 3-7643-3117-8
5 CINLAR/CHUNG/GETOOR. Seminar on Stochastic Processes, 1982
ISBN 3-7643-3131-3
6 BLOOMFIELD/STEIGER. Least Absolute Deviation
ISBN 3-7643-3157-7
7 CINLAR/CHUNG/GETOOR. Seminar on Stochastic Processes, 1983
ISBN 3-7643-3293-X
8 BOUGEROL/LACROIX. Products of Random Matrices with Application to Schrödinger Operator
ISBN 3-7643-3324-3
9 CINLAR/CHUNG/GETOOR. Seminar on Stochastic Processes, 1984
ISBN 3-7643-3327-8
10 KIFER. Ergodic Theory of Random Transformations
ISBN 3-7643-3319-7
11 EBERLEIN/TAQQU. Dependence in Probability and Statistics
ISBN 3-7643-3323-5
12 CINLAR/CHUNG/GETOOR. Seminar on Stochastic Processes, 1985
ISBN 3-7643-3331-6

GPSR Compliance

The European Union's (EU) General Product Safety Regulation (GPSR) is a set of rules that requires consumer products to be safe and our obligations to ensure this.

If you have any concerns about our products, you can contact us on ProductSafety@springernature.com

In case Publisher is established outside the EU, the EU authorized representative is:

Springer Nature Customer Service Center GmbH
Europaplatz 3
69115 Heidelberg, Germany

Batch number: 10577896

Printed by Printforce, the Netherlands